Ecological Solidarity and the Kurdish Freedom Movement

Environment and Society

Series Editor: Douglas Vakoch

As scholars examine the environmental challenges facing humanity, they increasingly recognize that solutions require a focus on the human causes and consequences of these threats, and not merely a focus on the scientific and technical issues. To meet this need, the Environment and Society series explores a broad range of topics in environmental studies from the perspectives of the social sciences and humanities. Books in this series help the reader understand contemporary environmental concerns, while offering concrete steps to address these problems.

Books in this series include both monographs and edited volumes that are grounded in the realities of ecological issues identified by the natural sciences. Our authors and contributors come from disciplines including but not limited to anthropology, architecture, area studies, communication studies, economics, ethics, gender studies, geography, history, law, pedagogy, philosophy, political science, psychology, religious studies, sociology, and theology. To foster a constructive dialogue between these researchers and environmental scientists, the Environment and Society series publishes work that is relevant to those engaged in environmental studies, while also being of interest to scholars from the author's primary discipline.

Recent Titles in the Series

Ecological Solidarity and the Kurdish Freedom Movement: Thought, Practice, Challenges, and Opportunities, edited by Stephen E. Hunt

Ecomobilities: Driving the Anthropocene in Popular Cinema, by Michael W. Pesses

Embodied Memories, Embedded Healing: New Ecological Perspectives from East Asia, edited by Xinmin Liu and Peter I-min Huang

Wetlands and Western Cultures: Denigration to Conservation, by Rod Giblett

Sustainable Engineering for Life Tomorrow, edited by Jacqueline A. Stagner and David S. K. Ting

Nuclear Weapons and the Environment: An Ecological Case for Nonproliferation, by John Perry

Portland's Good Life: Sustainability and Hope in an American City, by R. Bruce Stephenson

Conservation, Sustainability, and Environmental Justice in India, edited by Alok Gupta

Environmental and Animal Abuse Denial: Averting Our Gaze, edited by Tomaž Grušovnik, Reingard Spannring, and Karen Lykke Syse

Living Deep Ecology: A Bioregional Journey by Bill Devall, edited with an introduction by Sing C. Chew

Environmental and Animal Abuse Denial: Averting Our Gaze, edited by Tomaž Grušovnik, Reingard Spannring, and Karen Lykke Syse

Secular Discourse on Sin in the Anthropocene: What's Wrong With the World?, by Ernst M. Conradie

Ecological Solidarity and the Kurdish Freedom Movement

Thought, Practice, Challenges, and Opportunities

Edited by
Stephen E. Hunt

Foreword by John P. Clark

LEXINGTON BOOKS
Lanham • Boulder • New York • London

Published by Lexington Books
An imprint of The Rowman & Littlefield Publishing Group, Inc.
4501 Forbes Boulevard, Suite 200, Lanham, Maryland 20706
www.rowman.com

6 Tinworth Street, London SE11 5AL, United Kingdom

British Library Cataloguing in Publication Information Available

Library of Congress Cataloging-in-Publication Data

Names: Hunt, Stephen E., editor. | Clark, John P., 1945- writer of foreword.
Title: Ecological solidarity and the Kurdish freedom movement : thought,
    practice, challenges, and opportunities / edited by Stephen E. Hunt ; foreword by
    John P. Clark.
Other titles: Environment and society (Lanham, Md.)
Description: Lanham, Maryland : Lexington Books, 2021. | Series: Environment and
    society | Includes bibliographical references and index.
Identifiers: LCCN 2021034437 (print) | LCCN 2021034438 (ebook) |
    ISBN 9781793633842 (cloth) | ISBN 9781793633859 (epub) |
    ISBN 9781793633866 (pbk)
Subjects: LCSH: Social ecology—Middle East. | Political ecology—Middle East. |
    Environmentalism—Middle East. | Women, Kurdish—Political activity—Middle East. |
    Environmental protection—Middle East. | Environmental policy—Middle East. |
    Middle East—Environmental conditions.
Classification: LCC GE199.M628 E26 2021 (print) | LCC GE199.M628 (ebook) |
    DDC 304.209566/7—dc23
LC record available at https://lccn.loc.gov/2021034437
LC ebook record available at https://lccn.loc.gov/2021034438

Contents

List of Figures and Tables

FIGURES

TABLES

Foreword

John P. Clark

Ecological Solidarity and the Kurdish Freedom Movement is a very significant work, a source of insight, understanding, inspiration, and hope. It constitutes an important advance in the study of Kurdish society and politics in relation to the theory and practice of social ecology, and it will serve as an essential resource for any future research on ecological issues in Kurdistan. It will also be of considerable interest to pro-Kurdish activists and to those who have a strong connection to Kurdistan and hopes for its future. Most significantly, it is a work that explores cultural and political dimensions of Kurdish society that have been generally neglected and shows that many aspects of both traditional Kurdish culture and contemporary revolutionary Kurdistan have urgently needed lessons to teach the world.

There is much that could be said about the virtues of this work, but I would like to focus on how it makes a major contribution in three areas. First, it goes far beyond most of the literature on the Kurdish experience in posing critical questions regarding social ecological theory and practice. Second, it demonstrates strikingly the profound significance of the revolutionary feminism of the Kurdish Freedom Movement (KFM). And third, its analysis of practices and institutions in Kurdistan contributes greatly to our understanding of the meaning of an ecological worldview and ethos, something that is so desperately needed in our age of fateful global crisis.

SOCIAL ECOLOGICAL THEORY AND PRACTICE

Throughout the book, the contributors not only reveal the practical and theoretical importance of the Kurdish struggle but also raise crucial questions about the nature of that struggle and its future. For example, Stephen Hunt points out that Abdullah Öcalan, the revered leader and visionary philosopher of the movement, transformed it fundamentally with his inspiring ideal of a free ecological society, but that this vision left many of the economic and technological details to be worked out by the autonomous communities and their confederations. As Hunt correctly observes, the necessary reconciliation of ideal and reality poses daunting challenges, since, for example, an ecological perspective recognizes that "a rapid break with the global fossil-fuel based economy is a precondition for geopolitical progress," while at the same time military attacks, embargoes, and other factors create formidable obstacles to such an ecological transition.

Another example of such acute analysis is Federico Venturini's astute observation that in the Kurdish case, as in the Zapatista experience that in many ways parallels it, great revolutionary achievements have been the result not only of "years of consistent work" but also of "unique events and spatial conditions." In both of these cases, there was a "vacuum in state power," and each occurred in a relatively "remote area" of the world system. Venturini concludes that from a social ecological perspective, this implies the necessity for flexibility in using such experiences as models. Thus, participation in municipal elections might be seen less as "a prescription" than as a powerful strategy that can be exercised to varying degrees according to "the specific context of the struggle." That context may sometimes imply that strategies, such as grassroots organizing, development of affinity groups, or direct action, may play a relatively greater role.

Ercan Ayboğa makes an important and closely related point concerning the mode of operation of the structures that the KFM organized in North Kurdistan through the Democratic Society Congress (KCD). He notes that the movement has faced the problem that "taking over state power at the local level offers advantages, but also leads to risks such as growing hierarchy, alienation, and corruption," particularly in locales where the Turkish government still appoints officials to important positions. This reinforces Venturini's emphasis on the need for carefully contextualized analysis in assessing the potential in any given area for effective action through municipal government.

Another analysis that is welcome and much needed for Western readers in particular is Cihad Hammy's discussion of Öcalan's efforts to decolonize and de-Eurocentrize the dialectic. As Hammy notes, "dialectical interpretations are abundant in the wisdom of the East." Western social ecologists need to recognize not only the longevity of 2,500 years of Asian dialectical (and other theoretical) traditions but also the depth and radicality of much that can be found there. A closely related issue is the manner in which the Kurdish social ecological struggle and its ideals are rooted in indigenous cultural traditions, as will be discussed below. In both these areas, it is important for Western observers to recognize the ways in which they can learn from and interact dialectically with non-Western thought and practice rather than merely subsuming the latter under Western theoretical categories.

Finally, one might mention Hunt's absolutely crucial point in the Introduction that social and ecological liberation cannot succeed without an effective connection between local, regional, and global organization. As he states, the Kurdish revolutionary struggle implies a "responsibility of the rest of the world . . . to take up, embrace, adapt, and advance complementary visions for democracy, social justice, and ecological solidarity." Later in the book, Hunt points out the global importance of the La Via Campesina, the enormous global association of peasants that is founded on a commitment to radical democracy, agroecology, and food sovereignty and which espouses principles of democratic confederation, participatory democracy, gender equality, and ecological sustainability that are much in accord with the aspirations of the KFM.

It is questionable whether La Via Campesina can ever become an organization that could occupy both the revolutionary imaginary space and the revolutionary political space that the International once did. Yet, the implication of such lines of analysis is that some such successor will be needed if the KFM's project of social ecological revolution is not only to be defended successfully but also universalized globally and particularized everywhere.

ANTI-PATRIARCHAL REVOLUTION

Although it is true, as the issues just raised indicate, that the KFM faces momentous challenges, much of the book documents the ways in which it is already meeting those challenges and demonstrates the degree to which it already constitutes a major world-historical achievement. It is this success that leads Engin Sustam to conclude, with great insight, that "we see 'a molecular revolution' at the heart of the process of a new Kurdish subjectivity coming into being." This is true nowhere more than in the feminist revolution that has been at the core of the KFM.

Michel P. Pimbert makes the important point that women's cooperatives, a key part of this revolution, are an important means by which women can create a sphere of production free from male dominance, an evil that tends to be carried over even in institutions that are formally committed to sexual equality. The development of such a cooperative sphere produces "a care economy" that "values women's productive and reproductive labor as well as the labor of nature." Very similar experiences of the liberatory and transformative role of women's cooperatives have been reported in the Zapatista revolution in Chiapas. This illustrates very well Venturini's point that an openness to diverse strategies will be necessary for achieving various revolutionary goals in the most effective way.

One of the most important and most inspiring chapters of the book is Fabiana Cioni and Domenico Patassini's analysis of Jinwar, the "women's eco-village." They present a moving depiction of this community's many-sided project for "a progressive liberation from the patriarchal and hierarchical mentality" of the past. Men are allowed to enter the village only if they are willing to "challenge patriarchal hierarchical structures, male privilege, the use of language as an expression of power and competition." The community transforms into everyday practice Öcalan's famous slogan about the necessity to "kill the dominant male," which means in effect, the dissolution of any remnants of the domineering, all-consuming ego that has developed under patriarchal civilization.

The practices of the community are a model for the creation of a cooperative community. These include not only the institution of formal structures such as "a collective decision-making process with direct democracy" but also a determined focus on transformation of the personal and subjective dimension through liberatory, communitarian practices such as "empathic listening," true dialogue, and the establishment of an "organic time," which makes possible, for example, "opportunities for expression during the stages of the assembly."

A notable element of communal practice in the women's village is *Tekmîl* or "revolutionary constructive criticism," which is based on "the ability to question oneself." Contrary to self-questioning being seen as a weakness (as in masculinist ideology), it is shown to be an expression of personal strength and confidence and a measure of the depth of one's commitment to the community. This practice attests once more to the reality of the "molecular revolution" at work within the KFM.

Furthermore, the members of the community "agree to live communally, sharing everything" rather than merely committing themselves to true communality as some distant ideal. The women's village thus becomes a powerful, and indeed invaluable, example of how the ideal can be embodied *here and now* in the lives of the members. Such embodiment and revolutionary immanence is almost entirely absent in the vast majority of political movements today, with predictable results.

Azize Aslan's chapter on "Women's Subjectivity and the Ecological and Communal Economy" further demonstrates the profoundly revolutionary character of the radical feminism of the KFM. Aslan shows that Kurdish revolutionary women reject the ideological assumption that women will actually "have the right to participate freely" merely because they have this right in principle. They point out that even in a putatively democratic assembly, there can exist such destructive masculinist phenomena as "strong (loud) speech," the "personalization of discussion," and a "failure to care for the participation" of all.

This makes the important point that until the poisonous legacy of patriarchy is entirely destroyed, there will always be a gap between the supposed *forms of freedom* and the liberatory *substance of freedom*. As the analysis in the chapter shows, when women have an effective rather than merely formal equality of participation, there is a substantive difference in the topics discussed in the assembly. Significantly, there is less about "profit and production" and more about "life and reproduction."

AN ECOLOGICAL ETHOS

This last point is closely related to, and indeed merges into, the third area in which the book makes a major contribution. This lies in its account of the profoundly ecological and communal worldview that is inherent in segments of traditional Kurdish culture and which is being further developed through contemporary social ecological revolutionary practice.

An extremely powerful dimension of the book is its extensive account of the traditional nature-oriented worldview and ethics, particularly within the Kurdish Alevi community. To discover ecological wisdom in its deepest and most culturally integral form, one can do no better than look to the ecological sensibility, spirituality, and ethos of this community, as described in several chapters of this collection.

The Alevi community holds the land to be in a very powerful sense sacred and sees it as connected, as Dilşa Deniz points out, to "the main holy power, the unity of the whole, the main form of a creator." This concept of creator is associated with cosmic feminine power and evokes nature's powers of giving birth. In addition, the community's "system of sacred grand law," the Dersim Rêya Heqî, is seen as the expression of a system of mutual interdependence between humanity and nature, between humans and other beings in nature, between particular human communities, and between one human being and another. To this should be added Ahmet Kerim Gültekin's description of the rituals at sacred protected sites called "jiares," which include "trees, forests, mountains, rocks, caves, rivers, lakes, fountains, fire, soil, wild animals, or the sun and moon." All of this evidence attests to the vast scope of Alevi nature spirituality.

Deniz's account of the Gola Buyêr myth shows it to be another striking expression of the Alevi ecological worldview. The theme of this sacred text is the operation of the fundamental spiritual and ethical system of nature that protects life. It expresses the sacred law against harming or killing the mother, the concern for protection of future generations and the preservation of species, the importance of numinous places, and the reverence for water as the source of life. It is an extraordinary expression of a powerful ecological imaginary.

This and other accounts of traditional Kurdish culture should in no way be looked upon as a mere exercise in exoticism. Instead, they must be seen as a powerful challenge to anyone who believes in the need for "an ecological society" to think deeply about what a truly ecological form of association must be and how it must shape modes of thinking, feeling, acting, and imagining. Moreover, as Deniz points out, the traditional ethos has an inherently anti-capitalist dimension. It forbade the treatment of the land as a commodity subject to market transactions. The land was not to be bought or sold, except in cases of extreme necessity, in which it could be sold only to another with "the same genealogy" and "kinship priority," in effect, to someone charged with the same responsibility of caring for it.

We find within the KFM a creative, revolutionary dialectic in which new meaning is infused into age-old nature-affirming values by contemporary social and ecological struggles. Gültekin quotes Bilgin's observation that a "new understanding of nature is being forged in the Kurdish Alevis' struggles against incursions by dam projects, mining companies, tourism policies, and other threats." As Gültekin observes, in these struggles, the Kurds' confrontation with the long-standing threat of *genocide* is being expanded into a profound social ecological understanding of the threat to both the land and people posed by *ecocide*.

SOCIAL ECOLOGICAL SOLIDARITY

In sum, this remarkable collection demonstrates that the KFM has a very rich social ecological conception of the meaning of freedom for Kurdistan, for humanity, and for the Earth. It shows what can be achieved when a society begins to confront seriously all the forces of domination, from the broadly institutional to the intimately personal, that stand in the way of the liberation of humanity and nature.

It shows that to be effective, the revolutionary project must encompass a spectrum of strategies, including, for example, the creation of strong counter-institutions, the use of what is best in the traditional ethos to help create a new revolutionary one, the critique of dominant ideologies, the creation of a new liberatory worldview based on liberty, equality, solidarity and communal autonomy, and the generation of a powerful social imaginary rooted in the flourishing of the Earth, the land, the people.

It suggests that anyone who truly appreciates both the achievements and the liberatory potential of the Kurdish Freedom Movement should be moved not only to admire it greatly but also to emulate it and to find ways of acting effectively in solidarity with it.

Acknowledgments

In addition to the chapter contributors, the editor would like to extend warm thanks to John P. Clark for supplying the foreword and the external reviewer Dr Shannon Brincat. He would also like to extend thanks to Fiona Scott, Kate Aydin, Jeremy Anbleyth-Evans, and Hazhar Jamali for their generous help and support when preparing this collection.

Chapter 2: The author thanks Debbie Bookchin for her generous support and her valuable comments on an earlier draft.

Chapter 6: The editor would like to thank Menekşe Kizildere, HDP Ecology Commission Co-Spokesperson, for granting this interview.

Chapter 9: The article is related to PhD research undertaken at the Università Iuav di Venezia, conducted in the field by Fabiana Cioni with the supervision of Professor Domenico Patassini. The authors would like to acknowledge the collaboration of Necibe Qaradaxi (from the Bruxelles Jineolojî Center) and Jinwar's Council.

Chapter 10: The editor would like to thank Karen Tiedtke for translating this chapter from the original Spanish.

Chapter 11: The author would like to thank Mansour Sohrabi, the interviewees, those mentioned by name and those who chose to remain anonymous, and the editor Stephen Hunt, for their contribution to this research.

Chapter 18: This project has received funding from the European Union's Horizon 2020 research and innovation program under the Marie Skłodowska-Curie grant agreement No. 796086.

Introduction

Ecology in the Kurdish Paradigm

Stephen E. Hunt

Kurdish culture has an intimate relationship to the living world, with the mountainous region geographically defined by the Tigris and Euphrates river basins forming its heartlands. While rich in beauty and biodiversity, this natural environment is, like elsewhere, under severe and increasing threat. *Ecological Solidarity and the Kurdish Freedom Movement* will evaluate some positive endeavors to address this threat. The collection includes perspectives from disciplines including political theory, environmental humanities, law, anthropology, and cultural studies to examine the basis for the Kurdish freedom movement's (KFM) foregrounding of ecology as a pillar of its strategy for change. Building upon campaigns against dam construction during the 1990s, in the early twenty-first century the movement integrated "an ecological model of society" into its 2005 Declaration of Democratic Confederalism (Öcalan 2005), where it became a core part of a revolutionary paradigm that has international significance. It is heartening that ecology has been included as a priority, essential within an integrated plan for social transformation. The embrace of social ecology as a theoretical approach represents advances upon both the reformist mentality of the mainstream Western environmental movement and, historically, much of the organized Left's marginalizing of ecological consciousness. This introduction will signpost the chapters that follow, consider the ecological challenges for the Kurdish region (divided between the Turkish, Syrian, Iraqi, and Iranian parts), and introduce the conceptual framework of social ecology and democratic confederalism.

THEORY, PRACTICE, AND ACTIVISM

While present ecological challenges are formidable, it is timely to examine the KFM's audacious efforts to address such issues. The following chapters explore the theoretical and practical issues encountered in implementing solutions-focused policies.

Part 1 sets out the theoretical underpinning, using social ecology and communalism to frame our understanding of the political experiments underway in the Kurdish region. Federico Venturini makes the case for the value of social ecology, the philosophy introduced by the political ecologist and communalist, Murray Bookchin, as an explanatory and strategic thinking tool for both understanding the entwined social and ecological crises that confront humanity and as a means to address them. For Venturini and Cihad Hammy, social ecology provides a deep critique of capitalist expansionism in tandem with a practical vision of ecologically sustainable social reconstruction. Hammy examines the KFM's imprisoned figurehead and prominent theoretician, Abdullah Öcalan's interpretation of Bookchin's ideas and their further development. Engin Sustam positions the KFM's ecological pillar within the context of international post-capitalist resistance, situating the ecological aspirations of the Kurdish revolutionaries within the wider program of emancipatory communalism. Nicholas Hildyard elaborates the critical distinction between a revolutionary ecological paradigm embracing social justice, and the shallow, even reactionary approach of much mainstream Western environmentalism. As Hildyard argues, there has often been a failure to recognize the deep roots of ecological destruction in social injustice, while philosophical nature-culture dualism has often led the living world to be treated as matter to be commodified for human benefit.

Part 2 seeks to explore and explain some of the practical ways that ecological solidarity complements the Kurdish model of democratic autonomy. Ercan Ayboğa (co-founder of the Mesopotamia Ecology Movement) provides insight into the way that ecological principles are embedded within the movement's democratic structures. In Turkey, the MEM developed ecology assemblies in several Kurdish districts, including Amed, Dersim, Antep, Urfa, Van, Batman, and Mardin, together with smaller assemblies at the neighborhood level (Öztürk 2018). The HDP, the broad-based left-wing political party with strongholds in the Kurdish southeast, has an Ecology Commission. It has recently supported campaigns including water rights and opposition to climate change, deforestation, and environmentally destructive quarrying, and seeking to promote ecological objectives throughout Turkey. Presently, in 2021, the AKP government is undertaking legal efforts to ban the HDP, which obviously makes it difficult for the party to advance its policy priorities. Chapters by Clémence Scalbert-Yücel and Michel Pimbert reflect upon efforts to extend food sovereignty through agroecology and the diversification of production. Contributions by Fabiana Cioni and Domenico Patassini, and by Azize Aslan, turn to the connections between ecology and the Kurdish women's movement in Rojava and with *jineolojî* (the science of women). The former finds inspiration in fieldwork undertaken at Jinwar, the women's village which has been designed with ecological principles, while Aslan focuses on the role of the solidarity economy in furthering sustainability.

The chapters in part 3 feature further aspects of Kurdish ecology as a social movement and its relationship to other social movements. Allan Hassaniyan's work here and elsewhere (2020) rebalances attention to other parts of Kurdistan by revealing the significant challenges that confront the environmental movement in the predominantly Kurdish area of Iran, known as Rojhelat. This area in northwestern Iran suffers from exclusion, underdevelopment, and significant environmental degradation as water sources are redirected to central regions, and the oil industry and dam building have negative impacts. In this context, Kurdish environmentalists have been subject to considerable state oppression. Kurdish and other Iranian environmental campaigners have been imprisoned and threatened with execution, while several activists have become known as "martyrs of ecology" since they have died in suspicious circumstances (Kamal Soleimani, "Rojhelat," Scottish Solidarity with Kurdistan webinar, August 25, 2020; Sanford 2019). Communication about the Kurdish Iranian environmental movement is therefore problematic. Hazhar Jamali, living outside of the country, reports that currently "because of heavy restriction in Iran, there is no possibility for organizations to promote democratic confederalism. They usually discuss DC in private and with the people who they trust" and so there is "no possibility for them to freely discuss and promote ecology" (Interview by e-mail with Stephen Hunt, June 27, 2020). Hassaniyan researches this central aspect of Iranian Kurdish cultural identity and environmental resilience with an analysis of the Chya Green Association.

The 2013 Gezi Park Protests, the most broad-based and striking series of mobilizations against Erdoğan's government, began as a campaign by environmental campaigners to defend Istanbul's historic urban park against the construction of a shopping mall. It spiraled into a nationwide expression of disaffection against the AKP's autocratic, neoliberal policies. Kumru Toktamiş and Isabel David analyze the response of the Kurdish movement, which after some initial tactical uncertainties in navigating the political implications of the protests, finally experienced the resistance as a moment of revitalization linked to the emergence of the HDP.

The anti-dam campaigns were a major development in the emergence of the regional environmental movement alongside the KFM. Laurent Dissard examines the complicated relationship between the two movements and considers how the campaigns to save Hasankeyf, the most prominent of such mobilizations, influenced the shift toward environmental issues in the Kurdish struggle. Sadly, the settlement of Hasankeyf, a world heritage site and area of natural beauty, was lost when it was inundated by the Ilisu Dam in 2020.

In their thought-provoking chapter "The Kurdish Ecology Movement and Human Rights," Marlene Payva Almonte and Thomas James Phillips explore the contradictory role of human rights legal discourse for social movements, such as the KFM, both problematically framing ecological justice tactically within the context of a statist structure, yet also holding the potential to enable powerful strategic mobilizations for the far-reaching transformation necessary to make a more democratic and ecologically sustainable future possible. In the final chapter of part 3, Hunt highlights the contribution of "Make Rojava Green Again," an initiative of internationalist volunteers.

Part 4 examines Kurdish Alevi nature sympathy, an influence often overlooked in Western accounts. Sakine Cansiz, one of the PKK cofounders raised in Alevi culture in Turkey's Dersim Province (see Figure 1.3), was intensely aware of the beauty that illuminated and tragedy that haunted, her homeland, a source of commitment that led her to reject the notion that internationalism implied the "denial of one's identity" (2018, 117). Anthropologists Ahmet Kerim Gültekin and Dilşa Deniz reveal the deep connections between culturally specific notions of sacred land and reverence for the natural world and animals in the Kurdish Alevi, or "Rêya Heqî," cosmology and popular mobilizations. Drawing upon adapted, and sometimes perhaps invented, traditions, anti-dam protests began in Dersim in the 1990s, while the first Munzur Culture and Nature Festival was held in 2000. This shows Kurdish ecological activism was established before Öcalan began to read Bookchin's works. While the latter is not, therefore, the origin of the movement's involvement in ecological activism, Kurdish Alevi heritage and preexisting campaigns to protect the environment may in part explain why the movement readily adopted and embraced Bookchin's philosophy of social ecology. Defense of the natural world was prominent in the "Dersim awakening" and so integral to the Kurdish Alevi population's belief system, that nearly the entire population of Dersim joined a march against dam construction in 2014, making it the largest environmental protest in Turkey to date (Gültekin 2019, 18).

The two chapters in part 5 reflect the argument that war is both a cause and agent of environmental destruction. Pinar Dinc's research on "Forest fires in Dersim and Şırnak," informed by primary data and media discourse analysis, reveals correlations between forest fires and conflict in Turkey's Kurdish region. This raises questions as to whether forests have been destroyed as part of counterinsurgency measures, as they were historically. Ceri Gibbons discusses the Campaign Against Arms Trade's recent revelations about the complicity of the international arms industry in supply-ing military hardware, with reference to weaponized drone technology, used to invade Rojava and to target opponents of NATO-member Turkey's poli-cies. Gibbons's case study on the role of EDO, one of the United Kingdom's key suppliers of arms to Turkey, tracks the West's complicity in supposedly indigenous Turkish drone production. This reveals ways in which, far from being a distinct regional issue, the supply of arms has global connections and ramifications. Here, the specific underlying drivers of realpolitik and corporate interests are made apparent as supply chains for Unmanned Aerial Vehicles (UAVs) are identified, exposed, and challenged. Such work high-lights the role of researcher activists in raising awareness of Western state and corporate interventions and revealing the arms trade's role in regional destabilization and insecurity as well as destructive impacts upon the physi-cal environment.

Concluding the present work, Hunt's chapters consider the prospects for developing democratic confederalism, with its ecological dimension, along-side and beyond the KFM.

WHAT ARE THE ECOLOGICAL CHALLENGES?

As in other parts of the world, the effects of pollution, waste disposal, defor-estation, and soil erosion are widespread, as well as more recent threats, such as hydraulic fracturing, already advancing in Iraqi Kurdistan. The following identifies the major ecological harms.

Air Pollution

Air pollution is a major challenge to well-being throughout the Kurdish region. In addition to its harmful impact on the natural environment and contribution to climate change, dangerously high levels of air-borne con-taminants have negative consequences for physical and mental health (includ-ing respiratory illnesses such as lung cancer and asthma, viruses including coronaviruses, and also depression) that go against health and educational objectives. In Rojava, such social consequences are themselves outcomes of an economic situation where, particularly in the transport and energy sectors, historic modes of development and present embargoes work against cleaner alternatives. Oil and diesel are plentiful, and presently necessary for revenue, yet the capacity to refine them is rudimentary. Such abundance reduces incen-tives to seek renewable alternatives, despite dependence upon polluting diesel generators to produce electricity and heavy road traffic due to limited public transport infrastructure (Knapp, Flach, and Ayboğa 2016, 219). Janet Biehl was dismayed by the environmental degradation that she experienced first-hand when visiting Rojava in 2019.

> I met Sero Hinde, a Kurdish intellectual, a co-head of the Rojava film commune, who asked me what I saw as the biggest problem in the revolution. I said, the air pollution. He responded: "Everyone says that." Sure enough, I came upon a middle-aged woman in a souk in Qamislo and asked her what were the problems of the revolution. She immediately said, "the pollution." It's terrible, she said. People get lung cancer and go to the clinics. They get chemotherapy but it doesn't help. What's the second biggest problem? I asked her. There is no second-biggest, she said. This is the problem. (Biehl, e-mail to Stephen Hunt, April 4, 2020)

Poor air quality is a concern in all Kurdish cities. In the oil-based economy of the Kurdistan Region of Iraq, significant emissions of carbon dioxide and other greenhouse gases from road traffic occur alongside substantial and increasing levels produced by diesel generators, cement production, and other heavy industry. Studies in the major cities have revealed alarming concentrations of particulates, toxic, acidic, and carcinogenic combinations of lead and other heavy metals, and also sulfur and nitrogen oxides and volatile organic compounds (Meena and Omar 2015; Majid 2016; Shilani 2019).

Across the border in predominantly Kurdish parts of Iran, similar problems persist. Iran was rated as the world's seventh most polluted country in 2018, and while Tehran's air pollution problems are notorious, there are threats to public health in other major centers of population too, with measurable increases in premature mortality attributable to this cause (Karimi, Shokrinezhad and Samadi 2019, 438). Studies have found serious effects upon physical and mental health, and general well-being, in majority Kurdish cities, such as Samandaj, Marabad, and Urmia, where air pollution due to traffic exhaust and other human causes are compounded by particles from dust storms and sources such as desiccating lakes (Nadrian et al. 2019; Mohammadi et al. 2019). A detailed study found the cities of southeastern Turkey, where there is the largest Kurdish population, to be consistently among those with the highest levels of air pollution, in this case attributed to "heavy usage of coal due to the lack of natural gas infrastructure" (Arı, Özköse, and Gencer 2016).

Climate Change

Climate change is a major global threat, a defining characteristic of the Anthropocene, the geological era so-called since it is proposed that the impact of human activities has come to determine both the Earth's atmospheric and geological conditions and its ecosystems. Climate change constitutes one of the nine principal planetary boundaries identified by the Stockholm Resilience Centre's sustainability assessment. This team of international environmental scientists' influential research study warns that under current trends even a temperature increase set at a "guardrail" of 2°C above pre-industrial levels is in danger of being breached, with potentially catastrophic outcomes for life on Earth (Rockström et al. 2009). By the end of the twenty-first century, it is projected that the Middle East will experience "significant warming" and reduced rainfall (Bucchignani et al. 2018). There are, moreover, specific issues pertinent to the Kurdish region. In Rojava, for example, large-scale oil extraction and loss of tree cover from deforestation are significant contributory factors (Knapp et al. 2016, 212 and 219–20). Findings from a longitudinal research study of the Khabour River basin in Iraqi Kurdistan predicted that the increases in mean temperature and decreasing precipitation since 1980 will continue, with reduced streamflow leading to depleted rivers and other water sources, such as lakes, reservoirs, and aquifers (Abbas, Wasimi, and Al-Ansari 2016). Such meteorological changes have related consequences for biodiversity, aridification, and soil erosion and hence agricultural production.

Conflict

There are close links between warfare and climate change and other forms of environmental degradation. Ecological destruction, potentially, can cause, be an instrument of, and an outcome of military conflict.

Extractivism and ecological destruction can be conflict multipliers as rivalry for water, agricultural land, and other natural resources increase regional tensions. The stand-off between Turkey and Greece over the exploitation of gas reserves in the East Mediterranean is a case in point, although to date hostilities have been limited to a war of words. Climate change and drought further exacerbate poverty and instability, impacts that include the displacement of affected communities and populations, in turn potentially feeding hostilities and conflict. The exact nature and extent of causal links between climate change and specific conflicts, however, is a contested issue (Selby et al. 2017; Gleick 2017; Dalby 2018; Read 2019; Ash and Obradovich 2020). The case has been made that climate change, by aggravating economic instability, contributed to the conditions that led to the outbreak of the Syrian Civil War in 2011, the ongoing conflict that has affected not only the population within Syria's borders, but which has involved state and non-state actors, including from neighboring countries Turkey, Iran, and Iraq, notoriously ISIS, and become a proxy war representing a clash of competing geopolitical interests on the part of global superpowers, with complex political-ecological causes and consequences (Gürcan 2019).

Concerningly, environmental destruction is also weaponized as a military strategy, causing the loss of livelihoods and forcing the displacement of local communities. A recent example was the destruction of agricultural production, especially olive groves, to coerce civilian populations as a prelude to the 2018 invasion of the Kurdish canton of Afrîn, now part of the Turkish occupation of northern Syria, together with ongoing disruption of water supplies. Dinc's chapter in this collection notes the alleged complicity of armed forces in the deliberate destruction of forested areas by fire to destroy tree cover as a counterinsurgency measure in Turkey's Kurdish region, while attacks on Kurdish dissidents have also been cited in Hassaniyan's chapter as an explanation for unchecked forest fires in Iran.

Finally, conflict has grave humanitarian and ecological outcomes. Negative humanitarian consequences include direct harm to civilians and combatants, displacement due to ethnic cleansing, landmines, and coercion to make way for military infrastructure, and also collateral or intentional damage to agricultural production and archaeological heritage. In addition to the direct effects of weaponry, outbreaks of hostilities clearly inflict damage upon local environments in other ways, through the firing of oil wells, the contamination of watercourses, and deteriorating air quality (in extreme cases due to the use of chemical and biological weapons), the legacy of toxic materials and debris—including unexploded ordnance scattered upon the landscape—the consumption of resources in military operations, and the energy use embedded in the processes of reconstruction required to replace ruined infrastructure. Another specific ecological impact of war is violence against other species, directly or through the destruction of contiguous habitats, leading to reduced biodiversity.

The prospects for the Kurdish ecology initiatives, therefore, cannot be separated from security issues affecting the KFM. Direct conflict continues, particularly linked to the oppression of Kurdish populations in Turkey and Syria, and the bombing of PKK positions based in Iraq's Qandil Mountains. The conflict between the Turkish state and the PKK resumed in 2016 after the breakdown of peace negotiations. As of 2021, the Syrian Democratic Forces are defending AANES from Turkish forces and Turkish-backed factions of the Free Syrian Army, Salafist Jihadist, and other Islamist militias, including clandestine cells of ISIS fighters, and, also, Ba'athists loyal to Bashar al-Assad's regime.

Turkey, the hostile neighboring Kurdistan Regional Government in Iraq (the Barzani administration is a political rival of the PYD), Syria's Ba'athist regime, and Iran all further coerce the population of the AANES through the imposition of economic embargoes. Such blockades not only prevent people from crossing borders and impede trade in basic commodities but also extend to medical supplies and other humanitarian aid, reconstruction materials, renewable energy technology, such as solar panels, and even equipment for the removal of landmines to protect civilians from the lethal legacy of ISIS operations (Nasrat 2017, 46–47). The importation of perishable goods, including foodstuffs, and hardware, is also hindered by the fluctuating but decreasing exchange value of the local currency, the Syrian Lira. The withholding of support by the United Nations and mainstream NGOs, due in part to their reluctance to deal directly with the non-state administration and to concerns regarding regional insecurity, also undermines social well-being for the area's estimated current population of 4–5 million residents (Rojava Information Center 2019, 53 and 13). In turn, conflict has led to significant upheaval, forcing Kurds and other ethnic minorities to either migrate from the region or to seek refuge within the AANES, their physical displacement often accompanied by psychological trauma.

In all these ways, the ecological situation in the Kurdish region and the Middle East is intimately connected to wider geopolitical factors. The role of external state actors and corporate interests in fueling military conflict through the supply of arms is particularly consequential. In this respect, the complicity of Western nations, corporations, and the shortcomings of organizations of the international community jeopardizes the future of the Kurdish-led democratic-confederalist approaches that defeated ISIS. The current conflicts in the Kurdish region are rarely adequately or accurately framed in the mainstream Western media; indeed, even the Syrian War is no longer regularly reported. Despite this absence, it is essential to position the current circumstances of the KFM, including its ecological dimension, within the global context. At the time of writing, the Syrian Democratic Forces have in part been holding their ground due to their success in attracting fighters from not only Kurdish communities but also other ethnic groups, including, significantly, Arab militias.

It is helpful to understand a perspective that is representative of the PYD, the dominant political grouping since the eco-socialist revolution in north Syria in 2012. Saleh Moslem, one-time co-chair of the PYD and current member of the Party's Co-Presidency Council, makes the case that the AANES is able to repel an attempted ground invasion by Turkish armed forces and the allied militias and mercenaries that make up the Turkish-backed Free Syrian Army but is imperiled by the use of air attack and UAVs or "drones," because Rojava has no air defense system. Observing that seventy-three planes were used to undertake the invasion of Afrîn, the implementation of a no-fly zone is, therefore, an urgent priority. Moslem acknowledges that such a move may seem unlikely in the context of political support for, and substantial arms sales to, Turkey, NATO's second-largest army. He argues, however, that measures to curb the Turkish state's ambitions are in the international community's strategic interests, since Turkey's influence is a destabilizing influence across the wider region of the Middle East, North Africa, and the Caucasus, making it "dangerous for everybody," given its current interventions in Libya, the East Mediterranean, and in exploiting tensions between Armenia and Azerbaijan.[1]

Alongside the low volume of mainstream commentary on Kurdish issues in the West, researchers, such as Andrew Feinstein, and anti-arms trade activists, including members of the Kurdish diaspora, have sought to amplify awareness of the arms industry's role in destabilizing international relations, supporting internal repression of populations, and proliferating corruption across the globe. Demonstrations, such as those at London's annual Defence and Security Equipment International (DSEI) Arms Fair, one of the largest events of its kind in the world, endeavor to stimulate debate about, the role of an industry otherwise largely absolved of significant public scrutiny. The supply of arms to Turkey is a particular focus of attention.[2]

Dams and Water Scarcity

The reduction of available water supplies is a critical social and ecological problem, leading to reduced harvests and with consequences extending to regional instability. Water shortage is attributed to the Turkish dam-building program and control of the Euphrates River leading to reduced flow. It has been suggested that the disruption of the agricultural economy, due to reduced water supplies, outweighed the impact of climate change as a causal factor in the Syrian conflict (Karnieli et al. 2019, 1564). Kurdish sources have reported news of repeated attacks on water infrastructure as part of ongoing disputes, even during the coronavirus pandemic (Wilgenburg 2020). In addition to such weaponizing of water, forebodings about the sustainability of water supplies and usage are one of the leading natural resource concerns throughout the Middle East.

Soil Degradation

Soil degradation is a problem across the semi-arid and arid region of the East Mediterranean and west Asia. Unsustainable agricultural methods and other human activities that threaten soil health include increased erosion due to irrigation practices and deforestation, disruption of the nutrient cycle, and increased salinity. The intensive application of chemical fertilizers and pesticides is also accompanied by harmful ecological consequences, such as the contamination of groundwater sources through runoff, and the increased release of greenhouse gases, such as nitrous oxide. Given its fractured political administration, it is not of course possible to assess the issue of soil degradation in the Kurdish region as a discrete geographical entity. Nevertheless, relevant research relating to the wider area shows the challenges involved in the conservation of the soil upon which all human livelihoods, and the lives of other species, depend.

Negative ecological impacts have recently been evidenced in data from detailed studies of cultivation on the Harran Plain, an area spanning part of southeastern Turkey and reaching down to the Syrian border (Bilgili et al. 2018; Darama, Yılmaz, and Melek 2021). Research by Bilgili et al. documents that increased irrigation and chemical fertilizers initially brought about increases in the yields of wheat and cash crops, such as cotton. Nevertheless, it became apparent that such plenitude came at the cost of longer term viability. Yields began to reduce when soil quality diminished. In response, farmers applied increased quantities of chemical fertilizers to maintain nitrogen levels leading to further disturbance of the nitrogen cycle and contamination of groundwater sources caused by leaching (Bilgili et al. 2018, 11). The consumption of chemical fertilizers has continued to increase in Turkey over the past four decades, although overall inputs per hectare remain lower than those for Western nations (Motesharezadeh et al. 2017, 3). Bilgili et al. further suggest that earlier increases in agricultural output following irrigation by the GAP (Southeastern Anatolia Project) stimulated economic growth, accelerating development and urbanization which in turn reduced the land available for agriculture and heightened demand for water (2018, 11). Darama, Yılmaz, and Melek conclude that climate change also exacerbates droughts and is associated with the "serious matter" (2021, 12) of soil erosion on the Harran Plain.

The need for soil conservation is equally pressing on the Syrian side of the border. As Ayboğa, Pimbert, and Aslan observe in their chapters in the present collection, when the Ba'athist Syrian Arabic Republic controlled the Kurdish areas of north Syria, it imposed industrialized food production based on monoculture. This suited an economic and political agenda that prioritized development in other parts of the nation. Under this form of internal colonialism, food staples were also distributed to other parts of Syria and for export. There were severe restrictions on the cultivation of fruit, vegetables, and tree planting to prevent agricultural self-sufficiency that could support local autonomy (Knapp, Flach, and Ayboğa 2016, 192). According to Knapp, Flach, and Ayboğa, the area lost its forests even earlier in the early twentieth century, when trees were felled to supply timber for the Berlin to Baghdad Railway and to make way for agricultural production (2016, 212). In addition to this legacy, the economic embargo upon the current autonomous administration makes it difficult to import seeds and fertilizers. There are tensions between the objectives to diversify crop production and uphold principles of agroecology, while controlling prices so that basic foodstuffs are afford-able, and at the same time prioritizing the need to allocate many resources to social defense in the context of a war economy. Compromises include the domestic production of chemical fertilizers, justified as a means to achieve food sovereignty (Cemgil and Hoffman 2016, 71). Unfortunately, there has been a shrinkage of cultivated areas during the period of conflict throughout Syria, with a likelihood of "rapid and continued increase in erosion and deg-radation of agricultural soils," creating further social and ecological problems (Mohamed, Anders, and Schneider 2020, 22). More positively, research by Lanlani et al. suggests "significant reasons for hope and promise" (2018, 17), based on experiments with the implementation of conservation agricul-ture in Syria. These tentative findings are based on localized trials adopting agroecological solutions, such as intercropping and polyculture, low or no-tillage, and the use of organic mulches to conserve and improve soil quality. Jasim (2017) also finds enthusiasm for organic farming and "independent" agriculture, with reduced reliance upon chemical inputs and varieties of seeds supplied by multinational corporations in other areas of Syria which have contested regime control.

Loss of Biodiversity

Beyond concern for water and other natural resources to meet human needs, from an ecological perspective, it is also important to take into consideration the well-being and resilience of other species and to respect their intrinsic value. The "Principles and Objectives of the MEM" (Mesopotamian Ecology Movement 2016) is a document that sets out an expansive program for ecol-ogy in a society run according to social ecological principles. It recognizes the inherent worth of nonhuman species and their habitats to an extent that challenges the anthropocentrism in dominant attitudes to the natural world. The statement articulates an ecological view that rejects a rigid binary opposition between humanity and the rest of the biosphere, upholding the ecologically inclusive notion that "all other living creatures have also rights according to their species and variation." The document represents an impres-sive, and refreshing, aspiration to positively reconfigure the relationship between humanity and the living world according to the principles of social ecology. The egalitarian, non-speciesist articles set out in the "Principles and Objectives" are also reflected in the ethos of the ecology councils. Members of the Ecology Assembly in the city of Batman stated: "We don't just believe in the importance of human lives; we believe in preserving the existence of non-human life forms too" (Interview in Egret and Anderson 2016, 176). While species' decrease and loss is a concern throughout the Kurdish region, there are also contrasts between Turkey's Kurdish areas, which retain rich habitats to support biodiversity, and north Syria where ecological integrity is seriously depleted. Nevertheless, "Make Rojava Green Again" records the creation of the Hayaka and Mizgefta Nû nature reserves, supported by ambitious reforestation programs to which the campaign has been contributing (Internationalist Commune of Rojava 2018, 74–75).

Shifting patterns of land use have disrupted ecosystems through habitat loss and pollution in recent decades. The reconfiguration of land systems has resulted in damage to terrestrial habitat types, such as forests and wetlands, reducing ecological integrity and biodiversity. For example, a study by Azad Henareh Khalyani et al. into the driving forces behind decreasing tree cover in the Zagros Mountains, along the Iranian and Iraqi border, identifies the principal causes as the conversion of land for agriculture and mining to meet demand from urban populations, with climate change as a further contributory factor. Their findings indicate a 69 percent reduction in forest cover from 1972 to 2009 (Khalyani et al. 2013, 326). The situation of another species-rich environment, marshland, is also precarious. Research into the shrinking wetland habitat of the Mesopotamian Marshes in the Euphrates and Tigris River basin identifies drainage for farming and upstream dams as dominant causes, along with pollution from agricultural chemicals, industrial and domestic waste disposal, oil exploitation, and military operations as major sources of environmental degradation (Fawzi and Mahdi 2014, 8).

Data downloaded from the IUCN (International Union for Conservation of Nature) Red List indicates that of 1540 species listed for the broad Kurdish region (see Figure 1.1), 143 have been accorded a status of either "Critically Endangered" or "Endangered" internationally. This figure consists of four mammals, six birds, ten reptiles, thirty-three fish, four invertebrates (invertebrates are not extensively surveyed on the Red List), and eighty-six plant species.[3] It is striking that in nearly all cases, this is attributed to habitat loss and damage due to anthropogenic causes. For terrestrial species, the most common reason for decline is "human induced habitat loss and degradation caused by overgrazing." Other threats include road construction, deforestation and logging, mining, various forms of infrastructure development for tourism or the oil industry, climate change, and the poaching and "overcollection" of animals (for the meat and pet trades) and plants (such as rare bulbs for gardening), to be sold as commodities. For riverine and pelagic species, decreasing numbers are also attributed to human activity, identified as dam construction, pollution and water extraction, and game fishing. In this respect, it is salutary to note the multiple pressures upon a single species, the Euphrates soft-shelled turtle, since in its peril, it is sadly emblematic of the threats that the region faces: energy production and mining, over-harvesting of aquatic resources, war and civil unrest, dams and water management, and agricultural and other forms of pollution.

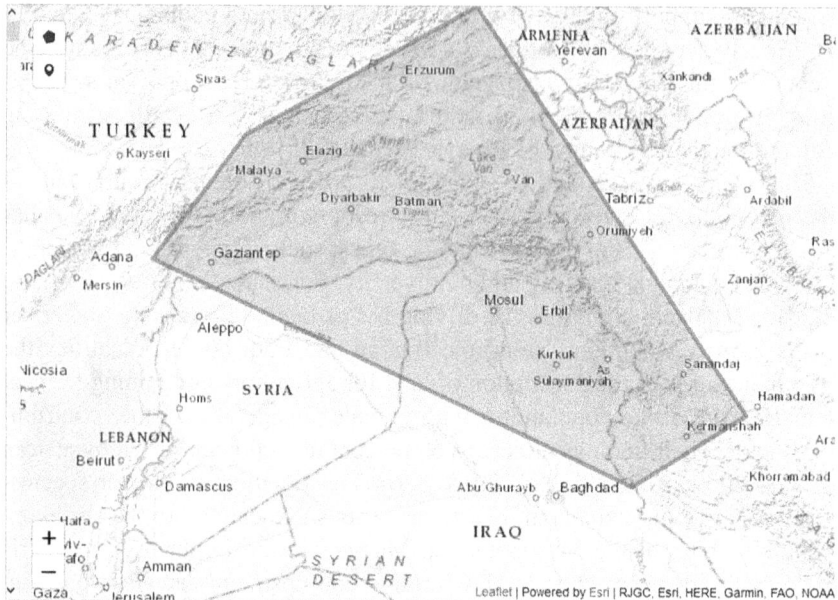

Figure I.1   Area Covered by Data Download from https://www.iucnredlist.org/ (March 3, 2021), approximately covering Kurdish majority region. Reference 2efa1ea8-8218-48a1-b1ba-e4f276159ded (approved March 17, 2021). *Source*: © IUCN Red List of Threatened Species. Generated by Stephen Hunt.

Mining

Of the four parts of the Kurdish region, mining activities are most established in Iran and Turkey. In Iran's Kurdish provinces, for example, copper, lead, and zinc are mined. Turkey (see Figure 1.2) holds the world's largest reserves of boron and thorium and is a major producer of copper, chromium, and gold. It continues to exploit reserves of highly polluting lignite (brown coal), with production concentrated in the country's western provinces, but also in the huge Elbistan coalfield in the west of the Kurdish region. As elsewhere in the world, mining operations can be contentious because of concerns about human well-being due to occupational diseases and industrial accidents and detrimental environmental impacts, such as the contamination of water sources and soil, and air pollution (Yesilnacar and Kadiragagil 2013). Opposition to mining can also be based on resistance to the reconfiguration of sites associated with sacredness, nature, and cultural memory as Gültekin discusses in the present collection, where he references the case of the campaign to protect Milli Köyü, a village with a Kurdish Alevi identity, from an expanding stone mine. Lignite also fuels the majority of Turkey's coal-fired power stations which provide more than a third of the nation's energy (more lignite plants

Figure I.2 Mining at Mardin in 2018. *Source*: © Mesopotamia Ecology Movement.

are planned) (European Association for Coal and Lignite 2021). The governing AKP has privatized and deregulated much of the former state-owned coal industry in recent years. Much extraction of metal ores and other minerals is controlled by Turkish companies with multinational links, such as Eti Bakır, and Ölmez Madencilik, which concentrates its activities in the Kurdish southeast. Nevertheless, it is estimated that smaller-scale or artisanal enterprises undertake 80 percent of mining activities (Demiral and Ertürk 2013). This sector is difficult to monitor for health and safety and ecological impacts, especially since it is documented that many of these smaller enterprises are either quasi-legal or operate outside the law (Kılıç, Özdemir, and Yavuz 2020).

The IUCN red list cites quarrying and mining activities as threats to the habitats of several wildlife species in the Kurdish region, including the critically endangered Bolland's blue butterfly and the vulnerable MacQueen's bustard. A revealing note from 2005 appended to the record for the Bolland's blue states that "an illegal chromium mine was established and did some damage in this area before it was stopped" (IUCN Red List data download, March 3, 2021). This demonstrates that the proliferation of unregulated mining can threaten particular endangered species, as well as adding to the general contamination of the environment.

THE ECOLOGICAL PILLAR OF DEMOCRATIC CONFEDERALISM

To understand the response to these multiple ecological threats, it is essential to take notice of the presence of the natural world in Kurdish culture and the integration of political ecology in the thought of Abdullah Öcalan and within the wider movement. Öcalan read several of Murray Bookchin's works, in around 2003, prompting him to contact the veteran philosopher of social ecology the following year (Ahmed 2015). While no correspondence could be sustained since Bookchin was already nearing the end of his life, Bookchin's ideas have been revived due to their influence upon thousands of movement activists who have since read them on Öcalan's recommendation (Ercan Ayboğa, Skype interview with Stephen Hunt, August 18, 2017). The PKK Assembly issued a statement, following Bookchin's death, asserting that he "broadened the consciousness of humanity: the ecological consciousness" and pledged to make him "live in our struggle" (PKK Assembly 2006).

While Murray Bookchin's ideas have been profoundly influential, this collection also challenges Eurocentric assumptions that attention to ecological matters only appeared in the Kurdish movement after Öcalan encountered Bookchin's texts and recommended them. Cihad Hammy, moreover, shows that Öcalan did not simply absorb and apply Bookchin's ideas, but had a far more dialectic relationship with his texts, developing social ecology and democratic confederalism for the twenty-first century by drawing upon a life of struggle dating back to the 1960s and personal study that has incorporated a broader range of philosophical ideas. Indeed, Havin Guneser observes that the Kurdish revolutionaries expressed concern for the living world a decade earlier, citing for instance Öcalan's report to the 5th Congress of 1994 (forthcoming in translation) where he talks about ecology (Guneser, "Öcalan's *Sociology of Freedom*," Scottish Solidarity with Kurdistan webinar, November 3, 2020). Nevertheless, Bookchin certainly supplied a coherent and advanced conceptual framework for political ecology.

The green shoots of Kurdish ecological awareness have also germinated from several other, often indigenous, seeds. Significant parts of Kurdish society have remained rural, retaining first-hand experience and knowledge of the living world, as Cioni and Patassini's chapter examines through horticulture and traditional construction techniques using natural materials as recently revived at Jinwar. PKK guerrillas have often of necessity lived within the natural world and relied upon it for much of their sustenance. In keeping with their revolutionary ideology, Kurdish ecology shares the insight, common to Bookchin and Öcalan's social ecology, that the problem of extreme environmental degradation is a systemic threat, an outcome of a paradigm characterized by hierarchy and domination, both within human society and over the living world. Since this accelerating threat can only be understood and addressed in that way, ecological well-being and social justice are united in the Kurdish emancipatory discourse.

The publication of Öcalan's *The Sociology of Freedom* in translation, in 2020, revealed to an English-speaking readership more about the extent to which ecological concerns ran as a sinuous green thread throughout his system of thought. The title echoes Bookchin's major text, *The Ecology of Freedom* (1983), a significant choice as a tribute that references and extends social ecological ideas, while contending other currents of revolutionary thought, including his own Marxist-Leninist heritage. For Öcalan, the processes of thought and accumulation that enabled emerging civilizations, manifest in the agricultural revolution and some of the earliest cities, in Sumer, Mesopotamia, were brought about at the cost of increasing exploitation of the biosphere, a severing of humanity's relationship to nature that was an originating source of the hierarchical domination and alienation, central to what he termed the "social problem." Such, he argues, are the deep roots of the present phase of advanced capitalism, with its monopoly of wealth and power, ultimately threatening humanity's future. For this reason, Öcalan stressed the natural environment's utmost theoretical and practical importance:

> I don't intend to add anything to the already existing disaster scenarios; but, according to our abilities, each of us must do and say what is necessary as responsible members of society. This is our responsibility and our moral and political duty, the very reason for our existence. (Öcalan 2020, 103)

Öcalan, in keeping with Bookchin, holds that the ecological crisis is the inexorable and inevitable outcome of the prevailing social and economic system of capitalist modernity, its deep civilizational roots having profound consequences for historical development up to the present precarious situation. A principal factor is the subject-object distinction which is at once the basis of the increasing disenchantment, domination, and exploitation of the natural world and of a hierarchical mentality that led to the exploitation of women—which, perhaps following Maria Mies's ideas, Öcalan terms "the oldest colonised group" (2013, 56)—and subjected peoples. Upon this foundation is built the accumulation and concentration of power and capital which, for Öcalan, are forces that both undermine social nature and hollow out the political sphere and destroy the integrity of the natural world, upon which real needs and livelihoods depend. To this end, the ideology of industrialism underpinning ongoing extractivism (leading to ecocide), colonialism, and hierarchical structures of domination and oppression (such as racism, classism, and sexism) and resulting in genocide and "societycide" (Öcalan 2020, 275) are inextricable, logical outcomes of capitalism's need for perpetual expansion within the framework of the nation-state.

As a counterbalance to the dynamic of capitalist modernity, which in order to accumulate also destroys, Öcalan identifies an alternative social current, which he terms democratic modernity. There are similarities between Öcalan's notion of democratic modernity and the Romantic anti-capitalism

theorized by Robert Sayre and Michael Löwy (2020). This is directly relevant since seminal environmental ideas appeared as part of the Romantic
revolt that began to question and counter the Industrial Revolution from the
eighteenth century onward, a focus of interest for Öcalan since it signified
what he considers to be the emergence of "the real crisis of urbanization"
(Öcalan 2020, 113). Perhaps unsurprisingly, Löwy explicitly recognizes that
the "romantic revolutionary culture" now ongoing in "Libertarian Kurdistan"
(2017) has importance for other social movements, which Öcalan identifies
as "anti-system forces" in *The Sociology of Freedom.* Such social movements most conspicuously appeared in the French Revolution and were
reinvigorated, for instance by the 1968 mobilizations from which emerged
the present-day movements for civil rights, women's liberation, ecology, and
indeed Kurdish freedom (and often understood as such by Kurdish freedom
sympathizers: Make Rojava Green Again 2019). Öcalan recognizes potential
for transformative change in these countercultures, yet acknowledges the
ever-present risk that capitalist neoliberalism can readily appropriate opposing forces if their supporters are not critically aware. He calls for a profound
philosophical shift, based on understanding that current existential problems,
including the development and proliferation of nuclear weaponry, "halting the rampant automobile madness," (206) climate change (100), and the
emergence of unsustainable mega-cities (115), share historic socio-economic
causes and cannot be addressed by technological innovation divorced from
a revolution in social power. At the same time, while Öcalan asserts that
"eco-economy and eco-industry would be taken into consideration in all
social activity" and are "one of the most fundamental dimensions of this
revolution" (2020, 253), he does not explain what these mean in practice.
This leaves us to speculate as to how we might define and assess what
technologies and modes of production would constitute "eco-industry." The
answer, perhaps, is to be found in Murray Bookchin's notion of "liberatory
technology" (1974), a description measured not so much by a particular technology's form considered in isolation but by a more holistic appraisal of its
socially embedded function regarding its emancipatory impact and ecological sustainability.

Öcalan stresses the importance of ecological thinking throughout his
writings from the mid-2000s, but does not (reasonably, considering the
circumstances of their composition in prison), provide a greenprint for a
transition to an ecological society, detailing how it would function and what
its technological basis might be. Thereafter, the KFM's adoption of ecological solidarity signaled a significant element of a paradigm shift in thinking.
Integrating the ecological dimension into the present-day Kurdish policies
and structures, however, is hugely challenging. As we have seen, the extent
of the ecological problems that confront all parts of the Kurdish region should

Figure I.3   Mountains of Dersim. *Source*: © Pinar Dinc.

not be underestimated. Yasin Duman, for example, a researcher undertaking extensive fieldwork interviews in Rojava, found that while the AANES has integrated ecological philosophy into its principles, ecology and eco-industry are less understood than other policies and difficult to adopt in practice (Duman 2019).

Nevertheless, without its ecological dimension, democratic confederalism would be an incomplete model. If any of the core elements of democratic confederalism were to be removed, the structural coherence of the paradigm would topple. Indeed, for a political model to be viable, it needs ecological sustainability, or more fittingly, in the mother tongue of green utterance, a permaculture, since the social economies upon which all livelihoods depend are reliant upon dynamic and abundant ecosystems. Current circumstances increasingly suggest that a rapid break with the global fossil-fuel-based economy is a precondition for geopolitical progress. Concerningly, world energy use continues to increase and, according to the industry's own figures, oil, gas, and coal still account for more than 84 percent of energy consumption (BP 2020, 4: table 1). The Kurdish paradigm and the courageous experiments underway in a region under heavy military attack and embargoes face constraints that are difficult to outmaneuver in isolation. While it offers way-markers for a different pathway in a time of crisis, it is also the responsibility of the rest of the world, therefore, to take up, embrace, adapt, and advance complementary visions for democracy, social justice, and ecological solidarity. The following chapters deal with some of the major issues facing the Kurdish region and beyond; since ecology is holistic, all are interconnected.

NOTES

1. All citations of Saleh Moslem's ideas in this paragraph are from "Your Freedom and Mine," Black Rose webinar, August 15, 2020. https://www.youtube.com/watch?v=Sj6AUMCnkFQ.
2. See also: https://caat.org.uk/resources/countries/turkey/uk-arms-sales-to-turkey/.
3. Data download from https://www.iucnredlist.org/ March 3, 2021. Reference 2efa1ea8-8218-48a1-b1ba-e4f276159ded (approved March 17, 2021).

REFERENCES

Abbas, Nahlah, Saleh A. Wasimi, and Nadhir AL-Ansari. 2016. "Assessment of Climate Impacts on Water Resources in Kurdistan, Iraq, Using SWAT Model." *Journal of Environmental Hydrology* 24, no. 10. http://www.hydroweb.com/journal-hydrology-2016-paper-10.html.

Ahmed, Akbar Shahid. 2015. "America's Best Allies against ISIS Are Inspired by a Bronx-born Libertarian Socialist." *Huffington Post*, updated December 18, 2017. https://www.huffingtonpost.co.uk/entry/syrian-kurds-murray-bookchin_n_5655e7e2e4b079b28189e3df?ri18n=true.

Arı, Emin Sertaç, Hakan Özköse, and Cevriye Gencer. 2016. "Ranking Turkish Cities and Regions for Air Quality Using a Multi-Criteria Decision-Making Method." *Polish Journal of Environmental Studies* 25, no. 5: 1823–30. https://doi.org/10.15244/pjoes/63172.

Ash, Konstantin, and Nick Obradovich. 2020. "Climate Stress, Internal Migration, and Syrian Civil War Onset." *Journal of Conflict Resolution*, 64, no. 1: 3–31. https://doi.org/10.1177/0022002719864140.

Bilgili, Ali Volkan, İrfan Yesilnacar, Kotera Akihiko, Takanori Nagano, Aydın Aydemir, Hüseyin Sefa Hızlı, and Ayşin Bilgili. 2018. "Post-irrigation Degradation of Land and Environmental Resources in the Harran Plain, Southeastern Turkey. *Environmental Monitoring and Assessment* 190: 660. https://doi.org/10.1007/s10661-018-7019-2.

Bookchin, Murray. 1974. "Towards a Liberatory Technology" [1965]. In *Post-Scarcity Anarchism*, 85–139. London: Wildwood House.

BP. 2020. *Statistical View of World Energy 2020*. 69th edition. https://www.bp.com/content/dam/bp/business-sites/en/global/corporate/pdfs/energy-economics/statistical-review/bp-stats-review-2020-full-report.pdf.

Bucchignani, Edoardo, Paola Mercogliano, Hans-Jürgen Panitz, and Myriam Montesarchio. 2018. "Climate Change Projections for the Middle East–North

Africa Domain with COSMO-CLM at Different Spatial Resolutions." *Advances in Climate Change Research* 9, no. 1 (March 2018): 66–80. https://doi.org/10.1016/j .accre.2018.01.004.

Cansiz, Sakine. 2018. *Sara: My Whole Life Was a Struggle: The Memoirs of a Kurdish Revolutionary.* Translated by Janet Biehl. London: Pluto.

Cemgil, Can, and Clemens Hoffmann. 2016. "The 'Rojava Revolution' in Syrian Kurdistan: A Model of Development for the Middle East?" *IDS Bulletin* 47, no. 3 (May 2016): 53–76. https://bulletin.ids.ac.uk/index.php/idsbo/article/view/2730/ HTML.

Dalby, Simon. 2018. "Climate Change and Environmental Conflicts." In *Routledge Handbook of Environmental Conflict and Peacebuilding*, edited by Ashok Swain and Joakim Öjendal, 42–53. Abingdon, UK: Routledge.

Darama, Yakup, Kutay Yılmaz, and A. Berhan Melek. 2021. "Land Degradation by Erosion Occurred after Irrigation Development in the Harran plain, Southeastern Turkey." *Environmental Earth Sciences* 80: 2011. https://doi.org/10.1007/s12665 -021-09372-5.

Demiral, Yücel, and Alpaslan Ertürk. 2013. "Safety and Health in Mining in Turkey." In *Occupational Safety and Health in Mining Anthology on the Situation in 16 Mining Countries,* edited by Kaj Elgstrand and Eva Vingård, 87–93. Gothenburg: Arbete och Hälsa, University of Gothenburg. https://core.ac.uk/download/pdf/163 36484.pdf#page=91.

Duman, Yasin. 2019. "Cultivating Autonomy in Rojava." Interview with Ashish Kothari, *Radical Ecological Democracy* website. December 15, 2019. https:// www.radicalecologicaldemocracy.org/redweb-conversations-series-cultivating- autonomy-in-rojava/.

Egret, Eliza, and Tom Anderson. 2016. *Struggles for Autonomy in Kurdistan & Corporate Complicity in the Repression of Social Movements in Rojava and Bakur.* London: Corporate Watch Cooperative.

European Association for Coal and Lignite. 2021. *Country Profile: Turkey.* Accessed May 13, 2021. https://euracoal.eu/info/country-profiles/turkey/.

Fawzi, Nadia Al-Mudaffar, and Bayan A. Mahdi. 2014. "Iraq's Inland Water Quality and their Impact on the North-Western Arabian Gulf." *Marsh Bulletin* 9, no. 1: 1–22. https://www.iasj.net/iasj/download/c3ccfed2a3968f78.

Feinstein, Andrew. 2012. *The Shadow World: Inside the Global Arms Trade.* London: Penguin.

Gleick, Peter H. 2017. "Climate, Water, and Conflict: Commentary on Selby et al. 2017." *Political Geography* 60: 248–50.

Gültekin, Ahmet Kerim. 2019. "Kurdish Alevism: Creating New Ways of Practicing the Religion." Working Paper Series of the HCAS "Multiple Secularities—Beyond the West, Beyond Modernities 18. Leipzig: Leipzig University.

Gürcan, Efe Can. 2019. "Extractivism, Neoliberalism, and the Environment: Revisiting the Syrian Conflict from an Ecological Justice Perspective." *Capitalism Nature Socialism* 30, no. 3: 91–109. https://doi.org/10.1080/10455752.2018.15 16794.

Hassaniyan, Allan. "Environmentalism in Iranian Kurdistan: Causes and Conditions for its Securitisation." *Conflict, Security and Development* 20, no 3 (June 8, 2020): 355–78. https://doi.org/10.1080/14678802.2020.1769344.

Hunt, Stephen E. [2017]. "Prospects for Kurdish Ecology Initiatives in Syria and Turkey: Democratic Confederalism and Social Ecology." *Capitalism Nature Socialism* 30, no. 3 (2019): 7–26. https://doi.org/10.1080/10455752.2017.1413120.

Internationalist Commune of Rojava. 2018. *Make Rojava Green Again: Building an Ecological Society*. London: Dog Section Press and Internationalist Commune of Rojava.

Jasim, Ansar. 2017. "Agriculture and Food Sovereignty in Syria." Heinrich-Böll-Stiftung website. Translated from German by Bernd Herrmann. October 4, 2017. https://lb.boell.org/en/2017/10/04/agriculture-and-food-sovereignty-syria.

Karimi, Behrooz, Behnosh Shokrinezhad, and Sadegh Samadi. 2019. "Mortality and Hospitalizations Due to Cardiovascular and Respiratory Diseases Associated with Air Pollution in Iran: A Systematic Review and Meta-Analysis." *Atmospheric Environment*, 198 (February 1, 2019): 438--47. https://doi.org/10.1016/j.atmosenv.2018.10.063.

Karnieli, Arnon et al. 2019. "Was Drought Really the Trigger Behind the Syrian Civil War in 2011?" *Water* 11, no. 8: 1564--75. https://doi.org/10.3390/w11081564.

Khalyani, Azad Henareh, Audrey L. Mayer, Michael J. Falkowski, and Daya Muralidharan. 2013. "Deforestation and Landscape Structure Changes Related to Socioeconomic Dynamics and Climate Change in Zagros Forests." *Journal of Land Use Science* 8, no. 3: 321--40. https://doi.org/10.1080/1747423X.2012.667451.

Kılıç, Sadık, Caner Özdemir, and Kemal Yavuz. 2020. "Reflections of Desynchronized Neoliberalism in Artisanal and Small-Scale Mining: Evidence from Zonguldak, Turkey." *The Extractive Industries and Society* 7, no. 4 (November 2020): 1385–91. https://doi.org/10.1016/j.exis.2020.08.008.

Knapp, Michael, Anja Flach, and Ercan Ayboğa. 2016. *Revolution in Rojava: Democratic Autonomy and Women's Liberation in Syrian Kurdistan*. Translated by Janet Biehl. London: Pluto.

Lalani, Baqir, Bassil Aleter, Shinan N. Kassam, Amyn Bapoo, and Amir Kassm. 2018. "Potential for Conservation Agriculture in the Dry Marginal Zone of Central Syria: A Preliminary Assessment." *Sustainability* 10, no. 2: 518. https://doi.org/10.3390/su10020518.

Löwy, Michael. 2017. "Libertarian Kurdistan: It Matters for Us Too!" *Verso Blog*, May 4, 2017, https://www.versobooks.com/blogs/3201-libertarian-kurdistan-it-matters-for-us-too.

Majid, Salih N. 2016. "Analysis of Organic Functional Groups and Some Trace Heavy Metals in the Settleable Dust Particles (Dustfall) of Sulaimani City / Kurdistan Region-Iraq, *Journal of Zankoy Sulaimani* 18, no. 2 (Part A, June 20): 333–47.

Make Rojava Green Again. 2019. "Interview with MRGA after the III. International Social Ecology Meeting." Make Rojava Green Again website, October 5, 2019.

https://makerojavagreenagain.org/2019/10/05/interview-with-mrga-after-the-iii-international-social-ecology-meeting/.

Meena, Bashdar Ismael, and Karzan Abdulkareem Omar. 2015. "Chemical Analysis of Rainwater and Study Effect of Vehicle Emissions on Human Health in Kurdistan Region." *Oriental Journal of Chemistry* 31, no.2: 689–95. http://dx.doi.org/10.13005/ojc/310210.

Mesopotamian Ecology Movement. 2016. "Principles and Objectives of the MEM, Approved at the 1st Conference of the Mesopotamian Ecology Movement (MEM) on April 23+24, 2016, in Wan." Mesopotamian Ecology Movement website, accessed December 20, 2016. http://www.mezopotamyaekolojihareketi.org/index/icerik/principles-and-objectives-of-the-mem/ (site discontinued).

Mohamed, Mohamed Ali, Julian Anders, and Christoph Schneider. 2020. "Monitoring of Changes in Land Use/Land Cover in Syria from 2010 to 2018 Using Multitemporal Landsat Imagery and GIS." *Land* 9, no. 7: 226. https://doi.org/10.3390/land9070226.

Mohammadi, Amir, et al. "Health Effects or Airborne Particulate Matter Related to Traffic in Urmia, North-West Iran." *Journal of Air Pollution and Health* 4, no. 2 (Spring 2019): 99–108. https://doi.org/10.18502/japh.v4i2.1234.

Motesharezadeh, Babak, Hassan Etesami, Sepideh Bagheri-Novair, Hormoz Amirmokri. 2017. "Fertilizer Consumption Trend in Developing Countries vs. Developed Countries." *Environmental Monitoring and Assessment* 189, no. 3: 103. https://link.springer.com/article/10.1007%2Fs10661-017-5812-y.

Nadrian, Haidar et al. 2019. "'I am Sick and Tired of this Congestion': Perceptions of Sanandaj Inhabitants on the Family Mental Health Impacts of Urban Traffic Jam." *Journal of Transport and Health* no. 14 (September). https://doi.org/10.1016/j.jth.2019.100587

Nasrat, Sayed. 2017. "Embargo and Humanitarian Aid in Rojava." *Seton Hall Journal of Diplomacy and International Relations* 18, no. 1 (Spring): 45–53.

Öcalan, Abdullah. 2005. "Declaration of Democratic Confederalism in Kurdistan." http://www.freemedialibrary.com/index.php/Declaration_of_Democratic_Confederalism_in_Kurdistan (site discontinued). Accessible via Wayback Machine, March 30, 2021. https://web.archive.org/web/20160929163726/; http://www.freemedialibrary.com/index.php/Declaration_of_Democratic_Confederalism_in_Kurdistan.

Öcalan, Abdullah. 2013. *Liberating Life: Women's Revolution*. Cologne: International Initiative Edition.

Öcalan, Abdullah. 2020. *The Sociology of Freedom: Manifesto of the Democratic Civilization, Volume III*. Translated by Havin Guneser. Oakland: PM Press.

Öztürk, Öner. 2018. "The Kurdish Movement: Ecology Practices in Bakur." Video contribution to "Ecosocialist Experiences in the World: A Critical Debate."

PKK Assembly. 2006. "2006: PKK's Salute to Bookchin." Posted by Janet Biehl on *Ecology or Catastrophe* (blog), August 27, 2015. https://www.biehlonbookchin.com/pkk-salute-bookchin/.

Read, Mark R. 2019. "Climate and the Syrian Civil War." In *The Environment-Conflict Nexus: Climate Change and the Emergent National Security Landscape*, edited by Francis Galgano, 167–76, Advances in Military Geosciences. Cham, Switzerland: Springer.

Rockström, Johan, Will Steffen, Kevin Noone, Åsa Persson, F. Stuart III Chapin, Eric Lambin, Timothy M. Lenton, *et al.* 2009. "Planetary Boundaries: Exploring the Safe Operating Space for Humanity." *Ecology and Society* 14, no. 2: 32. http://www.ecologyandsociety.org/vol14/iss2/art32/.

Rojava Information Center. 2019. *Beyond the Frontlines: The Building of the Democratic System in North and East Syria*, version 4 (December 19, 2020): https://rojavainformationcenter.com/storage/2019/12/Beyond-the-frontlines-The-building-of-the-democratic-system-in-North-and-East-Syria-Report-Rojava-Information-Center-December-2019-V4.pdf.

Sanford, Robert. 2019. "Environmental Degradation as a Form of Persecution Against the Iranian Kurds." Washington Kurdish Institute website (June 25, 2019). https://dckurd.org/2019/06/25/environmental-degradation-as-a-form-of-persecution/.

Sayre, Robert, and Michael Löwy. 2020. *Romantic Anti-Capitalism and Nature: The Enchanted Garden*, Routledge Explorations in Environmental Studies. Abingdon, UK: Routledge.

Selby, Jan, Omar S. Dahi, Christine Fröhlich, and Mike Hulme. 2017. "Climate Change and the Syrian Civil War Revisited." *Political Geography* 60 (September): 232–44. DOI: https://doi.org/10.1016/j.polgeo.2017.05.007.

Shilani, Hiwa. "Kurdistan Officials Warn of Alarming Levels of Air Pollution in Duhok." *Kurdistan 24* website, July 20, 2019, https://www.kurdistan24.net/en/news/34c679d8-bf5e-4dd9-adaf-06bd755b6094.

Wilgenburg, Wladimir van. 2020. "Turkish-backed Group Cuts Water Supply to Northeastern Syria for Fourth Time." *Kurdistan 24* website, March 30, 2020. https://www.kurdistan24.net/en/news/e2a45738-8bad-41a1-a5f9-b65c7ad62e43.

Yesilnacar, M. Irfan, and Zekiye Kadiragagil. 2013. "Effects of Acid Mine Drainage on Groundwater Quality: A Case Study from an Open-pit Copper Mine in Eastern Turkey." *Bulletin of Engineering Geology and the Environment* 72: 485–93. https://doi.org/10.1007/s10064-013-0512-5.

Part I

THEORY

Chapter 1

The Value of Social Ecology in the Struggles to Come

Federico Venturini

Memory and imagination are the only forces that can bring about real change. Remember and imagine! (Chodorkoff 2011, 338)

MURRAY BOOKCHIN'S SOCIAL ECOLOGY

Faced with deep social and ecological crises, Murray Bookchin, the founder of social ecology, expressed his concerns for the future of humanity: "If we do not do the impossible, we shall be faced with the unthinkable" (Bookchin 2005, 107). As a political ecology, social ecology argues that today's social and ecological crises are inextricably rooted in the domination of human by human which, in turn, gives rise to the notion of domination over nature. These two crises are exacerbated by the present hegemonic social, economic, and political system: capitalism.

Various theories using the term "social ecology," or similar, have been developed since the 1950s, spanning multiple fields and disciplines within the social sciences, and attempts have been made to merge such diverse strands of social ecology thought with Bookchin's social ecology (Crowe and Foley 2013; Krøvel 2014). In this piece, however, I focus on Bookchin's milieu that today stands as a coherent political philosophy on its own terms (Biehl 1999; Curran 2006; Marshall 2008; White 2008; Morris 2012). Prominent among those within Bookchin's tradition are Daniel Chodorkoff, Brian Tokar, Ynestra King, Chaia Heller, John Clark, Janel Biehl (although she has broken with social ecology: Biehl 2011), Dimitri Roussopoulos, Matt Hern, and Eleanor Finley.

Social ecology is more than mere diagnostic for today's crises. Permeated by dialectical naturalism, social ecology presents two important projects. On the one hand, it challenges the current capitalist system and all forms of oppression including racism, ethno-centrism, and patriarchy. On the other hand, social ecology offers a reconstructive and revolutionary vision for an ecological post-scarcity society. Social ecology considers current societal struggles, which surface in both urban and rural contexts, while also addressing central questions of nature, science, and technology which arise in these contexts. What is more, social ecology proposes ways in which to construct a new society, promoting prefigurative political organizing strategies, such as affinity groups, the formation of directly democratic social movements, as well as educational and political projects that include communalism. Moreover, social ecology provides an ethics of complementarity which lays at the foundation of struggles to promote sex/gender liberation, horizontalism, egalitarianism, mutual aid, self-determination, and decentralization. This is the power of social ecology: it offers a coherent and holistic theory that, while critiquing current social and ecological crises, provides a reconstructive vision and a theory of action aimed at achieving a free and ecological society (Bookchin 1986a). Social ecology is, thus, a grounding theory to foster social change and the development of social movements.

Bookchin has been acclaimed as an anarchist thinker since the 1970s. However, since 1987, he has been involved with two major ideological clashes: first with deep ecology (Bookchin 1987) and then with anarchism (Biehl 2007). These debates have been tarnished by *ad hominem* attacks on Bookchin, creating a caricature of his thought (Price 2012). White (2008) argues that in the last period of his life, Bookchin retreated in his positions, narrowing the debate on social ecology and making it a more rigid theory. In 1999, he broke with anarchism (Biehl 2007). However, since Bookchin's death, his concept of social ecology remains of inestimable value, leading to its revival, most visible in its vital influence on the Kurdish resistance in Syria and Turkey (Hammy and Finley 2015; Stanchev 2015; Hunt 2017) and upon the municipalist movement around the world (Mansilla 2017; Rubio-Pueyo 2017).

Social ecology has deeply influenced both the Kurdish and the municipalist movements, yet often this core influence is merely mentioned or unacknowledged. Indeed, given the renewed interest, and the frequent caricature of Bookchin's social ecology (Price 2012), it is essential to give an all-round vision, highlighting its key features and prospects. Social ecology is considered as a form of anarchist political ecology (Clark 2005 and 2012; Roussopoulos 2015), although some features distinguish it from other parts of this tradition (see White 2008). Social ecology's stress upon the connection between human and ecological crises (Robbins 2012), particularly links it to the political ecology tradition. More specifically, social ecology is a "political ecology with a libertarian and communitarian social perspective" (Clark 2005: 1569), an approach that marks it as the precursor of an anarchist political ecology.

The object of this chapter is twofold and is structured as follows. First, it highlights key ways that social ecology can contribute toward analyses and practices of social movements all over the world, such as the Kurdish freedom movement, in the struggles to come. To do this, it introduces key concepts of social ecology: its core concepts, dialectical naturalism, social ecology praxis, communalism and, finally, its idea of eco-communities. The second part of the chapter shows how to reconcile social ecology with the anarchist tradition. To this end, I focus on the relationship between anarchism and social ecology, before critically assessing communalism. Finally, I offer some conclusions, highlighting the necessity to build a culture of resistance.

BETWEEN FREEDOM AND DOMINATION: THE MOTOR OF SOCIAL ECOLOGY

Social ecology is based on challenging social hierarchy and social domination, with a theoretical elaboration that attempts to go beyond Marxist ideas about social class and the state. This elaboration is based on the idea that "the domination of nature by man stems from the very real domination of human by human" (Bookchin 2005, 65). Crucially, "nearly all our present ecological problems arise from deep-seated social problems" (Bookchin 1993, np): environment and society are inextricably linked to each other. The pressing planetary environmental problems (Rockström et al. 2009) that demand solutions can be addressed only by facing social problems. As stated in Price's recent reassessment of Bookchin's work (2012), social ecology is fundamentally grounded upon an understanding that hierarchy and domination as such "were in place *before* the emergence of the surplus. They are thus in the Bookchin programme *not* the inevitable by-product of the move to an economic world; they are not, perhaps most importantly, the by-product of a human project to dominate a stingy, harsh natural world" (157), but are born further back in human history with the emergence of gerontocracy. This understanding is the central "motor" of social ecology (Price 2012) that is derived from a re-examination of human civilization (Bookchin 2005).

This idea deviates from the classical Marxist tradition and uses concepts that help us move beyond an analysis based solely on economic relationships toward a deeper understanding of domination. For this reason, Bookchin

(1986a) stresses the importance of concentrating on domination and hierarchy and away from a focus on class and exploitation.

The organic world is defined as "an evolutionary process" (Bookchin 1995a, 17) in which two different kinds of nature, to which human beings belong, coexist: "first nature" and "second nature," where the former is constituted by mere biological evolution and the latter to human social evolution (Bookchin 1995a, 2005). Second nature developed from first nature, not in opposition to it, where humans are "nature rendered self-conscious" (White 2008: 109). Many Western thinkers offered a dichotomy between nonhuman nature and human society (Pepper 2003), while for social ecology humankind plays a multifaceted role within nature, being a unique expression of it, still part of it, but often acting in an antagonistic way toward it.

In this way, social ecology critiques both the tendencies of anthropocentrism (where humanity is considered superior to nature) and deep ecology (where humans have to return to nature), understanding the uniqueness of human progress in parallel with natural evolution and proposing an organic point of view in analyzing the problem (Bookchin 1996; Staudenmaier 2005).

Social ecology focuses on a rational analysis of "second nature" and the origins of social hierarchy and domination within it. This analysis starts with early human communities; what Bookchin calls "organic societies," "spontaneously formed, non-coercive and egalitarian" (White 2008, 36). Other important features of organic societies were the absence of the idea of human destiny to rule nature, the common usufruct of land, the presence of an irreducible minimum for everyone (e.g., in terms of food and shelter), the equality of the unequal, commitment to freedom, and cooperation. In these societies, humanity is thus viewed as part of nature, the concept of domination is not yet developed, and women and men lived in balance. In Bookchin's interpretation, the first form of domination emerged from organic societies: that of the elders over the young. Following this analysis, "the notion that all ecological problems are social problems, [. . .] stem[s] in the first instance from the emergence of hierarchy in its nascent form in the gerontocracy" (Price 2012, 157). On this point, Bookchin is clear, and he also stresses that the concept of domination of nature is a human construct that does not exist in first nature: "what we talk about when we speak of 'the domination of nature' is an ideology, not a fact. 'Nature' can no more be 'dominated' than an electron or an atom" (Bookchin 1995d, np). Moreover, according to this interpretation, social evolution is an effect of "changes in social forms and relationships and not because of the natural reaction to a harsh and necessitarian natural world; [thus] hierarchy as such is perhaps not the inevitable by-product of humanity's historical move through civilisation" (Price 2012, 157).

While Bookchin's anthropological accounts have been problematized by some authors (White 2008), it is important to consider, however, the value of social theories based on ethnographic or anthropological work aimed at outlining and describing alternative possibilities (Graeber 2004). In this regard, Price (2012), while acknowledging the necessity of clarifications, underlines the "coherence and unity" (195) of his social history.

Overcoming domination in social forms and relations is thus fundamental for building a new society based on freedom. This call emphasizes that changing our social structures can also dialectically change our relationship with nature. With this dialectical process, Bookchin goes beyond the classical Marxist concept that emphasizes the contrast between nature and human, or indeed between the rural and the urban. In his analysis, this contrast has shaped the development of our society through history, exacerbating exploitation. Morris (2012) adds that "Marx [. . .] was preoccupied with the preconditions of freedom (technological development, material abundance) not with the conditions of freedom (decentralization, the formation of communities, direct democracy, and technologies and urban life on a human scale)" (245) and social ecology aims to fulfill this gap, analyzing the conditions of freedom. To make a substantial change, for a real ecological society, a process of profound reconciliation between nature and human is necessary, pointing at the birth of a new nonhierarchical society based on concepts of freedom and cooperation, going beyond first and second nature (Bookchin 1996). The actualization of human potentialities can only be achieved after the relation between first and second nature has been established on a non-exploitative basis.

Bookchin's (1995b, 2005) analysis of history focuses on domination and freedom and highlights different times in which domination developed or aspects of freedom flourished. With these historical accounts, he not only underlines the various legacies of domination and freedom but also underscores its immanent emancipatory potentiality.

DIALECTICAL NATURALISM, RECONSTRUCTIVE ETHICS, AND LIBERATORY POWER

Dialectical naturalism is the grounding philosophy of social ecology. Skeptical of the epistemological turn of postmodernism, eco-mysticism, and primitivism, social ecology emphasizes the importance of human reason and proposes ways of thinking that "encompass processes, developments, and the unfolding of phenomena in which potentialities, like seeds, initiate the becoming of a given thing or condition" (Bookchin 1995a, 233). In particular,

building on Hegelian dialectical idealism and Marx's dialectical materialism, Bookchin proposes dialectical naturalism as an ecological thinking, as a
way of exploring reality and potentiality, that is, what it is and what should
be, building a "dialectical or organic way of thinking" (Morris 2012, 242).
Potentiality involves, in a call for action, "a sensitivity to the latent possibilities that inhere in a given constellation of phenomena, not a surrender to
predetermined inevitability" (Bookchin 1986a: 13).

In a domain in which false realism rules, ethics are ignored. In opposition to this, dialectical naturalism aims toward ethics that are "neither absolutist nor relativist, authoritarian nor chaotic, necessitarian nor arbitrary"
(Bookchin 1986a, 11). To offer this much-needed guidance, dialectical
naturalism explores the first nature and human evolution that brought us to
the development of second nature. It starts from the idea that first nature is
not just governed by a vulgar interpretation of Darwinian natural selection.
Bookchin, following the mutualistic naturalism of Kropotkin (Macauley
1998), understands natural selection to operate on the principles of diversity
and cooperation (Bookchin 1996, 2005). As summarized by Heller, first
nature is "a dialectical process of unfolding that is marked by tendencies
toward ever greater levels of differentiation, consciousness, and freedom"
(1999, 136).

For social ecology, despite first nature not being a locus wherein finding
inspiring examples for human behaviors is possible, "it is the 'matrix' for
an ethics, and ecology can be a 'source of values and ideals'" (Marshall
2008, 610-611). From first nature can be dialectically extrapolated principles for a nature-informed ethics (Bookchin 1996), an understanding
of ways in which human action can self-consciously develop human
potentialities.

Under this definition of ethics, "whether a society is 'good' or 'bad,'
moral or immoral, for example, can be objectively determined by whether
it has fulfilled its potentialities for rationality and morality" (Bookchin
1996, 24). Bookchin (1986a) stresses how key principles in dialectical
naturalism are participation—crucial in the natural world in building eco-
communities—and differentiation—linked to ecological stability and freedom. Without participation, the individual is disempowered and atomized;
with no differentiation, the individual becomes a homogenized thing. To
conclude, dialectical naturalism, by justifying an opposition to the unchal-
lenged ethical roots of domination, posits the basis for a realizable ecological utopia aiming to fulfill the full potentialities of first and second nature
(Bookchin 1986a).

Social ecology thus provides a new ground for the anarchist project, justified not from an anthropological perspective, as Graeber (2004) suggested,

but from a biological-ethical point of view. This approach can support the idea that, even if a legacy of hierarchy and domination has developed throughout human history, domination is not an innate aspect of the human project. There is a powerful legacy of freedom too, as expressed by continuous eruptions of freedom.

For social ecology, freedom is defined as "the autonomous individual's freedom to shape material life in a form that is neither ascetic nor hedonistic, but a blend of the best in both—one that is ecological, rational, and artistic" (Bookchin 2005, 300). To understand how freedom can be affirmed over domination, the issue of "power" should be addressed. In opposition to the Foucauldian argument, prevalent within anarchist theory, that power can be destroyed, Bookchin asserts "Power cannot be abolished; it is always a feature of social and political life. Power that is not in the hands of the masses must inevitably fall into the hands of their oppressors" (2015, 143). Power can, however, assume different forms, such as the "power to destroy and power to create" (Bookchin 1988), where the former is an expression of human domination and the latter is an expression of freedom. Power to create is the power to build an alternative—a free and democratic utopia—in the sense that Bookchin called "a liberatory use of power" (1995a, 183) and Clark later termed "power of self-determination" (2012, 513). What needs to be eliminated is "hierarchy, domination, and classes, [as well as] the use of power to force people to act against their will" (Bookchin 1995a, 183). Bookchin thus makes a clear distinction "between power held by state institutions and power claimed by popular institutions or between institutions that lead to tyranny and those that lead to freedom" (Bookchin 1995a, 184). This is a position that echoes the dichotomy between power from above versus power from below—or power of resistance (see Sharp et al. 2000). In this analysis, we can aspire to distribute power, to eliminate its abuses. However, this claim is disputed. For instance, Holloway (2010), coming from a nonorthodox Marxist point of view, states that all past emancipatory examples of taking power have failed. Also referring to the Zapatista experience, he hopes for a society built by people "who do not exploit and do not want to exploit, [. . .] who do not have power and do not want to have power" (203). Answering this, from a social ecological perspective, Bookchin stresses that the "power that does not belong to the people invariably belongs to the state and the exploitative interests it represents" (2007, 50). Similarly, Legard argues that power has a neutral connotation and "is always a mix in between 'power-to' and 'power-over'" (2010, 70). From the perspective of social ecology, "power-to" needs to be distributed and made accessible, while "power-over" must be avoided.

THE PRAXIS OF SOCIAL ECOLOGY: DIRECT ACTION

Linked to anarchist praxis, at the core of the social ecology tradition lays direct action (Bookchin 2004), both as tactic and strategy. Direct actions are diverse, and they can span from grassroots work to more confrontational and militant actions. In its tactical form, direct action is "a method of abolishing the state without recourse to state institutions and techniques" (Bookchin 2004, xi). As strategy, Bookchin describes direct action as: "A mode of praxis intended to promote the individuation of the 'masses'" (2004, xi). Doing so, he stresses the importance of direct action in empowering people (Bookchin 1988, 2004). Through a process whereby the identity of the particular is asserted within the framework of the general, the praxis of direct action frees people from a homogeneous mass and puts their future in their hands "mak[ing] people aware of themselves as individuals who can affect their own destiny" (Bookchin 2004: xi).

Direct action is thus a moment of empowerment, in which people are enabled to take back control of the future of the city, "learning how to manage every aspect of our lives from producing to organizing, from educating to printing" (Bookchin 1988, 53). Thus, direct actions are not just a tactic but embryonic modes for a different way of living today, a way that prefigures a different tomorrow and fights for it. Further than stressing its ability to empower, Bookchin (2005) also makes a strong connection between direct action and the need to reaffirm citizenship in its primary sense: "I must emphasize that direct democracy is ultimately the most advanced form of direct action" (438). The next section will explore the political project of social ecology that will enable direct democracy: communalism.

A POLITICAL PROJECT: COMMUNALISM

In the late 1980s, Bookchin, heavily inspired by the experience of ancient Athens and the town assemblies in New England, proposed a coherent project for a new political system: communalism (see Bookchin 1995b, 2007, 2015; Biehl 1998; Roussopoulos 2015). Communalism is the final, and definitive, chosen title of a project that Bookchin (2007) previously called libertarian municipalism. Communalism has been positively reviewed by Harvey (2012) and is currently implemented by the Kurdish population (Hunt 2017). Communalism is conceived as both a revolutionary strategy in the present and as potential societal organization in a liberated future.

Those who share a perspective from social ecology are aware of the importance of the city for human development. Urbanization and citification, which are usually accepted as synonyms, are perceived as antagonistic (Bookchin 1995b). Only citification presupposes the idea of society, while the notion of "urban" refers to an amorphous environment that absorbs all available space. Social life and civilization are developed in cities thanks to the proximity of marketplaces to living quarters that fosters everyday social interactions. Citification enhances human life in two different domains, everyday life and the political, which together form the truly social life. However, the concepts of "civitas" and citizenship are lost in the modern metropolis, and the city is becoming the space where the state affirms its power and control.

Today's urban belts and megalopolis cannot be considered "real cities" (Bookchin 1986b, 1995b). Urbanization is, paradoxically, destroying the very reason for the birth of cities: the potentiality for individuals to build communities that can fully develop society's possibilities. Moreover, urbanization is contaminating those values and institutions born from civic relationships, replacing them with values of anonymity, homogenization, and institutional gigantism (Bookchin 1995b). This is reflected by the increase of individualism, the dissolution of social or civic commitment to private life, the retreat of the intelligentsia into academia, and in the "consumption, not only production, [that] has become an end in itself" (Bookchin 1995b, 194). The possibility of a critical mass of people living together in cities created politics: the possibility to communally practice direct democracy.

Aware of the impossibility of achieving a completely self-reliant community, the solution is a confederation of self-governed municipalities, a "Commune of communes":

A network of administrative councils whose members or delegates are elected from popular face-to-face democratic assemblies, in the various villages, towns, and even neighborhoods of large cities. The members of these confederal councils are strictly mandated, recallable, and responsible to the assemblies that choose them for the purpose of coordinating and administering the policies formulated by the assemblies themselves. Their function is thus a purely administrative and practical one, not a policy-making one. (Bookchin 1995b, 253)

Bookchin introduces this new system to allow people to return to the heart of political debate, suggesting an organization that should encourage public participation and direct democracy. In developing this, he refers to Proudhon and Kropotkin's (Macauley 1998) idea of "communes," led by principles of self-management, complementarity, and mutual aid, and he defines "decentralization," "statelessness," "collective management," and "direct democracy" as the principal characteristics of communalism (Bookchin 1986b, 1995b). To maintain these features, the delegates at the confederal level must be under the control of their assemblies.

The aim of communalism is to "advance a perspective for extending local citizen-oriented power at the expense and ultimately the removal of the nation-state by village, town, and city confederation" (Bookchin 1995b, 1). In this way, communalism emerges as a "new" political project, a "fundamental duality of power" (Bookchin 1995b, 10) in which confederated municipalities acquire power at the expense of the nation-state. One of Bookchin's aims is to propose "a self-conscious practice in which confederal municipalists can engage in local electoral activity" (Bookchin 1995b, 9). From an economic point of view, this approach proposes a new form of economy that goes beyond nationalization or collectivization (Bookchin 1995b; Staudenmaier 2002), toward a moral economy (Bookchin 1986a): communalism proposes the "municipalization of the economy and its management by the community as part of a politics of public self-management" (Bookchin 1986b, 181).

Moreover, concerning the organization of the commune, it is important to remember the difference between policymaking, which is a people's duty, and administration, which addresses technical and logistical problems and in which assemblies' participation is not entirely necessary. Thus, the constitution of administrative bodies of municipalities is considered possible. However, Bookchin acknowledges that a complete autarchy would be harmful: "interdependence among communities is no less important than interdependence among individuals" (Bookchin 1995b, 237). Of course, singular decentralization, self-sufficiency, human-scale communities, and technology alone are not sufficient to create democratic social changes: only their combination ensures hope for a better future.

Local elections should also be used to foster dual power; a position that stimulated harsh debates (Biehl 2007). Classical anarchists have, for example, critiqued electoral participation which they reject as a form of statism, upholding principles of consensus decision-making over the majority system. The practice of communalism remains, nevertheless, always in tension with the state and suggests that the movement should avoid running candidates at the regional or national level because history has taught us that "state power is corruptive" (Bookchin 1995b, 11). However, Bookchin suggests participation in elections at the civic level as a strategy for implementing social change, believing in the possibility of intervening at a this level without being compromised by the central or local state. Elections at the municipal level could be seen differently: the municipality is the formal political arena closest to the people, and thus similar to the Greek polis.

ECO-COMMUNITY: TOWARD POST-SCARCITY AND NEW URBAN FUTURES

As explained, communalism is the political organization of the utopian eco-communities envisioned in social ecology and is by definition an ecological politics (Bookchin 1986a). Once their political structure is determined, it is also important to consider their feasibility in terms of resources. In this regard, Bookchin introduces the idea of "post-scarcity" (Bookchin 2004). Following the political ecology tradition, the current economic capitalist system is increasingly promoting worldwide social imbalances and environmental problems, causing a deep scarcity of resources (Swyngedouw 2004; Heynen et al. 2006). However, in Bookchin's project, the concrete availability of resources and a tremendous advancement in technics can contribute to building a different society, a post-scarcity society, where it is possible to imagine "the fulfilment of social and cultural potentialities" (Bookchin 2004, iv). Aware of the legacy of human domination over fellow humans and nature, social ecology is not calling for a technocratic approach. In its conceptualization, to reduce scarcity, deep social changes are required, rooted in the construction of a new relationship with nature in which humans are conscious of their impacts (Bookchin 1988; Hopkins 2008; Hern 2010). In this context, post-scarcity is not understood as merely a material status: the possibility of having a large enough quantity of goods for all people to survive at a decent level opens the way to a deeper possibility, the achievement of freedom (Bookchin 2004, xvi). The main requirements for building a post-scarcity society are autonomy, self-reliance, sustainability, and communality (Imboden 2011).

Having recognized the detrimental impacts on the environment of the employment of certain types of technology, Bookchin is, however, far from stating that technology is intrinsically "bad" or "good." Technology is indeed considered fundamental to reach positive and wide-ranging solutions to today's crises and to achieve a "balance between man [*sic*] and the natural world" (Bookchin 1965, 188). For social ecology, in the building of new communitarian social relations, moral technologies, conceived as the "good" technologies based on social ecology ethics (Bookchin 1996), will permit the existence of eco-communities, built on a renovated human scale (Bookchin 1995b).

ANARCHISM AND SOCIAL ECOLOGY

Both a social ecology perspective and social-movement practices place high importance upon direct democracy, direct action, and prefiguration, which are also key elements in the anarchist tradition. To support and analyze current social-movement practices, however, social ecology furthers its position in terms of consensus decision-making, militant direct action, and grassroots work. To do so, social ecology re-establishes a relationship with its anarchist roots.

Until the beginning of the 1990s, Bookchin indeed favored grassroots initiatives that emerged from social movements (Bookchin 1988, 2004), defining them as "emerging 'free space' for popular, often libertarian, civic entities, and the civic bases for a new body politic" (Bookchin 1988, 186). He also speaks about "alternative organizations, technologies, periodicals, food cooperatives, health and women's centers, schools, even barter-markets [and] local and regional political coalitions" (Bookchin 1986a, 152) as key examples that challenge capitalism. However, in his later work, Bookchin modified his approach and criticized the effectiveness of these experiences and concentrated his efforts on developing communalism (1995b, 2007, 2015).

Bookchin explicitly disagrees with communitarian and social experiments, such as cooperatives, social clubs, and neighborhood centers, common solutions posed to the capitalist system (Bookchin 1995b). His critique rests on the limitation of their success and the deterioration of their social dimensions caused by "the pressure of competition or simply greed, [which turns these initiatives] into corporations in their own right" (Bookchin 1995b, 2). He observes that every business seems to necessitate molding itself to the imperative to "grow or die," if it wants to survive in the current system. Bookchin believes that these experiments invariably disappear or are incorporated into the capitalist system. Another signal of his opposition to these solutions can be found in Bookchin's discussion (1995c) of so-called "lifestyle anarchism," which had especially spread in the United States in the 1990s, in contrast with what he called "social anarchism." Lifestyle anarchists, he argues, are those who dress in an anarchistic style or live in certain ways, but do not align their activities with the development of a revolutionary project. Additionally, he makes a connection between lifestyle and individualism: "individualist anarchism remained largely a bohemian lifestyle, most conspicuous in its demands for sexual freedom ('free love') and enamored of innovations in art, behavior, and clothing" (Bookchin 1995c, 8). He also opposes primitivist and postmodern forms of anarchism, especially these notions posed by anarchist philosophers, John Zerzan and Hakim Bey. Instead, Bookchin proposes a return to an anarchism with a strong "social framework." He furthermore, criticizes consensus decision-making processes. In the present day, however, practices of consensus are spreading (Marshall 2008; Trapese Collective 2007; Seeds for Change 2013), as demonstrated by social movements throughout the world (Sitrin and Azzellini 2014) and the Kurdish experiences (Knapp, Flach, and Ayboga 2016). It does not follow that, as Bookchin feared

(1995d), they lack internal organization. Despite some well-argued points, such as his justified critique of the emphasis on "lifestyle" present in the anarchism movement, this work is the weakest part of Bookchin's production (Price 2012).

However, in Bookchin's earlier work, there is a valuable analysis of the role of social movements that deserves to be recovered. Many anarchist thinkers try to articulate the tension between personal autonomy and social freedom: these two aspects cannot be separated and must be interconnected. Replying directly to Bookchin, Clark (2013) states that "the bridge [between social anarchism and lifestyle anarchism] is crossed many times each day by those who practice the anarchist ideal of communal individuality in their everyday lives" (192). Under this bridge, for social ecologists as well, social movements can be a privileged actor for social change. Indeed, social movements have expressed a need to inspire social transformation throughout history (Chodorkoff 1980, 2014). Underlining the importance of social movements in supporting and grounding urban struggles, Chodorkoff (2014) outlines strategies to move from the stagnant aftermath of the 2011 square occupations to the construction of "*permanent* autonomous zones" (175), to "solidify" the Occupy movement (Imboden 2012). For social ecology, the occupation of public squares represents an important example of direct democracy; however, we should consider that while "directly democratic processes are in the movement context they do not constitute direct democracy, they constitute movement democracy" (Chodorkoff 2014, 174–175). In other words, what was reached is not a "real" direct democracy, but rather an inspiring example, which instilled understanding and practice of deeper democracy into participants and witnesses. The development of this practice and understanding within public life creates a kind of proof that direct democracy is possible, at least within movement contexts; now is the moment to expand, to put this in practice in our everyday lives, and to build dual power.

CRITICALLY ASSESSING COMMUNALISM

As shown in the previous section, one reason for the rupture between anarchism and social ecology has been the debate on electoral participation. The political project of social ecology, communalism, makes use of electoral strategies. However, communalism, as a political project, goes far beyond a simple electoral strategy and shares with social movements the need to build direct democracy practices. Moreover, as Harvey (2012) underlines, communalism can be the potential vision for organizing a future society that today's social movements need to establish an alternative to the current capitalist system. Communalism is, for Harvey (2012), one of the most advanced political projects of the radical Left. This project has been developed both as a vision for future social organization and as a pathway toward this vision.

The first aspect, which proposes a more ideal form of social organization to come, is strongly theorized; it has direct links to the communalist-like experiences in Turkish and Syrian Kurdistan and could be linked to the Zapatistas' communities. Despite several critiques and questions that still need to be answered (Light 1998; Harvey 2012; Clark 2013), communalism contains a sufficient level of detail as a future vision to which to aspire; it is supported by the evidence of living examples that can be studied as practical alternatives and viable forms of social organization. However, despite the fact that the Kurdish and Zapatista autonomous experiences have taken place as a result of years of consistent work, they have also been facilitated by unique events and spatial conditions: both took advantage of a vacuum in state power (especially the Kurdish movement), and both are located in remote areas with difficult access (particularly the Zapatistas).

While the KFM and Zapatistas are indisputable examples of alternative autonomous systems and offer invaluable lessons, they are not able to provide a recipe for social change exportable tout court to other countries, especially core and semi-periphery countries where the state is strong and stable. Moreover, in Syrian Kurdistan and Chiapas, there were neither previous municipal elections nor the creation of popular assemblies as suggested by communalism. Assemblies acquired a key role only after initial communalist-like projects were initiated. Moreover, in Turkish Kurdistan, political parties, which are supporting a communalist-like project, also participate in national elections, departing from traditional communalist doctrine that allows for participation only in local elections. Furthermore, Öcalan, inspired by Bookchin, did not apply communalism per se, but a version of the social ecological political project adapted to the specific circumstances of the Middle East: democratic confederalism (Heider and Kontny 2004). This highlights the need to adapt communalism, creating popular knowledge geared toward social change specifically tailored to the needs of the locality.

Harvey (2012) suggests that communalism can support social movements in building social change. The question is how to build a transition to communalism in other countries, where the starting conditions are different from the communalist-like examples we currently have. How can today's Left construct a path toward this communalist vision? It is this second aspect of communalism, as a path to be followed, where the "instruments" Bookchin offered may be insufficiently sophisticated to be put into practice today. I wish to stress the potential role of social-movement practices in the construction of communalist-like pathways, a role that has, so far, been neglected. Social ecology needs to learn from them.

Bookchin referenced specific historical examples to ground his general and prescriptive political project: democracy in ancient Athens, medieval free cities, and the tradition of town meetings in New England. While not to diminish the importance of these examples, making them the main, or even exclusive, pillars of communalism creates difficulty for this project to be understood and adopted in different contexts. Indeed, we should recuperate the first theorization of communalism:

> Communalism is not a fixed electoral dogma that depends upon the state, in whatever form, to initiate municipal institutional changes. In practice, it will obviously vary from locality to locality and country to country. (Bookchin 1995b, 12)

As Souza (2010) suggests, communalism should acquire a more flexible stand. For communalism to become an effective strategy for social change, participation in municipal elections, which is now a prescription, should instead be considered a recommendation to follow or not depending on the specific context of the struggle. Social ecology should relax the prescription that elections are the main, or only, path. It is necessary to recover the role of grassroots work (Martin 2010) in preparing the terrain for the widespread practice of direct democracy (see in this context the ideas of social ecologists Chodorkoff 2014; Roussopoulos 2015), affinity groups, and the role of direct action. This applies, as well, to communalism's prescriptive requirement to build popular assemblies.

This more flexible communalism advances the political project and responds explicitly to Souza's critique:

> Bookchin's solution is both limited and risky. The Argentine and several other experiences suggest that the challenge of institutional struggle cannot be confined to the boundaries of "libertarian municipalism." The complexity of state/"civil society" relationships and the seductive power of (neo)populism, "participatory management and planning" ("participatory budgeting," for instance), and so on, seem to require both *cautiousness and creativity*, probably more than Bookchin suspected. (2012a, 26)

The relationship between social movements and the state is multifaceted and complex. The survival of the communalist project depends upon taking these phenomena into account and the importance of social movements' actions.

CONCLUSIONS: BUILDING A
CULTURE OF RESISTANCE

In this chapter, I first explore the value of social ecology theory in the struggles to come for social movements. Second, informed by experiences of existing social movements, I explored the main shortcomings of communalism, offering solutions to continue to develop social ecology as an anarchist political ecology.

To conclude this piece, it is important to stress the reconstructive vision (Chodorkoff 2014) of social ecology, as a way of creating "a new society that questions all the presuppositions of the present-day society, [. . .] its inherent ability to see the future in terms of radically new forms and values" (Bookchin 1988, 280). A strategy for change should be based on a renewed culture of resistance inspired by dialectical naturalism's ethics and developed on key principles of "self-management, mutual aid, horizontal organization, and the fight against all forms of oppression" (Kuhn 2017, 6), where educational, grassroots projects and dual power initiatives are crucial.

Building a future ecological society "must be a holistic process that integrates all facets of a community's life. Social, political, economic, artistic, ethical, and spiritual dimensions must all be seen as part of a whole" (Chodorkoff 2014, 21). Working against all forms of domination consists not only of "mass" liberation but also of an individual search for sustainable relationships in which reconstructive ethics are adopted in daily life practices. This is a utopian approach that goes beyond social relations to the depth of individual life and vice versa; seeds of utopia emerge "not only in the factory but also in the family, not only in the economy but also in the psyche, not only in the material conditions of life but also in the spiritual ones" (Bookchin 1988, 76).

It is necessary to repair the relationship with nature and produce a new urbanism "that combines the features of urban and rural life in a harmonized future society" (Bookchin 1986b, xi), examples being the practice of permaculture (Mollison 1990; Hemenway 2105) and the Homeless Workers' Movement (MTST) in Brazil (Souza 2012b). This approach recovers two important features: human scale and the communitarian dimension (Bookchin 1986b), toward "decentralized eco-communities, each carefully tailored to the natural ecosystem in which it is located" (Bookchin 1986b, 161).

As Catherine, a character in Chodorkoff's novel *Loisaida* (2011), insists in this chapter's epigraph: memory and imagination are key. Every day we face incredible challenges posed by the capitalist system, with its destruction of the environment and persistent social inequality. On the one hand, we must remember and analyze the past, what we did, what "they" did, and whether it worked or not. On the other, we should use our imagination, or what Bookchin calls "the *creativity* of life" (1986a: 26), to explore new and alternative forms for human and ecological liberation, deploying them differently depending on the times and contexts (see, on social ecology and degrowth, Vansintjan 2019; Finley 2018, and on social ecology and the right to the city, Venturini, Degirmenci, and Morales 2019). Applying, reinventing, decolonizing, debating, and questioning social ecology as an anarchist political ecology are thus necessary practices to continuously reimagine itself in different contexts and times. Social ecology, then, is an invaluable tool for the crises of our times and the struggles to come.

REFERENCES

Biehl, Janet. 1998. *The Politics of Social Ecology: Libertarian Municipalism.* Montréal: Black Rose Books.

Biehl, Janet, ed. 1999. *The Bookchin Reader.* Montréal: Black Rose Books.

Biehl, Janet. 2007. "Bookchin Breaks with Anarchism." *Communalism*, no. 12: 1–20.

Biehl, Janet. 2011. "Report from The Mesopotamian Social Forum." *New Compass*, http://new-compass.net/node/265/.

Blumenfeld, Jacob, Chiara Bottici, and Simon Critchley, eds. 2013. *The Anarchist Turn.* New York: Pluto Press.

Bookchin, Murray [Lewis Herber, pseud.]. 1965. *Crisis in Our Cities.* Englewood Cliffs, NJ: Prentice Hall.

Bookchin, Murray. 1986a. *The Modern Crisis.* Philadelphia: New Society.

Bookchin, Murray. 1986b. *The Limits of the City.* Montréal: Black Rose Books.

Bookchin, Murray. 1987. "Social Ecology Versus Deep Ecology: A Challenge for the Ecology Movement." *Anarchy Archives*: http://dwardmac.pitzer.edu/Anarchist_Archives/bookchin/socecovdeepeco.html/.

Bookchin, Murray. 1988. *Toward an Ecological Society.* Montréal: Black Rose Books.

Bookchin, Murray. 1990. *Remaking Society: Pathways to a Green Future.* Boston: South End Press.

Bookchin, Murray. 1993. "What is Social Ecolog?" *Anarchy Archives*: http://dwardmac.pitzer.edu/Anarchist_Archives/bookchin/socecol.html/.

Bookchin, Murray. 1995a. *Re-enchanting Humanity: A Defense of the Human Spirit Against Anti-Humanism, Misanthropy, Mysticism and Primitivism.* London: Cassell.

Bookchin, Murray. 1995b. *From Urbanization to Cities: Toward a New Politics of Citizenship.* New York: Cassell.

Bookchin, Murray. 1995c. *Social Anarchism or Lifestyle Anarchism: An Unbridgeable Chasm.* Oakland: AK Press.

Bookchin, Murray. 1995d. "Reply to Moore." *Social Anarchism*: http://www.social anarchism.org/mod/magazine/display/25/index.php/.

Bookchin, Murray. 1996. *The Philosophy of Social Ecology*. Montréal: Black Rose Books.

Bookchin, Murray. 2004. *Post-Scarcity Anarchism*. Oakland: AK Press.

Bookchin, Murray. 2005. *The Ecology of Freedom*. Oakland: AK Press.

Bookchin, Murray. 2007. *Social Ecology and Communalism*. Oakland: AK Press.

Bookchin, Murray. 2013. *Social Ecology and Communalism*. Oakland: AK Press.

Bookchin, Murray. 2015. *The Next Revolution*. London: Pluto Press.

Bray, Mark. 2013. *Translating Anarchy: The Anarchism of Occupy Wall Street*. Winchester, UK: Zero Books.

Castells, Manuel. 2015. "Neoanarquismo." *La Vanguardia*, (May 21).

Chodorkoff, Dan. 1980. "Un Milagro de Loisaida: Alternative Technology and Grassroots Efforts for Neighborhood Reconstruction on New York's Lower East Side." PhD thesis, New School for Social Research.

Chodorkoff, Dan. 2011. *Loisaida: A Novel*. Burlington: Fomite.

Chodorkoff, Dan. 2014. *The Anthropology of Utopia*. Porsgrunn: New Compass.

Clark, John. 2005. "Social Ecology." In *Encyclopedia of Religion and Nature*, edited by Taylor, Bron, 1569–1571 London: Continuum.

Clark, John. P. 2012. "Political Ecology." In *Encyclopedia of Applied Ethics,* edited by Ruth Chadwick, 505–516. San Diego: Academic Press.

Clark, John. 2013. *The Impossible Community: Realizing Communitarian Anarchism*. London: Bloomsbury.

Crowe, Philip. and Karen Foley. 2013. *The TURAS Project: Integrating Social-Ecological Resilience and Urban Planning*. Dublin: TURAS.

Curran, Giorel. 2006. *21st Century Dissent: Anarchism, Anti-globalization and Environmentalism*. New York: Palgrave Macmillan.

Finley, Eleanor. 2018. "Beyond the Limits of Nature: A Social-Ecological Perspective on Degrowth." *Capitalism, Nature, Socialism,* 29, no. 1: 244–50.

Graeber, David. 2004. *Fragments for an Anarchist Anthropology*. Chicago: Prickly Paradigm Press.

Graeber, David. 2013. *The Democracy Project: A History, a Crisis, a Movement*. London: Penguin Books.

Hammy, Jihad, and Eleanor Finley. 2015. "Bookchin and Öcalan: Fruits on The Tree of Humankind." *Kurdish Question*. http://kurdishquestion.com/oldarticle.php?aid =bookchin-oecalan-fruits-on-the-tree-of-mankind/ (site discontinued).

Harvey, David. 2012. *Rebel Cities: From the Right to the City to the Urban Revolution*. London: Verso Books.

Heider, Reimar, and Oliver Kontny. 2004. *Öcalan Intermediaries to Bookchin* (letter, April 6). https://www.documentcloud.org/documents/3284166-Ocalan-Intermediaries -to-Bookchin-Biehl-4-6-2004.html/.

Heller, Chaia. 1999. *The Ecology of Everyday Life: Rethinking the Desire for Nature*. Montréal: Black Rose Books.

Hemenway, Toby. 2015. *The Permaculture City: Regenerative Design for Urban, Suburban, and Town Resilience.* White River Junction, VT: Chelsea Green Publishing.

Hern, Matt. 2010. *Common Ground in a Liquid City: Essays in Defense of an Urban Future.* Oakland: AK Press.

Heynen, Nikolas Cunnington, Maria Kaika, and Erik Swyngedouw. 2006. *In the Nature of Cities: Urban Political Ecology and the Politics of Urban Metabolism.* London: Routledge.

Holloway, John. 2010. *Change the World Without Taking Power.* London: Pluto Press.

Hopkins, Rob. 2008. *The Transition Handbook: From Oil Dependency to Local Resilience.* Totnes, UK: Green Books.

Hunt, Stephen E. 2017. "Prospects for Kurdish Ecology Initiatives in Syria and Turkey: Democratic Confederalism and Social Ecology." *Capitalism Nature Socialism*, (2017): 1–20. DOI: 10.1080/10455752.2017.1413120.

Imboden, Charles. 2011. "Twenty-First Century Social Ecology: Toward a Social Ecology of Sonora." *Better Worlds, Brighter Futures.* http://socecology.files. wordpress.com/2011/08/imboden_21st_century_social_se1.pdf/.

Imboden, Charles. 2012. "Solidity Occupy." *New Compass*, (4 January). http://new -compass.net/articles/solidify-occupy/.

Krøvel, Roy. 2014. "Social Production of Community Resilience." *Resilience: International Policies, Practices and Discourses* 2, no. 1: 64–70. https://doi.org/10 .1080/21693293.2014.880600.

Kuhn, Gabriel. 2017. *Revolution Is More Than a Word: 23 Theses on Anarchism.* London: Active Distribution.

Legard, Sveinung. 2010. Change the World by Taking Power. *Communalism*, no. 2: 62–70.

Light, Andrew. 1998. *Social Ecology after Bookchin.* New York: Guilford Press.

Macauley, David. 1998. "Evolution and Revolution: The Ecological Anarchism of Kropotkin and Bookchin." In *Social Ecology after Bookchin*, edited by Andrew Light, 298–342. New York: Guilford Press.

Mansilla, José. A. ed. 2017. Nuevos municipalismos y conflicto urbano [special issue]. *Quaderns-e*, 22, no. 1.

Marshall, Peter. 2008. *Demanding the Impossible: A History of Anarchism.* Oakland: PM Press.

Martin, Ian. 2010. *Back to the Roots: Anarchists as Revolutionary Organizers.* Fordsburg, South Africa: Zabalaza Books.

Knapp, Michael, Anja Flach, and Ercan Ayboğa. 2016. *Revolution in Rojava: Democratic Autonomy and Women's Liberation in Syrian Kurdistan.* London: Pluto Press.

Mollison, Bill. 1990. *Permaculture, A Practical Guide for a Sustainable Future.* Island Press, Washington, DC.

Morris, Brian. 2012. *Pioneers of Ecological Humanism.* Brighton: Book Guild Publishing, 2012.-

Pepper, David. 2003. *Eco-socialism: from Deep Ecology to Social Justice*. London: Routledge.

Price, Andy. 2012. *Recovering Bookchin: Social Ecology and the Crisis of Our Time*. Porsgrunn: New Compass.

Robbins, Paul. 2012. *Political Ecology: A Critical Introduction*. Chichester, UK: John Wiley & Sons.

Rockström, Johan, Will Steffen, Kevin Noone, Åsa Persson, F. Stuart III Chapin, Eric Lambin, Timothy M. Lenton, et al. 2009. "Planetary Boundaries: Exploring the Safe Operating Space for Humanity." *Ecology and Society* 14, no. 2: 32. http://www.ecologyandsociety.org/vol14/iss2/art32/.

Roussopoulos, Dimitrios. 2013. "The Politics of Neo-Anarchism." In *Anarchism: A Documentary History of Libertarian Ideas* (Vol. 3), edited by Robert Graham, 31–42. Montréal: Black Rose Books.

Roussopoulos, Dimitrios. 2015. *The Politics of Social Ecology*. Porsgrunn: New Compass.

Rubio-Pueyo, Vicente. 2017. *Municipalism in Spain: From Barcelona to Madrid, and Beyond*. New York: Rosa Luxemburg Stiftung.

Seeds for Change. 2013. *A Consensus Handbook: Co-operative Decision-making for Activists, Co-ops and Communities*. Lancaster: Footprint Workers' Co-op.

Sharp, Joanne, Paul Routledge, Chris Philo, and Ronan Paddison, eds. 2000. *Entanglements of Power: Geographies of Domination/Resistance*. London: Routledge.

Sitrin, Marina, and Dario Azzellini. 2014. *They Can't Represent Us! Reinventing Democracy from Greece to Occupy*. London: Verso.

Souza, Marcelo L. 2010. "Which Right to Which City? In Defence of Political-Strategic Clarity." *Interface* 2, no. 1: 315–333.

Souza, Marcelo L. (2012a). "The City in Libertarian Thought." *City* 16, no. 1–2: 4–33.

Souza, Marcelo L. (2012b). Marxists, Libertarians and the City. *City* 16 no. 3: 315–331.

Stanchev, Petar. 2015. "From Chiapas to Rojava: Seas Divide Us, Autonomy Binds Us." *ROAR Magazine* (February 7, 2015). http://roarmag.org/2015/02/chiapas-rojava-zapatista-kurds/.

Staudenmaier, Peter. 2002. "Economics in a Social-Ecological Society." *Harbinger* 3, no 1. https://social-ecology.org/wp/2002/09/harbinger-vol-3-no-1-economics-in-a-social-ecological-society/.

Staudenmaier, Peter. 2005. "Ambiguities of Animal Rights." *Institute of Social Ecology*. http://www.social-ecology.org/2005/01/ambiguities-of-animal-rights/.

Swyngedouw, Erik. 2004. *Social Power and the Urbanization of Water: Flows of Power*. Oxford: Oxford University Press.

Trapese Collective. 2007. *Do it Yourself: A Handbook for Changing Our World*. London: Pluto Press.

Vansintjan, Aaron. 2019. "Urbanisation as the Death of Politics: Sketches of Degrowth Municipalism." In: *Housing for Degrowth: Principles, Models, Challenges and*

Opportunities, edited by Anitra Nelson and François Schneider, 196–209. London: Routledge.

Venturini, Federico, Emet Değirmenci, and Inés Morales-Bernardos, eds. 2019. *The Right to the City and Social Ecology: Towards Ecological and Democratic Cities.* Montréal: Black Rose Books.

White, Damian. F. 2008. *Bookchin: A Critical Appraisal.* London: Pluto Press.

Chapter 2

Social Ecology in Öcalan's Thinking

Cihad Hammy

It was during the heroic resistance against the Islamic State in the battle of Kobanî that the Rojava Revolution caught the attention of the left on a global scale. Since then, though not to a sufficient extent, the subject of the Rojava Revolution has gained some space both in academic research and in radical nonacademic outlets.

In these texts, the authors, in general, tend to reduce the political thinker and co-founder of the PKK (Partiya Karkerên Kurdistanê) Abdullah Öcalan's radical transformation more or less to his engagement with the historian and philosopher Murray Bookchin. The core argument in such literature is that Öcalan was a Marxist-Leninist revolutionary who then, in prison, became a radical libertarian socialist with a strong anarchist and communalist tendency after he familiarized himself with the works of Bookchin. As a response to that, some of the more narrowly nationalistic Kurdish academics and activists dismiss or belittle, in one way or another, Bookchin's influence on Öcalan. Neither approach does any good in illuminating Öcalan and Bookchin's connection philosophically and politically.

My argument is that Öcalan was simply receptive when he encountered Bookchin's works. In this chapter, I will show how Öcalan went through a gradual intellectual development from Marxism-Leninism toward libertarian socialism before reading Bookchin. In prison, he begins his "quest for the truth" (Öcalan 2012a). He was already on this quest for alternatives when he discovered Bookchin, who provided him with his answers.

It has to be acknowledged that there were many other philosophers and thinkers who significantly influenced Öcalan. To name a few: Immanuel Wallerstein, Fernand Braudel, André Gunder Frank, Samir Amin, Friedrich Nietzsche, Michel Foucault, Edward Said, Hannah Arendt, Marxist feminists, Rosa Luxemburg, and the Frankfurt School. In fact, each of these names or schools requires separate chapters to elaborate on how they contributed to the development of Öcalan's thinking. And while Bookchin and Öcalan would not agree on certain philosophical and theoretical issues, I believe they would share the same political ends.

Though there were other central questions—for example, women's liberation—which influenced Öcalan's transformation throughout his struggle, I will mainly focus in the first part on how Öcalan dialectically transformed his political and philosophical thinking such that it would later be pertinent to social ecology from the mid-1980s up to early 2002, the era just before he encountered Bookchin's works. In the second part, I will show how Öcalan incorporates the philosophical and political dimensions of social ecology.

ECOLOGICAL AWARENESS

In the process of his ongoing critique—from the mid-1980s and onward—of the philosophical basis of the Marxism-Leninism of the USSR, Öcalan saw the limitation of the narrow class analysis underpinned by the dualistic contradiction that Karl Marx called the "two great hostile camps" within the capitalist system: the bourgeoisie and the proletariat. Capitalism, according to Öcalan, not only inherently produces contradiction through the exploitation of the dispossessed working class (the proletariat) by the dominant property-owning class (the bourgeoisie) but also produces wider social contradictions which extend to the destruction of nature:

> It is no longer only the contradiction between classes that capitalism has launched against society, but against nature as well. This monster (capitalism) not only reinforces class division and colonialism, but also destroys nature. It has perforated sky and land, thereby turning the ecological balance upside-down and pushing society to the brink of extinction. (Öcalan 2003, 29)

In the same book, reflecting on the great danger capitalism has inflicted on nature and social life, Öcalan identified some imperative and urgent issues that are produced by capitalism, such as unprecedented population growth, the rise of pandemics, climate change, global warming, environmental pollution, and desertification. Öcalan evaluated these ecological problems as incredibly dangerous, more so than "the danger of class conflict by forty times." Moreover, he went on to describe capitalism as a cancer on the environment and social life: "[c]apitalism pollutes the environment and biosphere, and renders an incurable cancer upon society" (Öcalan 2003, 72).

Öcalan was not satisfied with the reformist solutions proposed by the green movement to deal with the ecological crisis. Instead, he saw "revolutionary approaches as indispensable to correctly analyze the human relation to nature" (Öcalan 2003, 40). Öcalan believed deeply that any revolutionary movement must overcome the limitations of the nineteenth-century paradigm—to use Immanuel Wallerstein's words—and henceforth he expanded his analysis to include "[s]truggling against the terrible destruction of the environment and consumption insanity" (quoted in Özcan 2006, 100). The new paradigm for socialism, he envisioned, must encompass ecology, utopianism, science, and ethics (Öcalan 2003).

Environmental issues were one of the central themes dealt with in his writings (Öcalan 1999). For example, in Öcalan's first major book authored in prison, he outlined a new program for the Kurdish movement which centered ecological issues. As he put it, "For a truly new beginning, environmental issues should be included in the program." He went on to stress the urgency of environmental issues: "The antagonism between man and environment can be as problematic as social issues" (Öcalan 2007, 255).

A few decades before Öcalan penned his own critiques, Bookchin expressed a similar dissatisfaction with the limitations of the Marxian analysis, which in his view confined the contradictions of capitalism to the social and economic sphere. For Bookchin, these contradictions within bureaucratic state capitalism not only permeate all hierarchical social forms, but threaten natural life as well: "This contradiction also opposes the exploitative organization of society to the natural world—a world that includes not only the natural environment, but also man's 'nature'" (Bookchin 2004, 5–6). He goes on to stress the importance of a radical approach in tackling these contradictions: "the contradiction between the exploitative organization of society and the natural environment is beyond co-optation: the atmosphere, the waterways, the soil and the ecology required for human survival are not redeemable by reforms, concessions, or modifications of strategic policy" (Bookchin 2004, 6).

TECHNOLOGY

Another question that Öcalan occupied himself with was technology. Debate over this question dominated the political left after World War II. The left critique of technology was dominated by Bookchin, the Frankfurt School (Herbert Marcuse, Max Horkheimer, and Theodor Adorno among others), and Martin Heidegger. These philosophers attacked technology for its destruction of nature and its role in the projection of control and domination over the advanced industrial societies.

Aware of the ecological destruction perpetuated by the capitalist system, Öcalan started to reflect on the question of technology—monopolized by capitalism—and its relation to control, science, and nature. He would lecture

PKK fighters that "nowadays, technology is a monster, destroying the environment and biosphere." Moreover, he regarded technology and its rule and structure as a means to disempower people from their will and to erode intellectual freedom (Öcalan 2003).

As a dialectical thinker, Öcalan conceived of two possible potentialities of technology. He conceived of technology at once as a means for liberation, when it is under the service of the people in a liberated society, yet also as oppressive, when it is monopolized by the capitalist class. "Technology, which for the first time in history could turn man's utopia into reality is [instead] abused by a minority uninterested in mankind's basic needs, turning man's reality into a living hell." He continued the metaphor: "Technology is becoming a monster, threatening to devour its creators" (Öcalan, 2007, 245). To solve this, he advocated the democratization of technology, in order that it might be used in an eco-friendly and rational way (Öcalan, 2007, 245).

The outlook of Öcalan is closer to that of Herbert Marcuse, inspired by Marx's *Grundrisse*, in which Marx contended that technology had the potential to produce abundance for everybody, and that this would eventually lead people to overcome the realm of necessity and enter the realm of freedom. As he put it, "Technics by itself can promote authoritarianism as well as liberty, scarcity as well as abundance, the extension as well as the abolition of toil" (Marcuse 1998, 41).

On the question of technology, Marcuse's colleagues Horkheimer and Adorno rooted the issue of the oppressive and negative aspects of technology in the capitalistic relations of production and capitalist political power (Horkheimer and Adorno 2002, 95). Like Horkheimer and Adorno, Öcalan did not deprecate technology per se, but rather the capitalist mode of production and its forms of political power that have monopoly over technology: "The blame for this irrational situation should be laid on the old mode of production, and the political power behind it" (Öcalan 2007, 244).

Bookchin also wrote about technology, beginning in the 1960s, when his work was a major influence on the burgeoning green left. In *Towards a Liberatory Technology*, Bookchin talked about the potentiality of technology in eliminating toil, material insecurity, and centralized economic control (Bookchin 2004). He also stressed how technology—under industrialism, urbanization, and state capitalism—exacerbated the alienation of humanity from nature, as well as from her or himself. As he put it, "technology is transformed into a force above man, orchestrating his life according to a score contrived by an industrial bureaucracy" (Bookchin 2004, 78). Like many of the thinkers above, Bookchin did not see technology as harmful per se; "to view technological advances as intrinsically harmful . . . is as shortsighted as it is arrogant" (Bookchin 1995a, 238). Therefore, he held that "[i]gnoring technology, of course, is no solution" (Bookchin 2004, 79). He emphasized that technology, in a liberated society, could expand human formation and local possibilities, enrich creativity, and bring about reharmonization between nature and society. The use of renewable technology, such as solar and wind energy, would enable a free and decentralized society to metabolize with nature in harmony: "To bring the sun, the wind, the earth, indeed the world of life, back into technology, into the means of human survival, would be a revolutionary renewal of man's ties to nature" (Bookchin 2004, 76). Furthermore, he wrote about the role of alternative, ecological technology in "reawaken[ing] man's sense of dependence upon the environment" (Bookchin 2004, 64).

POLITICAL TRANSFORMATION

In the 1990s, Öcalan started to pose important questions concerning the bourgeois-class dominated and undemocratic nature of the state and abandoned the idea of an independent Kurdish nation-state as a solution for the Kurdish question in Turkey. As an alternative, he offered decentralized federalism and autonomy in a democratic republic. Truthfully, he was not intellectually prepared to radically move beyond the critique of the state within the Marxist traditions he had previously followed. Up until 2002, despite seriously attempting to understand the nature of the state and its historical roots (which he traced back to the Sumerian civilization, influenced by the work of Vere Gordon Childe and Samuel Noah Kramer), he always took an ambivalent stance toward the state. In *The Roots of Civilisation*, Öcalan took a similar position on the problem of the state to Marx, who did not conceive the state as a neutral political organ but rather as a political apparatus to secure the class interests of the bourgeoisie. In contrast to that he also appreciated the administrative role of the state (Eagleton 2011, 197–99).

What is crucial about this approach is that it paved the way for Öcalan to begin seriously working on a critique of the state and to think of political alternatives within the sphere of statelessness. It was precisely this that led him in 2002 to adopt the Gramscian theory of the "organic" civil society as an area for "democratic possibilities," thereby questing for a new way to escape the state (Bar-on, 2015). Despite his serious efforts, he was still working within the boundaries of the Marxian tradition (which sees the state as a necessary step for socialism) and did not venture to take the further step of moving beyond the Marxist tradition.

Similarly to the revolutionary martyr and Marxist thinker Rosa Luxemburg's critique of Lenin's rigid hierarchy and highly centralized and bureaucratic tendencies in the party, Öcalan, starting from the mid-1980s, launched a staunch critique of the USSR with regard to its heavily bureaucratic state apparatus, its "elimination" of the freedom of the individual, and its complete lack of meaningful democracy. Thus, in the mid-1980s, Öcalan began to reflect on the nature of democracy, the party, and the humanist question. He then reached the conclusion that socialism cannot be realized if it is not democratic and humanistic. "[. . .]To insist on socialism means to insist on being human," he would stress in the 1990s. He also further emphasized that "our goal is the liberation of the human who has been destroyed by the capitalist-imperialist system" (Öcalan 1998).

This critique was not simply an academic theorization of "real" socialism and the crimes of state communism in the USSR but rather a relentless self-critique against himself and the ideological foundation of the Kurdish movement at the time. Through this critique of "real" socialism, Öcalan was trying to transform the Kurdish movement so that it would embrace a more humanistic and democratic stance, rooted in the people. This was made concrete in the critique of the centralized, monopolized, and bureaucratic party structure and its cadres as an apparatus over society. To resolve this crisis, Öcalan stressed the importance of democracy and the people's participation in the Kurdish movement. As he clearly put it:

> We are in favour of giving way to [the] people [to rule] and of protecting this [right]. But there are many party cadres who make decisions on behalf of the people, who do not even give [the] people the right to talk. . . . The people will talk, the people will decide, the people will make their choice, the people will make changes. Our understanding of democracy is precisely this; it should not be distorted. (Quoted in Özcan 2006, 141–42)

The critique of Leninism did not remain there but was extended to the very core of the PKK's ideology of national liberation and self-determination: the right to decide their own destiny. Starting from the 1970s, the PKK defined Kurdistan as an international colony. That meant it needed liberation from the colonizers: the Turkish state and the imperialist powers behind it. At the time, the Leninist principle of self-determination (which understood the apogee of the decolonial process as the independent nation-state) was prevalent among the left in the third world, and the PKK was no exception (Jongerden 2016).

Based on the above critique of the USSR's socialism, Öcalan started to rearticulate his approach to the question of national liberation and the Leninist approach to self-determination. During his earlier years in prison and negotiations with the Turkish state in 1993, he offered federative autonomy as a form of self-determination, thereby abandoning the Marxist-Leninist paradigm of the independent nation-state as the end goal of self-determination. This approach is very similar to that taken by two founders of Négritude, the Martinican Aimé Césaire and the Senegalese Léopold

Sédar Senghor, who, like Öcalan, approached the issue on the basis of decolonization, therefore rethinking the traditional paradigm of self-determination. Both thinkers rejected the ideal of nation-state sovereignty, and instead proposed democratic and decentralized federalism based on popular participation and cooperatives as a program of self-determination (Wilder 2015).

While Öcalan was making these critiques, Bookchin's 1993 essay, "Nationalism and the 'National Question,'" touched on the same themes that Öcalan was struggling with. Bookchin was very critical of national liberation movements and the forms of self-determination they advocated, which were based on statism and narrow nationalism: "Nationalisms that only a generation ago might have been regarded as national liberation struggles are more clearly seen today . . . as little more than social nightmares and decivilizing blights" (Bookchin 2015, 133). Moreover, he saw that such movements eventually reproduced the very same forms of authoritarianism and dictatorship that they initially fought against. Bookchin observed that the success of those movements "has had the effect of creating politically independent statist regimes that are nonetheless as manipulable by the forces of international capitalism as were the old, generally obtuse imperialist ones." In order to overcome those pitfalls, Bookchin proposed that such movements must be grounded on democratic and universalist bases, and should be, just as Öcalan was trying to achieve, "universalistic in [their] view of humanity, cooperative in [their] view of human relationships on all levels of life, and egalitarian in [their] idea of social relations" (Bookchin 2015, 131). Such an outlook was also clearly expressed by Öcalan: "We will not fall into the swamp of narrow nationalism. We will not allow the question of borders to determine us. We will not discuss how much land belongs to whom. If we create the freedom of peoples, we have also created the most significant internationalism" (Öcalan 1998).

SOCIAL ECOLOGY IN ÖCALAN'S THINKING

In 1964, Bookchin employed the term "social ecology" for the first time in his early essay titled "Ecology and Revolutionary Thought," in which he emphasized that the idea of dominating nature has its origins in the domination of humans by humans. This meant that to solve the ecological crisis, all forms of institutionalized hierarchy and domination that exist within society must be dissolved and replaced by nonhierarchical, free, and rational institutions. In doing so, Bookchin gave ecology, which required a far-reaching political transformation, a sharp revolutionary edge. While Bookchin advanced a revolutionary outlook to ecology, the left, in majority, was oblivious to ecological issues and regarded them as a "petty bourgeois" deflection from the struggle against capitalism.

In 2002, Öcalan recommended that *From Urbanization to Cities* and *The Ecology of Freedom* should be read throughout Kurdistan to mayors, local officials of Kurdish-led towns, and the PKK (Jongerden and Akkay 2013). From 2003, the terminology of social ecology begins to appear in Öcalan's writings, with terms such as democratic and ecological society, communes, councils, and assemblies in frequent use (Öcalan 2005). From that point, as I will show, he adapted more or less the entire body of social ecology into his own work. In the following section, I will discuss in-depth the impact of the philosophical and political dimensions of social ecology on Öcalan's writings.

THE PHILOSOPHY OF SOCIAL ECOLOGY

As discussed previously, at this point (from the 1980s until early 2002), Öcalan was already deeply aware of ecology and the ecological crisis inflicted on society by capitalism. However, this awareness was deepened, conceptualized, and revolutionized philosophically and politically when he began to engage with the works of Bookchin. Öcalan started to view the ecological crisis as a central issue and saw that it was a product of social breakdown caused by hierarchy, domination, exploitation, and slavery: "when man began to enslave his brother, he also began to enslave nature" (Öcalan 2004, 184).[1] Moreover, Öcalan contended, "These extraordinary ecological problems are a consequence of a broken society" (Öcalan 2004, 81). One can clearly see that when Öcalan penned these sentences, the central tenet of social ecology was in his mind that "the very notion of the domination of nature by man stems from the very real domination of human by human" (Bookchin 2005, 1).

Thus, to solve our ecological crisis and to create an ecological society, Öcalan proposed a political program which would not "replicate the old hierarchical system or the traditional state-based system, nor of course the system of slavery, where society is oppressed and exploited" (Öcalan 2004, 54). Relying on Bookchin (see Bookchin 2007, 46), he instead advanced a nonhierarchical and ethical system based on a dialectical relationship with nature and direct democracy:

> We will build a morality-based system that involves sustainable dialectical relations with nature, a system that does not rely on internal power structures, and a system where common welfare is achieved by the means of direct democracy. (Öcalan 2004, 54)

Furthermore, as a true social ecologist (Bookchin 2015,142), Öcalan does not approach the ecological crisis with piecemeal measures. Indeed, he advocates sweeping changes that foster reharmonization with nature in a holistic way, which means establishing a democratic and socialist society:

> Awareness of, and caring for, the natural environment does not only entail caring for clean air and clean water. It means to live entirely with nature, to return from our present reductionism, from dividing nature into small parts, to a holistic view of nature. This too is the meaning of building a democratic and socialist society. (Öcalan 2004, 185)

Influenced by Bookchin, Öcalan considered that the ecological crisis was caused by the rise of hierarchy and domination in social life. Disillusioned with the limitations of Marxism's unit of analysis, "class" and economic reductionism, Bookchin located his argument within the anarchist tradition, advocating "hierarchy and domination" as a unit of analysis to understand the root of all social and ecological problems. For Bookchin, hierarchy is "the cultural, traditional and psychological systems of obedience and command, not merely the economic and political systems" (Bookchin 2005, 4). Bookchin goes on to identify the forms of domination as "the domination of the young by the old, of women by men, of one ethnic group by another, of 'masses' by bureaucrats who profess to speak in their 'higher social interests,' of countryside by town, and in a more subtle psychological sense, of body by mind, of spirit by a shallow instrumental rationality, and of nature by society and technology" (Bookchin 2005, 4). However, this does not mean that he abandoned class as a unit of analysis entirely. In fact, he incorporated "class and exploitation" into the larger and more comprehensive unit of analysis "hierarchy and domination" (Bookchin, 2001). This aided Öcalan to move beyond the narrow class analysis that he was criticizing.

Inspired by Bookchin's magnum opus *The Ecology of Freedom*, Öcalan turned back to "deep history" to uncover how hierarchy and domination originally emerged. In doing so, he aimed to understand the origins of hierarchy to find a way to dismantle it, as Öcalan pointed out: "When looking for the roots of the ecological crisis, which is exacerbating in parallel to the crisis of the system, it seems to make sense to start the search at the beginning of civilization" (Ocalan 2004, 119). Though Öcalan used the long narrative of the history of class-civilization (which he traced back to the Sumerian civilization) in *The Roots of Civilization*, he did not yet discuss the rise of hierarchy which preceded and gave rise to the state and class society and advanced democratic civilization (which he described as resistance and libertarian traditions) against class civilization. Reading *The Ecology of Freedom* provided Öcalan with the anthropological tools to deeply investigate the Neolithic era and the rise of hierarchy and domination in a more detailed and sophisticated way. Bookchin always stressed that even in classless and stateless societies, one could still find forms of hierarchy and domination (Bookchin 2005, 4). Formulated differently, Öcalan wrote, "The development of power and hierarchy even before class society emerged is a significant turning point in history" (Öcalan 2004, 15), and moreover that "the state is the representative of permanently institutionalized hierarchical structures" (Öcalan 2017, 50).

Both thinkers also discussed the era before the rise of hierarchy. Bookchin called it "organic society," while Öcalan named it "natural society," and they described this society as female oriented and structured by communal, non-hierarchical, peaceful, cooperative, and solidarity-based social relationships (Bookchin 2005; Öcalan, 2015). However, out of this "organic society," hierarchy started to emerge. Since then, the dialectic of the legacy of freedom and domination began to unfold (Bookchin 2005). For Öcalan, the initial form of hierarchy was the male alliance (shaman, wise elder, and strong man) against women: "The older matriarchal system, which relied on natural authority, was the first victim of hierarchical society. Women may have been the first social group to suffer oppression" (Öcalan 2004, 18). Taking the same line as the historian and feminist Gerda Lerner in her book *The Creation of Patriarchy* (1986), Öcalan argued that "all other forms of slavery and servitude developed because of the enslavement of women." It is worth noting that in dealing with the narrative of the rise of hierarchy, Bookchin saw gerontocracy as the initial form of hierarchy (Bookchin 1990, 57–58; Biehl 1991, 46), while for Öcalan this was patriarchy.

To ground social ecology within a solid and coherent philosophical body, Bookchin also voiced similar critiques of the absolutism inherent in Hegel's dialectical idealism and the crude materialism and scientism of Marxism's dialectical materialism. Perhaps Bookchin was the only philosopher after Hegel and Marx to have reinterpreted and developed the dialectic to explain the relationship between nature and society as what he termed dialectical naturalism. Simply put, dialectical naturalism is an organic reasoning that reflects on symbiotic relationships between "first nature" (biological) and "second nature" (human social life). It seems that second nature emerged from first nature, while potentially retaining the objective ethics of first nature without losing its specificity and integrity. However, due to the rise of hierarchy and domination within second nature, the process of alienation of humans from themselves and from nature unfolds. By overcoming hierarchy and domination in second nature, humanity reaches, what Bookchin calls, third nature (free nature); reharmonization of the first nature with the second but in a richer, freer, and rational form. In third nature, as Bookchin claims, "second nature would thus become first nature rendered self-reflexive, a thinking nature that would know itself and guide its own evolution" (Bookchin 1996,

136). This dialectical process, latent with potentiality, always challenges what merely is (the dominant social order, capitalism), and strives toward what ought to be (free and ecological society), an ethical notion inherent in natural and social development toward self-reflexivity, reason, and freedom.

Before reading the works of Bookchin, Öcalan criticized the "progressive" approach and scientism of the fundamental philosophical basis of Marxism—dialectical and historical materialism (Öcalan 2007). However, he could not find alternative pathways for his critique. Dialectical naturalism gave Öcalan the means to reject the Hegelian and Marxian interpretations of dialectics for their "very unpleasant and destructive consequences" (Öcalan 2020, 24). The core concept of Hegel's sublation (*Aufhebung*) in dialectics is development that does not annihilate the earlier stage but rather retains it in a new, potentially higher, and richer form. Following Hegel's concept of sublation, Bookchin used the term "cumulative" to describe, in his dialectic, the relationship between nature and society (Bookchin 1996, 89). Inspired by Bookchin, Öcalan employed a "constructive and developmental dialectic" in locating the human place in nature and the universe. Öcalan suggests that the followers of Hegel and Marx misinterpreted dialectics, yet adds that "It would, however, be more correct to look for the errors made by those who misinterpreted these dialectics in major ways and not in Marx or Hegel" (Öcalan 2020, 24). In particular, in the last two volumes in the series "Manifesto of the Democratic Civilization," one can sense that Öcalan has a strong affinity for Hegel's works. While Bookchin's dialectical naturalism rests on the dialectical legacy in Western philosophy, and Öcalan acknowledges the significant contributions of ancient Greece and Europe during the Enlightenment, Öcalan also seeks to decolonize and de-Eurocentrize the dialectic: "Dialectical interpretations are abundant in the wisdom of the East" (Öcalan 2020, 24).

Bookchin was always a fierce opponent of deep ecology and other ecological tendencies such as misanthropy, biocentrism, and anthropocentrism (Bookchin 1995b, 6; Bookchin, 1987). Drawing from Bookchin, Öcalan also made similar critiques (2017, 43–44). Applying dialectical naturalism to the relationship between nature and society as a critique of biocentrism, Öcalan contends that "In a way, society, as second nature, is a higher level, a reflection of the first nature" (Öcalan 2017, 45; Öcalan 2020. 21). Furthermore, dialectical naturalism helped Öcalan to overcome the dualism that pits society against nature in the philosophical form as subject (society) and object (nature) (Öcalan 2017, 45). Following Bookchin, Öcalan conceived of nature as a cumulative evolutionary process latent with potentiality, developing from the simplest of life forms toward increasingly and ever-greater degrees of differentiation, diversity, complexity, choice, subjectivity, and freedom (Öcalan, 2020; 277; Bookchin 1996). Indeed, Bookchin saw these principles as an objective ethics in the (biological) first nature; furthermore, he sought an objective knowledge of nature to reduce ecological ethics in first nature, which could then guide the ethical compass of second nature (human social life) (Bookchin, 1996, 89). In other words, this potentiality of first nature led to the emergence of social development, which retained the same ethical telos as first nature (Öcalan 2020; 277).

Another major aspect of the ecological crisis that Bookchin occupied himself with was urbanization. Bookchin sharply distinguished the city from urbanization. For Bookchin, urbanization erodes the potentiality of the city as an authentic arena of political life and citizenship. Also, it colonizes both the city and countryside, thereby exacerbating the ecological crisis further (Bookchin 1995a). Öcalan sees that the capitalist urban paradigm, alongside the old hierarchical civilizations which are based on distorted city, class, and state built cancerous structures in society, leads us to the brink of "societycide" and pushes the ecological crisis to the extremes by devouring the agrarian landscape. Following Bookchin, Öcalan conceives that "reason has developed in a close relationship with the city. The city is where human beings began to recognize the breadth of their capacity [potentiality]" (Öcalan 2020, 110). In fact, to overcome the problem of urbanization, he seeks to reclaim the free and true potentiality of the city: "We should consider how to rescue the limited remaining beauty, morality, and reason within the city" (Öcalan 2020, 116).

THE POLITICS OF SOCIAL ECOLOGY

To build an ecological and democratic society, the hierarchy, domination, and exploitation embodied in the capitalist nation-state system must be overcome. To achieve that, Bookchin advanced his program of communalism and its concrete political dimension, libertarian municipalism. Drawing on the best aspects of Marxism and the rich traditions of libertarian socialism, Bookchin transcended into a synthesis (communalism) to radically challenge and transcend the system of capitalism and the nation-state (Bookchin 2015). Communalism aims to engage people in democratic and civic assemblies to exercise face-to-face politics and make decisions concerning their public affairs. In doing so, it reclaims "real politics" from the state—that is, the full engagement of all citizens in public affairs (Bookchin 2015).

After the collapse of the USSR, Öcalan seriously reflected on the dangerous nature of the state, with the aim of understanding how to empower the people to escape it. In encountering Bookchin's work, he found a viable alternative for the Kurdish question and a blueprint for a free and democratic Middle East. Inspired by communalism, or "libertarian municipalism"

(Bookchin 1995a), Öcalan developed the anti-capitalist and stateless project he called "democratic confederalism," which rests on three pillars: social ecology, radical democracy, and women's liberation. As a non-state social paradigm, democratic confederalism aims to empower the people to run their affairs through grassroots assemblies, combined with a confederation of democratized municipalities and councils, elected by the people through direct face-to-face democracy and democratic politics. Drawing on Bookchin (1995a), Öcalan rearticulated and radicalized his vision of politics as self-management of the community through directly democratic popular assemblies. Following Bookchin, for Öcalan, the presence of the state above society entails the fundamental denial of "politics" in the wider Aristotelian and Arendtian sense of the word: "In any case, power and the state only come into existence when the denial of social politics is ensured. Wherever politics comes to an end, power and state structures are at work." He called anti-state politics "democratic politics" (Öcalan 2020, 31; Hammy 2016). In its highest iteration, democratic confederalism attempts to overcome the nation-state and capitalism.

While the anarchist tradition runs strong in Öcalan's writings in regard to his critique of the state, he cannot easily dismiss the state entirely within his program, as the Turkish state is the indispensable part of the negotiation, without which the Kurdish question cannot be solved. Rather, he attempts a democratization of the state, and the creation of what he called a "democratic republic," which stands at odds with the nation-state model and is receptive to a democratic system, basing itself on a democratic society (Öcalan 2012b). Both democratic confederalism and the democratic republic constitute what Öcalan later conceptualized as a Democratic Nation (Öcalan 2016).

As mentioned earlier, before and during his first years in prison, Öcalan started to rearticulate the concept of "self-determination," by abandoning the Kurdish nation-state project and instead advocating decentralized federalism. Under the influence of Bookchin's libertarian municipalism or communalism, he then radicalized his approach to self-determination as radical democracy against the state (Öcalan 2011; Matthews and Miley 2018) Politically, Öcalan takes a two-pronged approach: he aims to democratize the state in the form of the "democratic republic," while also empowering the people so that they can manage their affairs outside of the sphere of statecraft, through the democratic confederalist project. This is similar to Bookchin's slogan "Democratize the republic and radicalize the democracy."

Since 2005, the Kurdish movement has adopted democratic confederalism and the PKK and all affiliated organizations have restructured themselves under the name of the KCK (*Koma Civakên Kurdistan*, the Association of Communities in Kurdistan) (Jongerden and Akkay 2013). Since 2005 and 2006, this project has been practiced both in Bakûr (North Kurdistan, in

Turkey) (TATORT Kurdistan 2011) and, as is well known, in Rojava since 2012. Following the Rojava Revolution, several articles have been written on the influence of the political dimensions of social ecology on Öcalan and the Kurdish movement (Bookchin 2018; Gerber and Brincat 2018).

Politically, social ecology is both the institutionalized body and foundational basis of Öcalan's thinking. Yet, all in all, Öcalan cannot be intellectually reduced to Bookchin. At the same time, the great significance of social ecology cannot be denied in the impetus it gave Öcalan to embrace a more radical and democratic stance. Dialectically, Öcalan retained the entire body of social ecology as one of the foundational pillars of democratic modernity, thereby nurturing, developing, and enriching it within a very different social and political environment to that of Bookchin.

NOTE

1. From Abdullah Öcalan's *Bir Halkı Savunmak* (2004) (In Defense of the People, unpublished). I use some quotes from the unfinished English translation by the International Initiative Freedom for Öcalan, Peace in Kurdistan. This book will be published as "Beyond State, Power, and Violence" by PM Press. This book is heavily influenced by Bookchin's works, and includes a chapter titled "Social Ecology."

REFERENCES

Bar-On, Tamir. 2015. "From Marxism and Nationalism to Radical Democracy: Abdullah Öcalan's Synthesis for the 21st Century." In *Challenging Capitalist Modernity II: Dissecting Capitalist Modernity—Building Democratic Confederalism*, edited by Network for an Alternative Quest, 77–88. Neuss: Mezopotamya Publishing House.
Biehl, Janet. 1991. *Rethinking Ecofeminist Politics*. Boston: South End Press.
Bookchin, Debbie. 2018. "How My Father's Ideas Helped the Kurds Create a New Democracy." *New York Review*, June 15, 2018. https://www.nybooks.com/daily/2018/06/15/how-my-fathers-ideas-helped-the-kurds-create-a-new-democracy/.
Bookchin, Murray. 1987. "Social Ecology Versus Deep Ecology: A Challenge for the Ecology Movement." *Anarchy Archives* (website). http://dwardmac.pitzer.edu/anarchist_archives/bookchin/socecovdeepeco.html.
Bookchin, Murray. 1990. *Remaking Society: Pathways to a Green Future*. Boston: South End Press.
Bookchin, Murray. 1995a. *From Urbanization to Cities: Toward a New Politics of Citizenship*. New York: Cassell.
Bookchin, Murray. 1995b. *Re-enchanting Humanity: A Defense of the Human Spirit Against Anti-Humanism, Misanthropy, Mysticism and Primitivism*. London: Cassell.

Bookchin, Murray. 1996. *The Philosophy of Social Ecology*. Montréal: Black Rose Books.

Bookchin. Murray. 2001. *Anarchism, Marxism and the Future of the Left: Interviews and Essays, 1993–1998*. Edinburgh: AK Press.

Bookchin, Murray. 2004. *Post-Scarcity Anarchism*. Oakland: AK Press.

Bookchin, Murray. 2005. *The Ecology of Freedom*. Oakland: AK Press.

Bookchin, Murray. 2007. *Social Ecology and Communalism*. Oakland: AK Press.

Bookchin, Murray. 2015. *The Next Revolution*. London: Pluto Press.

Eagleton, Terry. 2011. *Why Marx Was Right*. New Haven: Yale University Press.

Gerber, Damian, and Shannon Brincat. 2018. "When Öcalan met Bookchin: The Kurdish Freedom Movement and the Political Theory of Democratic Confederalism." *Geopolitics* (October 2018): 1–25. https://doi.org/10.1080/1 4650045.2018.1508016.

Hammy, Cihad. 2016. "Two Visions of Politics in Turkey: Authoritarian and Revolutionary." *OpenDemocracy* website. August 26, 2016. https://www. opendemocracy.net/en/north-africa-west-asia/two-visions-of-politics-in-turkey-authoritarian-and-revolutionary/.

Horkheimer, Max, and Theodor W. Adorno. 2002. *Dialectic of Enlightenment*. Stanford: Stanford University Press.

Jongerden, Joost P., and Ahmet Hamdi Akkaya. 2013. "Democratic Confederalism as a Kurdish Spring: The PKK and the Quest for Radical Democracy." In *The Kurdish Spring: Geopolitical Changes and the Kurds,* edited by M.M.A. Ahmet and M.M. Gunter, 163–185. (Bibliotheca Iranica: Kurdish Studies Series; No. 12). Costa Mesa, CA: Mazda Publishers.

Jongerden, Joost P. 2016. "Colonialism, Self-Determination and Independence: The New PKK Paradigm." In *Kurdish Issues: Essays in Honor of Robert W. Olson*, edited by M. Gunter, 106–121. Costa Mesa, CA: Mazda Publishers.

Lerner, Gerda. 1986. *The Creation of Patriarchy*. Oxford: Oxford University Press.

Marcuse, Herbert. 1998. *Technology, War and Fascism: Collected Papers of Herbert Marcuse*. London: Taylor and Francis.

Matthews, Donald H., and Thomas Jeffrey Miley. 2018. "Review of Abdullah Öcalan's Manifesto for a Democratic Civilization." In *Your Freedom and Mine: Abdullah Öcalan and the Kurdish Question in Erdogan's Turkey*, edited by Thomas Jeffrey Miley and Federico Venturini, 331–356. Montréal: Black Rose Books.

Öcalan, Abdullah. 1998. "Our goal is the liberation of the human," a text from "Sosyalizmde Israr—Insan Olmakta Isrardir" (To insist on socialism means to insist on being human). Translated and published by Komun Academy, 2020: https ://komun-academy.com/2020/05/01/our-goal-is-the-liberation-of-the-human/.

Öcalan, Abdullah. 1999. *Declaration on the Democratic Solution of the Kurdish Question*. Translated by Kurdish Information Centre. London: Mesopotamian Publishers.

Öcalan, Abdullah. 2003. "Sosyalizm Israr—Insan Olmaktan Isrardir" (To insist on socialism means to insist on being human) Arabic translation, Democratic Enlightenment Union publication. الإصرار على الاشتراكية إصرار على "بناء الإنسان." منشورات اتحاد التنوير الديمقراطي٢٠٠٢

Öcalan, Abdullah. 2004. *Bir Halkı Savunmak.* Çetin Yayınları, Istanbul.

Öcalan, Abdullah. 2005. *Plädoyer für den Freien Menschen.* Mezopotamien Verlags- und Vertriebs- gesellschaft.

Öcalan, Abdullah. 2007. *Prison Writings I: The Roots of Civilisation.* Translated by Klaus Happel. London: Pluto Press.

Öcalan, Abdullah. 2011. *Democratic Confederalism.* International Initiative: Cologne, and Mesopotamian Publishers: Neuss.

Öcalan, Abdullah. 2012a. "Seeker of Truth." In *Challenging Capitalist Modernity: Alternative Concepts and the Kurdish Quest: Documentation of the 2012 Conference,* edited by Network for an Alternative Quest, 25–34. Bonn: Pahl-Rugenstein.

Öcalan, Abdullah. 2012b. *Prison Writings III. The Road Map to Negotiations.* Cologne: International Initiative.

Öcalan, Abdullah. 2015. *Manifesto for a Democratic Civilization: The Age of Masked Gods and Disguised Kings,* Vol. 1. Porsgrunn: New Compass.

Öcalan, Abdullah. 2016. *Democratic Nation.* Cologne: Mezopotamien Verlag.

Öcalan, Abdullah. 2017. *Manifesto for a Democratic Civilization: Capitalism,* Vol. 2. Porsgrunn: New Compass.

Öcalan, Abdullah. 2020. *Manifesto of the Democratic Civilization: The Sociology of Freedom Volume III.* Oakland: PM Press.

Özcan, Ali Kemal. 2006. *Turkey's Kurds: A Theoretical Analysis of the PKK and Abdullah Öcalan.* London: Routledge.

TATORT. 2013. *Democratic Autonomy in North Kurdistan: The Council Movement, Gender Liberation, and Ecology—in Practice; A Reconnaissance into Southeastern Turkey.* Translated by Janet Biehl. Porsgrunn: New Compass.

Wilder, Gary. 2015. *Freedom Time: Negritude, Decolonization, and the Future of the World.* Durham, NC: Duke University Press.

Chapter 3

Ecological Self-Governmentality in Kurdish Space at a Time of Neoliberal Authoritarianism

Engin Sustam

The Kurdish question in the Middle East currently expresses itself in a puzzle of political and social ecology. This chapter examines this and theorizes a change of political values in Kurdish life. We will speak of the micropolitical ecological emancipation at the center of Turkey's Kurdish region (Bakûr) and Syria (Rojava). It has taken the shape of a heterogeneous movement which is challenging the crisis of colonial society. It also struggles for the environment, feminism, and the emancipation of the Kurdish people. This chapter offers an analysis of Kurdish space during a time of new uprisings and global authoritarianism. The topical analysis of Kurdish spaces of resistance crosses over with other movements, in which new and unique Kurdish subjectivities create emancipatory experiences, an important concern for wider alternative politics. This defines the chapter's milestones in a revolution that is embedded in the era of world capitalism. These novel spaces are analyzed alongside the concept of "Kurdish communalism." This discussion is informed by Michel Foucault's concept of "governmentality" (2004b, 2008 and 1994, 642) which contributes an analysis of the change in neoliberal powers toward an emerging governmental and global authoritarianism, in which the state and its institutional arrangements are characterized by a specific mode of macro-power. It is essential to underline that the goal of the Rojava Revolution is completely counter to such authoritarian power, based on policies of control of the population. The Foucauldian analysis, therefore, will give us a fresh perspective upon this Kurdish practice.

It is important to consider the Kurdish political movement within the context of the wider political transformation, of global "insurrectionary" social movements. This influenced the revolution's thinker, Abdullah Öcalan, who transformed a Marxist-Leninist movement into an autonomist-libertarian movement (2013b). The theory of social ecology, in particular, has a rhetoric and outlook that is compatible with the priorities of the Kurdish political movement (Üstündağ 2018; Cooperativa Integral Catalana 2016; Rojava Information Center 2020). The Kurdish political movement has adopted a discourse centering on an alternative political system, characterized by ecological priorities and communalism and based on micro-identities. The movement's aspiration is to apply this system to territory, which is contiguous, but separated by nation-state borders across the Kurdish majority region. Since the 1990s (from the first experience of HADEP's thirty-seven municipalities), Kurdish municipalities organized workshops that developed the theoretical idea of an alternative economy in Turkey's Kurdish areas. This economy, however, remains dependent on the monetary system of the state and international corporations. In November 2016, the Kurdish political movement worked with ecology activists to organize a conference in the city of Van with the slogan: "Let's communalize our land, our water and our energy, let's build a democratic, free life!" This micropolitical, communal vision of a free territory is embodied by the Rojava Revolution post-2012. At the Van conference, attendees discussed the necessity to construct an alternative economy centered on social, "humanitarian," and environmental benefits, and the emancipation of women, one that would avoid "individualist" or statist, for-profit approaches. This perspective reflects the emerging principles of radical democracy, communalism, ecologism, the emancipation of women, equality of sexual identities, the eradication of poverty, and solidarity. The resolutions taken at the conference in Van aimed at communalizing the cultivation of land, as well as work against precarity in the Kurdish space. The brutality of war and the "necropolitics" (Mbembe 2003) of Turkish state violence had exacerbated all inequalities. These effects do not limit themselves to stimulating the micronational dialectics of the movement, but also express a sort of politics of "dissensus" (Rancière 2009) in the Kurdish regions. The ecological movement aims to be both political and practical, basing itself on material conditions, with its practice informed by a theoretical toolbox.

We will discuss the expression of emancipation in the public space in relation to the culture of urban insurrection in the Kurdish region. The complexity of the space of revolt engenders a new political perception of the revolution in the Kurdish cantons by means of counterpower and countercultural reproduction, which makes itself visible, transcending any conventional ideological behavior in the Middle East. This micro-revolutionary tendency encompasses heterogeneous realities. It formulates politics in the Kurdish regions in a perspective close to that developed by Félix Guattari (2012, 2014): we see "a molecular revolution" at the heart of the process of a new Kurdish subjectivity coming into being. In parallel, this chapter will examine the thought processes relating to ecology. The questions that impress themselves are: What are the current criticisms concerning the social consequences of the ecological crises in Kurdistan in times of conflict and war? What are the propositions for alternative projects as part of the Kurdish revolution in urban and rural spaces?

FROM THE ARAB SPRING TO THE ROJAVA REVOLUTION IN THE TIME OF GLOBAL UPRISING

At a time of global crisis, neoliberalism aims to take total control over the life of every citizen of the world. We are on the way to a new level of global governmentality based on the surveillance of everyday life and of dissidence (Sustam 2020). By contrast, amid conflict, racism, and violence, the Rojava experience invites us to reflect on what we had not seen coming: the unforeseen insurgencies and a molecular revolution in life and free territory. Kurdish practice helps us to analyze the new spaces of uprisings in the world and the crisis of capitalism in the twenty-first century. The emerging Kurdish space is made up of complex themes, to be theoretically understood through framing concepts such as pedagogy, social ecology, uprising, the social structure of emancipation, and insurrection. In this context, our objective is, therefore, to understand the conditions and the factors which favor the appearance of this advanced countercultural language of Kurdish revolts and communalism, against the hegemonic construction of the neoliberal (in the Middle East and the world) and state apparatus in the space of conflict. This chapter intends to examine the subversive and creative subjectivity of emancipatory spaces and uprisings in the making. It situates the pedagogy of new ontological forms of these micropolitical spaces on a world scale, where uprisings are traced globally. In this context, since July 19, 2012, the revolution in Rojava has been a striking and outstanding example among several others, notably: Tahrir Square, Istanbul's Gezi Park, the Diyarbakır Hevsel Gardens resistance, Place Maiden, Nuit Debout, the Zapatistas' autonomous administration in Chiapas, Brazil's Landless Movement (MST), ZAD, and the *Gilets Jaunes*, and also revolts in Chile, Bolivia, Hong Kong, Iraq, Lebanon, Iran, France, and elsewhere (see Sustam 2020).

Turkey's Kurdish region became the site of the new urban *Serhildan* (Kurdish popular urban uprising, literally meaning "to raise the head," from *ser*, meaning "head," and *hildan*, meaning to rise) against the state apparatus, with the state following a new strategy that entails establishing special security zones in the cities. This is a generation born during the war of the 1990s that is now behind the barricades. The emergence of this new generation (who formed the youth movement YDG-H/K: Movement of the Patriotic Revolutionary Youth/Women, that became YPS: Civil Protection

Units) shows a reflexive-strategic break with the 1990s (Collectif Ne Var
Ne Yok: 2016, 7–27). They have become the principal actors of the counter-
violence of the Kurdish intifada known as the Serhildan. After the uprising in
the areas that proclaimed themselves autonomous, the government installed
special security zones, thus engaging in an undeclared war that interrupted
the peace process. The intensification of the Serhildan in 2014–2016 was
connected to this breakdown in peace talks, and Daesh's recent genocide of
Yazidis, and attacks on Kobanî at this time. There were several manifestations of Serhildan, in Turkey's Kurdish areas, which echoed the Kurdish
uprising of the 1990s. It was during the period, when Daesh attacked Kobanî,
that Turkey wanted to conquer Rojava through such paramilitary groups.
Kurds in Turkey therefore use the term "Serhildan" as a powerful expression of anger at state violence and colonial denial. At the same time, some
ecological movements engaged in urban ecopolitics in diverse spaces, such
as the Hevsel Gardens resistance movement (See Erbay 2017 on the Amed
Ecology Platform and Evrensel 2014) in the center of Diyarbakır, against
gentrification, the expansion of the urban area, privatization of the city and
social exclusion.

To understand the political character of the Kurdish political movement, it
is necessary to concentrate on the Rojava Revolution and the resistance at the
barricades (*hendek*) in Kurdish cities. Self-governance is a new strategy for
the micro-power to reproduce itself when facing the oppression of the state,
which is militarizing the Kurdish region, and the domination of the "necropolitics" (Mbembe 2003) of the jihadist movements in the Middle East.
It is important to stress that the discourse on self-governance of the Kurdish
political movement in the Middle East proposes a diversified approach in the
elaboration of the social project. The counter-violence (against state, patriarchal, and jihadist violence) has created something that looks like non-state
organization in Kurdistan (Tatort Kurdistan 2013). Its self-organized movement is elaborating a resilient critique of the traditional armed struggle with
its hierarchies.

In this context, the objective is to understand the conditions and factors
that favor the transformation of the Kurdish space in Syria and the emergence of this insurrectionary language of revolts in the face of the hegemonic
construction of the neoliberal economy and state. During the last decade, the
Kurdish space has seen the emergence of what could be called a new political subjectivity, a perspective critical of political readings of Kurdicity as
based on the idealization of the nation-state. This changed criticism concerns
mainly the practical foundations of a whole space of revolt and urban insurrection facing the colonization of the dominant nation-states. In this context,
the concept and practice of self-management lead to a regional, complex, and
cross-border approach among Kurds. They operate a double shift from a state

government and the classic landmarks of the nation-state: to a revolution (in Rojava) based on the principles of social ecology and the idea of democratic confederalism.

In keeping with the ideas that Hardt and Negri explore in "Assembly" (2017), these uprisings present subjectivities and raise the potential to create a micro-revolutionary process. This represents a deviation from the transnationalized monetary system which, with its increasing domination of the potentially heterogeneous "people to come" able to realize collective subjectivities liberated from capitalism (Deleuze 1993, 15; Deleuze and Guattari 1991; Comité invisible 2007), threatens to bring bringing poverty, precariousness, and insecurity to societies across the planet (Lazzarato 2004; 2008). The revolutionary impetus and the emancipation of a colonized people carry the weight of the possibility of hope for a better life, ecology, and freedom. We also see new forms of "debt dependency" (Lazzarato and Negri 1991; Lazzarato, 2004 and 2008) and the kind of widespread surveillance and security measures that assert the future of international companies, creating the potential to get rid of state systems. On the one hand, global neoliberalism imposes a system of control upon society using state apparatuses, yet at the same time, it strives to free itself from state rule to suit its economic and transnational interests. In this respect, David Graeber's discussion of the debt mechanism of neoliberalism is relevant. Graeber highlights the role of debt in causing poverty, human misery, and ultimately the destruction of the planet, observing that "consumer debt is the lifeblood of our economy. All modern nation-states are built on deficit spending. Debt has come to be the central issue of international politics. But nobody seems to know exactly what it is, or how to think about it" (2011, 5).

Counter to this model of global capitalism, the cooperatives in Rojava, often led by women, provide an alternative example, one grounded in the system of democratic confederalism and inspired by indigenous people in the part of Chiapas administered by the Zapatistas (Baschet 2019). Following his visit to Rojava, Graeber thus underlined the political transformation of the Kurdish political movement toward a more libertarian practice in Kurdish northeastern Syria (Graeber 2017; Schaepelynck and Sustam 2018). Yet a question remains: What differentiates the Rojava Revolution and the internationalist uprising, allowing the Kurdish people to assert their existence with other peoples, in a context where the autonomous institutions are neither virtual nor imaginary but represent a real possibility to create a future "together"? How should we account for and read the concept of the communal revolution among the Kurds, as an instance of solidarity and radical democracy directly counter to the Middle East's totalitarian regimes? The following will try to analyze these questions.

KURDISH SELF-GOVERNMENT:
AN EXPERIMENT IN A STATELESS SOCIETY

It should be recalled that the Kurds generally claim an autonomous government in each part of the Kurdish region (Turkey, Iraq, Iran, Syria). Indeed,
we are talking about four systems with completely different components.
Kurds in Iraq have an already established experience of federal government
in the form of the Kurdistan Regional Government of Iraq, in existence since
1992. The autonomy of the Iraqi Kurds in a representative and parliamentary
democracy is based on a neoliberal, financialized oil economy. In Iran, it has
only been possible for Kurdish people to develop limited cultural autonomy,
due to the colonial domination of the despotic regime in power since the 1979
Islamic Revolution. The focus here will be upon Rojava, the self-government
and confederal system in northeastern Syria, and Turkish Kurdistan, or
Bakûr, as a municipal experience (quasi "self-government") based on the
political theory of radical and representative democracy dependent on the
Turkish neoliberal economy (Aslan 2016, 93-98).

The Kurdish political movement (that of the PKK, PYD, and others) uses
two relative terms according to the regional political needs: those of *Xweserî*
(autonomy) and *Xwesêrîya demokratîk* (democratic autonomy) and thus
Demokratîk Konfêderalîzm (democratic confederalism) (Bance: 2017; 2020).
The Kurds have established self-governing institutions such as those of
TEV-DEM (Movement for a Democratic Society) to manage the administration and the socio-ecological project of the Rojava Revolution. Democratic
autonomy is a proposal of the Kurdish political movement, coordinated by
the Union of Kurdistan Communities (KCK) from 2005, alongside civil
society groups in the DTK (Democratic Society Congress, formed in 2017
with a horizontal structure, without "hegemony of state power"). It is in this
environment that the transformation of the new Kurdish subjectivity in the
Middle East begins. For the Kurdish political movement, to put ecology, the
emancipation of women, and self-government at the center of the process of
social change is to question the capitalist system and to occupy a position for
the transformation of Kurdish society.

The first thing to say is that the Rojava Revolution promotes democratic
self-governance based on ideas of libertarian anarchism. A communal system is being built using Öcalan's theories (2011a, 2013a and 2015), which
in turn are based on the concept of social ecology developed by Murray
Bookchin (Bookchin 1995, 1998, 2006) and are a departure from Marxist
"orthodoxy." We can also recognize the legacy of historic self-governance,
such as anarchist governance in the Spanish Revolution, or experiences,
such as the Zapatista movement in Chiapas (Baronnet 2013; Baschet and
Goutte 2015). In the Kurdish regions, a hermeneutic decoding of political

and ecological philosophy (grounded in the ideas of thinkers, such as Marx, Bakunin, Fanon, Foucault, Deleuze, Bookchin, Wallerstein, and Negri) has been begun. Generally, the definition of a stateless society entails a transformation of political conditions through social struggle. Sharing, comradeship, knowledge, relationships, desire, and emotion, all these constitute material and immaterial kinds of work, which allow for mobility in political organization made up of interconnected networks. TEV-DEM, for example, is driven by a powerful representation of women in governmentality (Duman 2016, 79–115). Much has been said about the place of women and minority society at the center of political action in the Rojava Revolution. However, what is at stake is less a political struggle against patriarchy alone than a revolution with a self-governance approach, which goes beyond the form of nation-state.

The concept of "molecular revolution" functions as a helpful tool for the analysis of the ecological revolution in Rojava. For Guattari, the concept is key as a form of political criticism for understanding the institutional practice of a revolution that refers to a transformation beyond notions of national revolution or to a class recovering state power (Guattari 2012, 54–76, 199–204, 218, 266, and 371). This seems an important critical tool, especially for the Kurdish political movement, which transformed its vision from the Soviet model toward a new libertarian and ecological policy. Guattari observes precisely how the institutional dynamics of the state, with its old left structure and practices of class struggle, internalized the bureaucratic and despotic aspects of its political vision (Guattari 2007). Likewise, in the period following the collapse of the Soviet bloc, the Kurdish political movement critiqued the despotic institutions of the Soviet system (Öcalan 1993, 2004, and 2011b). This then links analysis of the revolution to the micropolitical struggle for "identity" emancipation. For the Kurdish political movement, overthrowing power for the working class is not enough. It focuses on the unfinished liberation of sexes, genders, social class, and ecology (Öcalan: 2011a). In this way, the mentality of the movement decolonizes the territory and decentralizes the question of the struggle toward communalism.

Despite the war, the "co-chairs" of northeastern Syria are implementing a self-administered society and establishing a micro-economic system regarding the alternative economy, ecology, and land ownership as provided by Rojava's constitution, the "Social Contract." Rojava, now AANES, adopted the "Social Contract," on January 6, 2014,[1] as the foundation of a new political model based on libertarian municipalism and rethinking education, the health system, security, and the local micro-economy. Minorities are equally represented in the municipalities, since the "Social Contract" also advances the political integration of all peoples, whether Kurds, Yazidis, Alevis, Arabs, Armenians, Assyrians, Christians, or other micro-ethnicities. For several years, workshops and conferences were being organized that developed the idea of an alternative economy. The cooperative economy, the political economy of the revolution, has become concrete practice in Rojava, where it takes the form of regional cooperative and participatory societies. It creates companies which belong to the employees, who elect their manager; the cooperative economy is also associated with networks of consumers and producers in the free zone. The cooperatives that make up the solidarity economy in Rojava are based on the cultivation of the land and primary agricultural products, such as olives, and also oil (Lebsky 2016; Shilton 2019; Madra 2016). It is important to add that Rojava's economy remains something of a working laboratory, not a final outcome (Küçük and Özselçuk 2016; Stefani and Ruge 2019).

The declaration of democratic confederalism reveals a reconfiguration of priorities in the public sphere. When speaking of the micro-economic alternative, the "molecular revolution," in the urban and rural spaces in relation to Kurdish resistance, some political activists underline ecology as an important expression of self-governance and the emancipation of the colonized identity. Political ecology in Kurdistan is influenced by the theories of Bookchin and Öcalan, who analyzes social ecology, together with the historical accumulation of knowledge, the nation-state, sexism, religion, the hegemonic state apparatus, in *Democratic Confederalism* (Öcalan 2011a, 9–22). The Kurdish movement connects democratic confederalism (Öcalan 2011a, 21–35) to environmental issues by grasping ecological practice as a political lever for territorial and social emancipation that incorporates a global critique of the despotic model of capitalist global companies and nation-states. For example, the demonstrators of the Hevsel Gardens resistance in Diyarbakır, subject to "Special Project Area" measures (Emeç 2017), used the type of resistance from the barricades or "Hendek" against an extensive gentrification project. The Kurdish movement takes a clear stance in criticizing capitalism as a form of colonialism. For the Kurdish ecologists, the governmentality of the war and conflict does not only base itself on the ideological colonization of identity and the Kurdish territory, it also physically, and by force, colonizes the geography and nature of Kurdistan. This is the reason why the ecology movement in Kurdistan draws upon the social values and mythical history of Mesopotamia in its theory and has a constitutive vision of ecology as a rejection of capitalism's takeover of micro-territory. In this Kurdish context, social ecology, based on Bookchin's libertarian municipalism (Bookchin 1998), seeks to criticize the system of "capitalist modernity" (Network for an Alternative Quest 2012 and 2015); not only its most flamboyant excesses but also its state legal system. According to Öcalan's definition, the term "capitalist modernity" is used to redefine the globalization era, the crisis of contemporary society, and the diversion of capitalist wealth.

The dynamic of Kurdish ecological ideas, in the form of the first ecology movements that emerged during the Mesopotamian Social Forum (2011, in Diyarbakır's Sumer Park), has been influenced by Bookchin's theory for "a libertarian ecology" (Leverink 2015; De Long 2015), adapting its anti-colonial critique to the specificity of the Kurdish space, opposing masculine domination and colonialist war in the case of Turkey. Bookchin's conception of social ecology underscores the adjective "social" in ecological matters to problematize the profound social change applied to the institutions of the capitalist system (Bookchin 1982, 8). Bookchin's conception of nature beyond the system of world capitalism is interpreted and adapted in Öcalan's thinking about an ecological society as part of the new political proposal for cultural, sexual, and feminist emancipation (Öcalan 2004, 79–173). The Kurdish political movement focuses on three propositions: a sociohistorical theory of colonized territory, social ecology, and gender equality (an emancipatory project of women), within the libertarian municipalism of the system of democratic confederalism. This perception is based on the need for a transformation of understanding in a gendered ecopolitical way in the face of colonialist patriarchal culture. It must be emphasized that it is the municipality which has also been the central form of the Kurdish space since 1991, within the political movement and the legal context. The experience of the Kurdish municipality has begun to materialize since the local election of 1999 (well before the Rojava Revolution in 2012).

How does such a patchwork of micro-identities that constitutes the Kurdish space connect between the dynamic of the resistance and of peace and connect them to the practice of ecologic emancipation? Social ecology rejects colonial domination, basing itself on a definition of eco-geography as a decolonized perception of Kurdish identity in a subaltern culture within the framework of the revolt for self-governance. It mobilizes many types of activist networks at the heart of the political question with its dynamics, tensions, and confrontations and is a significant part of the new momentum of the heterotopic Kurdish spaces. In Rojava, the population is organized into assemblies, which include ecology assemblies, alongside neighborhood assemblies, and those constituted for women, religions, energy, youth, and others. The current strategy is to consider the cantonal municipality as autonomous of the state executive power. According to the "Social Contract," the autonomy of municipalities is structured from below. In this vision, the Öcalan-proposed democratic confederal system is one that rejects the nation, patriarchy, positivist scientism, hegemony, state administration, capitalism, and Fordist or post-Fordist industrialism and is the place of democratic autonomy, a social and alternative ecology (Bouquin, Court, Hond: 2017). Indeed, we have another example of an alternative institution. The University of Rojava is an important experiment to observe, since it identifies itself as a self-managing educational institution that is completely at odds with the statist Ba'athist educational system resulting from the culture of pan-Arab power. Another example of an initiative in civil society is "Jinwar," the women's self-managed ecological village in the canton of Cizîrê, near the city of Amûdê in Rojava (Oke 2017). This is an initiative where the women are subjects of this free autonomous space, organizing their living space according to their own decisions. Influenced by the feminism of *Jineolojî*, the science of women and free life (see Kurdistan au féminin 2017; and l'OCL 2015), the village is also based on the principles of local self-reliance and women's labor and aims to empower women to meet their basic needs. In this free zone in the north of Syria, the residents take up permaculture and design organic systems of agriculture inspired by Mesopotamian cultural heritage and nature traditions. The village upholds direct democracy, gender equality, and ethnic and linguistic pluralism through female subjectivity under the influence of Jineolojî.

THE SOCIAL FORUM AND THE ECOLOGICAL MOVEMENT IN BAKÛR

I will mostly speak here of three sites of sociopolitical ecology in Kurdistan: two physical ecological movements, those of *Tevgera Ekolojiyê ya Mezopotamyayê* (Mesopotamia Ecological Movement, hereafter MEM), *Jîngeh* (Space of Life), and an internet portal for new journalism on Kurdish social ecology called *Jîyana Ekolojîk-Dengê Xwezayê* (Ecological life/the sound of nature).

The MEM (Mesopotamian Ecology Movement 2016a), was founded during the international Mesopotamian Social Forum in Diyarbakır in 2011. It created self-governed, regional ecological assemblies which tackle problems resulting from the war and the politics of the Kurdish question. There is solidarity with the ecological movements of western Turkey, and they work together to oppose nuclear and hydroelectric power plants on the coasts of the Mediterranean and the Black Sea. Such cement constructions treat the earth and the forest like goods to be used and alienate the local population. The movement for democratic confederalism organized a conference on April 23–24, 2016, in Van. The suffocating destruction capitalism inflicts on society and the depredations inflicted on nature were raised, together with the response of self-government as a form of resistance against the state. The movement goes further than the history of Kurdish anti-colonial resistance. It also criticizes poverty, precarity, unemployment, and the unhealthy food choices imposed by industrial agriculture and genetic modification. Another talking point was the enormously destructive project of the hydroelectric dams, started by Turgut Özal's government at the beginning of the 1990s, as well as the GAP project in south Anatolia and the Kurdish region. The AKP government continues to devastate the region with programs inherited from this era, such as dams on the Munzur and the Ilisu dam at Hasankeyf. The MEM positions itself against the nation-state and its capitalist capacity. According to the movement, the mobilization of an ecological resistance gives crucial importance to a culture of sharing. To fight against the socio-political destruction of the government and war, the movement proposes to communalize the earth, water, and electricity. As was declared at their last conference:

Our struggle is an important contribution to the liberation of people and nature on our planet. We strive to attain a truly natural society, the fundamental justification of our existence. [. . .] We announce the 21st century to be a brilliant age built by new generations. We will see a radical, democratic society of free women. The ecological struggle is larger than a single-issue struggle, it incorporates the vital essence of the paradigm of free life itself. Without ecology, society cannot exist, and without humanity and nature, the ecology cannot exist. Ecology is the essence of a millenarian dialectic of generation and regeneration, it connects all natural, interdependent elements like the rings of a chain. [. . .] It can only develop in a sociopolitical movement and through a struggle for liberation that takes a position against the system that puts nature, society and the individual into peril in the interest of profit, capitalism, and the hegemonic state. [. . .] In the Middle East, the history of ecology has not yet been written. To arrive at the liberation of women, it was necessary to learn the history of women. In the same manner, in order to arrive at an ecological society, it's necessary to understand the history of ecology. We can spread consciousness by opening ecology academies. Ecology will be an essential component of the study programs in all social spheres and all university programs. Spreading ecological consciousness and sensibility in the social sphere and in educational institutions is as vital as organizing our own self-governed assemblies. (*Sources*: Eko-teknoloji çalistayi (January 22, 2016, Amed), Enerji calistayi (January 9, 2016, Urfa), Orman çalistayi (January 10, 2016, Dersim), Su çalistayi (December 12–13, 2015, Wan), Ekolojik Kentler çalistayi (December 12, 2015), Çevre saglik çalistayi (December 26, 2015, Antep sonuç bildirgeleri)

Concluding Declaration of the 1st Conference of the MEM (see Aslan 2014):

• An intellectual, organizational strategy must be put into place, and coordination with national and international ecology movements must be assured in order to improve the discussions and the communal actions against destruction and ecological exploitation.

- Mental, physical, and ideological destruction must be fought, and the topics of energy, water, forests, earth, cities, seeds of agriculture, and technology must be addressed. We must mobilize for struggle to construct a new way of life on the basis of the politics of the Ecology Movement of Mesopotamia that we have discussed.
- We must fight against a system which demolishes the urban agglomerations and burns the forests in Kurdistan. We must make the ecological devastation that happened in Kurdistan known and chart a map of the destruction the war has caused.
- Actions must be planned in coordination with other ecological movements. Our actions must also address the destruction of Kurdish cities, and we must actively participate in solidarity platforms that have been established in those cities.
- The struggles to preserve the cultural and natural sites of Kurdistan have to continue. There are many sites that face extinction, such as Hasankeyf, Diyarbakır-Sur, the Munzur Valley, and "Gele Goderne." They are under threat because of the politics of energy and security.
- An ecological model adapted to Kurdistan must be developed.
- We must work toward a bigger and more regular presence in the print and digital media, and an ecology academy must be established.
- Legal battles currently running parallel to actions and campaigns must be brought to a successful end.
- Organizational structures everywhere in Kurdistan and the Middle East must be developed (translated from the Turkish original, Mesopotamian Ecology Movement 2016b).

Since its inception, the MEM has brought many projects to fruition. Together with the Kurdish municipalities, it has organized workshops that broach environmental issues such as water, forest, fields, agriculture, technology, ecological buildings, health, communal economy, and poverty. These workshops were organized in towns like Mardin, Van, Diyarbakır, Urfa, Dersim, Antep, and Batman. These big cities were chosen for their political history that positions them at the heart of the Kurdish resistance.

Concerning ecological groups in Turkey's Kurdish region, Jîngeh (space of life, the collective ecology) is an antiauthoritarian, decolonial anarchist movement based in Diyarbakır and Van that is part of the MEM. Jîngeh activists have undertaken a campaign tour (from Dersîm, to Hewlêr in the KRI) to attract attention to the forest fires in Turkey's Kurdish region. The Turkish state's burning of forests and villages in the 1990s marked the collective memory, and it is one reason why the Kurdish political movement makes ecology a central issue in the micropolitics of the Kurdish identity (Yeşil Gazete 2020). Jîngeh raises awareness about social ecology and biodiversity in Kurdistan, taking an interest in environmental repercussions and criticizing the territorial framework of the nation-state (Gazete Karinca 2021; Jingeh).[2] They explore and share knowledge about relevant subjects, for example, organizing evening events to discuss ecology or different micropolitical theories, including libertarian, anarchist theories.

Finally, it is important to mention the portal, Jiyana Ekolojî, an online and social media forum for environmental journalism that covers ecological questions and resistance in Kurdistan from an anti-colonial perspective. Jiyana Ekolojîk (meaning "ecological life") is a platform for the journalistic voice of the ecology movements in Kurdish spaces and is also in contact with ecologists from western Turkey. Jiyana Ekolojîk aims to revive traditional Kurdish practices and values. Nevertheless, this is not a movement, but a website gathering material to expand the internal and external relations of the ecology movements (Jiyana Ekolojîk 2015). The portal uses social media creatively to illustrate collective life and representing resistance at the everyday, micropolitical level. Last, in Turkey's Kurdish region, it incorporates self-government organizations in an autonomist archive, covering areas such as *Tevgera Ekolojiyê ya Mezopotamyayê* (MEM), *Rêveberiya Xweser a Demokratîk* (Democratic Self-Government), *Avedanî û Bajarvanî* (Environment and Urbanism), *Ziman, Çand û Bawerî* (Language, Culture, and Religion), Jîngeh (Space of life), *Ekolojî û Şîngeh* (Ecology and Shelter), *Geşkirina Aboriya Xwecihî* (Local Economic Growth), *Tenduristiya Gel* (Public Health), and *Heyberên Çandî û Dîrokî û Turîzm* (Cultural, Historical, and Tourist sources). The critique of capitalism as a global system is specifically situated here, as a theory and discourse appertaining to the Kurdish space. The ecology movement defines nature as oppressed, like the Kurdish identity. Its central focus is on the war, which has a double impact on nature and the population, relentlessly exposed without their own state. The ecology movement takes the question of nature, opposing the big industries that run agriculture, as a starting point, but it ultimately forces us to rethink the entire emancipation struggle of the Kurds and reformulate it as part of a desire for liberation that faces a hegemonic system. To a great degree, taking into consideration the specificity of colonial Turkey, the liberation of the Kurds makes ecological critique its symbol, positioning itself outside of modern, capitalist civilization, as much as outside of colonial power. The Kurds see ecological practice and life as intrinsic to their revolt against Turkish colonialism and capitalism in the Middle East.

This new transformation of Kurdish space also constitutes a dissolution of the state military apparatus in the region, an extension of the realm to political relationships. The Kurdish zone then becomes a factory of autonomist political action, since it is transnationalized by various effects of struggle for emancipation and actor-networks. The dynamics of ecology in Kurdistan, from the appearance of the first ecology movements in the Social Forum of Mesopotamia at Bakûr (for "an ecology of freedom"), is based on an anti-colonial critique in relation to the nature of the Kurdish space toward the patriarchal and colonialist domination of the war in Turkey. And Kurdish actors export this experience in their own context in Rojava after the 2012 Revolution. Kurdish ideas about social ecology, especially as applied in Bakûr and Rojava, forms the new challenge of a geopolitical approach which sets aside orthodox left politics (the nation-state, Marxism, Leninism, and Maoism) and colonial intervention and thus positions itself as a way to reflect decolonized anti-capitalism. According to the ecologists' approach, the governmentality of war and conflict does not only colonize Kurdish identity and territory, it is also a mechanism for colonizing geography and nature in Kurdistan. This is why the thesis of the ecology movement in Kurdistan interprets social values, the mythical history of Mesopotamia, the constitutive vision of ecology as a perspective of micro-territoriality against colonialism and capitalism.

In addition, it was decided, for example, to incorporate articles on animal rights and their protection (animal liberation) and a conscientious objection against the call to compulsory military service (civil disobedience and anti-militarism) into Rojava's "Social Contract." The "Social Contract" defends the collective rights of societies, rejects patriarchy, and favors a self-managed economy to advance politically toward the emancipation of women and society. The cantons continue to "reinforce" their autonomist goals despite the demands of wartime. The mobilization of ecological resistance in Kurdish space also creates a crucial sharing and commonality within the colonized society in Kurdistan, proposes a struggle of social ecology against the socio-political destruction of the necropolitics of violence and war that eliminate the areas of freedom for peoples and rural life and the environment. That is why the movement proposes to communalize the land, water, and energy, setting up a free, democratic life against the nation-state and capitalism. The system of "capitalist modernity" prevents the creation of ecological cities, alternative energies, and a sharing free society due to the monopoly of the big industries over agriculture, villages, and other aspects of rural space.

CONCLUSION

In summary, we have a critical approach on three levels. The Kurdish space comes with an urban micropolitical patchwork of identity affiliations, an emancipation embodying the politics of social ecology. This emancipation comes into being through the performance of a heterogeneous movement putting up resistance against the war. We approach it dialectically from the angle of counterpower and countercultural reproduction which transcends any conventional ideological behavior (especially the ideology of the state) in the Middle East. These revolts must be analyzed as a new micropolitics which takes strong positions in relation to the environment, micro-identities, the crisis of society and, on the opposite side, necropolitics and, to use Foucault's term, biopolitical governmentality, where subversive violence is in action. We also try to grasp the new codes of collective pronouncement that were created in the revolts. This Kurdish space became the manufacturing center for political libertarian action as a result of having been singularized and transnationalized by various actors since the Rojava Revolution, as I have argued elsewhere, and as Graeber suggests in an interview given in 2017 (Sustam 2016; Graeber 2017).

To conclude, social ecology rejects colonial domination and builds on the concept of eco-geography defined as an anti-colonial geography of the Kurdish minority identity. It is part of a larger framework of self-governed revolt that mobilizes networks and actors that, in their turn, are at the center of a political question with its own dynamics, tensions, and confrontations. We pose the following question: how can a patchwork of micro-identities connect the dynamic of resistance, peace, and the practice of ecological emancipation? Therefore, there is a need to broaden the debate on building radical and direct democracy in the Kurdish space for the future, which after the era of uprising shows the communalist social imaginary based on the "constituent power" approach. This Kurdish communalism is a cross-border experience and expresses the capacity of Rojava's "Social Contract" to represent democratic and autonomist opposition against colonialist and state sovereignty in the Middle East. The administration of the Kurdish municipalities of Rojava is democratic and semi-decentralized; the local administrations of several settlements and communes give people autonomy and control in making decisions that may affect their lives. All that is significant to observe after the experience of democratic municipality in Bakûr and the ecological revolution in Rojava, the Kurds also created their own notion regarding the struggle for emancipation and ecology, such as "jîngeh, xweserî, jînwar, jinêolojî," xweseriya ekolojî, and çalakî (for definitions of Kurdish ecological terms see Ecomark 2019). The destruction of the habitat in Kurdistan threatens many species. The Kurdish political movement has also embarked on cultural and administrative initiatives on land and health. Among all these intergenerational transformations, the most emblematic subjects dealt with are nature, ecology, municipality, and gender within the struggle for emancipation. This political change from the anti-colonial politics of the 1990s is integrated into the new decolonial position of the movement and the institutional application through the organizations of the municipality, local government, and other aspects of the political framework.

In the period following the Arab Spring, the Kurds had the historic opportunity to experience liberation for the first time in more than a century in Syria, a moment of emancipation which has evolved into new practices informed by social ecology and communalist life. This experience also gave the Kurds opportunity to decolonize their territory and reverse previous policies. Kurdish communalism and libertarian municipalism represent a historic, democratic form of political organization based on the recognition of ecology and women's freedom and on the defense of the autonomy of multiple communities against the "repressive state" and despotic regimes. As we have seen over the last decade, democratic forces operating in this political space include not only libertarian perspectives but the solidarity economy and women's spaces (such as Jinwar) and thus ecological production zones. Writing at the dystopian time of the COVID-19 pandemic, it seems to me that we must have debates based on the rejection of corporate capitalism and the global factory by creating a new ecological perspective and an alternative democratic society in opposition to the society of control and consumption.

NOTES

1. "Charter of the Social Contract: Self-Rule in Rojava." January 29, 2014. *Peace in Kurdistan* website. https://peaceinkurdistancampaign.com/charter-of-the-social-contract/.

2. Jîngeh on Facebook: https://www.facebook.com/J%C3%AEngeh-584222668356661/.

REFERENCES

Aslan, Azize. 2014. "Democratic Economy Conference in Wan." [November 8–9, 2014]. Published by the Institute for Social, Political and Economic Studies, *Co-operation in Mesopotamia* website. https://mesopotamia.coop/democratic-economy-conference-in-wan/.

Aslan, Azize. 2016. "Economic Self-governance in Democratic Autonomy: The Example of Bakûr (Turkish Kurdistan)." *Birikim* no 325 (May): 93–98.

Bance, Pierre. 2017. *Un autre futur pour le Kurdistan? Municipalisme libertaire et confédéralisme démocratique.* Paris: Noir et Rouge.

Bance, Pierre. 2020. *La fascinante démocratie du Rojava: le contrat social de la Fédération de la Syrie du Nord.* Paris: Noir et Rouge.

Baronnet, Bruno. 2013. "L'Expérience D'Education Zapatiste au Chiapas: Entre Pratiques Politiques et Imaginaires Autochtones à L'école." *Cahiers de la Recherche sur L'Education et les Savoirs* 12: 133–52. https://journals.openedition.org/cres/2341.

Baschet, Jérôme. 2019. La rébellion zapatiste: Insurrection indienne et résistance planétaire, Paris: Flammarion.

Baschet, Jérôme and Guillaume Goutte. 2015. *Enseignements D'une Rébellion: La Petite école Zapatiste.* [S.l.] : Éditions de l'Escargot.

Bookchin, Murray. 1982. *The Ecology of Freedom: The Emergence and Dissolution of Hierarchy*, Palo Alto, CA: Cheshire Books.

Bookchin, Murray. 1995. *Social Anarchism or Lifestyle Anarchism, An Unbridgeable Chasm.* Edinburgh: AK Press.

Bookchin, Murray. 1998. *The Politics of Social Ecology: Libertarian Municipalism.* Montréal: Black Rose Books.

Bookchin, Murray. 2006. *Social Ecology and Communalism.* Oakland: AK Press.

Bookchin, Murray. 2011. *Une société à refaire: Vers une écologie de la liberté*, Montréal: Éditions Écosociété.

Bookchin, Murray. 2015. *Au-delà de la rareté.* Montréal: Éditions Ecosociété.

Bouquin, Stephen, Mirielle Court, and Chris den Hond, eds. 2017. *La Commune du Rojava: l'alternative kurde à l'état nation.* Brussels: Coédition Critica and Paris: Editions Syllepse.

Collectif Ne var Ne Yok. 2016. *"Serhildan," Le soulèvement au Kurdistan: Paroles de celles et ceux qui luttent pour l'autonomie,* [Marseille]: Niet Editions.

Comité invisible. 2007. *L'insurrection qui vient.* Paris: La Fabrique.

Cooperativa Integral Catalana. 2016. "Economic Self-Governance in Democratic Autonomy: The Example of Bakûr (Turkish Kurdistan)." *Cooperativa Integral Catalana* website. July 9, 2016. http://cooperativa.cat/en/economic-self-governance -in-democratic-autonomy-the-example-of-bakur-turkish-kurdistan/.

De Long, Alex. 2015. "Kurdish Autonomy Between Dream and Reality." [interview with Joost Jongerden], *Roar Magazine* (June 4, 2015): https://roarmag.org/essays/ kurdish-autonomy-jongerden-interview/.

Deleuze, Gilles. 1993. *Critique et clinique.* Paris: Éditions de Minuit.

Deleuze, Gilles, and Félix Guattari. 1991. *Qu'est-ce que la philosophie?* Paris: Éditions de Minuit.

Duman, Yasin. 2016. *Rojava, Bir Demokratik Özerklik Deneyimi.* Istanbul: Iletişim.

Ecomark. 2019. "Termên Ekolojî." *Ecomark* website. https://www.ekoloji.com/ku/ ekoloji/ekoloji-terimleri/, with accompanying English language webpage.

Emeç, Hülya. 2017. "Dicle Valley and Hevsel Gardens Open for Pillaging with Statutory Decree." *ANF News* website. January 17, 2017. https://anfenglish.com/ku rdistan/dicle-valley-and-hevsel-gardens-open-for-pillaging-with-statutory-decree -18053.

Erbay, Vecdi. 2017. "Amed Ekoloji Platformu: Hevsel Bahçeleri Talan Ediliyor." *Gazete Duvar,* March 17, 2017. https://www.gazeteduvar.com.tr/gundem/2017/03 /17/amed-ekoloji-platformu-hevsel-bahceleri-talan-ediliyor.

Evrensel. 2014. "Hevsel Direnişi Güçlenerek Sürecek," *Evrensel* website, March 11, 2014. https://www.evrensel.net/haber/80019/hevsel-direnisi-guclenerek-surecek.

Foucault, Michel. 1994. *Dits et écrits, Tome III: 1976–1979.* Paris: Gallimard.

Foucault, Michel. 2004a. *Naissance de la Biopolitique.* Paris: Gallimard.

Foucault, Michel. 2004b. *Sécurité, territoire, population.* Paris: Editions du Seuil.

Foucault, Michel. 2008. *Le Gouvernement de soi, et des Autres*. Hautes Etudes, Gallimard, Seuil.

Gazete Karinca. 2021. "Mezopotamya Ekoloji Hareketi." *Gazete Karinca* website (reports relating to MEM). https://gazetekarinca.com/tag/mezopotamya-ekoloji-hareketi/.

Graeber, David. 2011. *Debt: The First 5,000 Years*. Brooklyn, NY: Melville House.

Graeber, David. 2017. "Syria, Anarchism and Visiting Rojava." *Cooperation in Mesopotamia* website. July 5, 2017. https://mesopotamia.coop/david-graeber-syria-anarchism-and-visiting-rojava/.

Guattari, Félix. 2012. *La Révolution moléculaire*. Paris: Les Prairies Ordinaires.

Guattari, Félix. 2014. *Qu'est-ce que l'écosophie?* Paris: Éditions Lignes.

Guattari, Félix, and Suely Rolnik. 2007. *Micropolitiques*. Paris: Empêcheurs de Penser en Rond.

Günes, Cengiz. 2012. *The Kurdish National Movement in Turkey: From Protest to Resistance*. London: Routledge.

Hardt, Michael, and Antonio Negri. 2017. *Assembly*. New York: Oxford University Press.

Jiyana Ekolojîk. 2015. [Homepage of Jiyana Ekolojîk]. http://www.jiyanaekolojik.org/ (site discontinued). Accessible via Wayback Machine, March 7, 2021. https://web.archive.org/web/20160303134011/http://www.jiyanaekolojik.org/.

Knapp, Michael, Anja Flach, and Ercan Ayboğa. 2016. *Revolution in Rojava: Democratic Autonomy and Women's Liberation in Syrian Kurdistan*. London: Pluto Press, 2016.

Küçük, Bülent, and Ceren Özselçuk. 2016. "The Rojava Experience: Possibilities and Challenges of Building a Democratic Life." *South Atlantic Quarterly* 115, no. 1: 184–196. https://doi.org/10.1215/00382876-3425013.

Kurdistan au féminin. 2017. "Qu'est-ce que la Jineolojî, une science des femmes?" *Jineolojî* website. December 24, 2020. http://jineoloji.org/fr/quest-ce-que-la-jineoloji-une-science-des-femmes/.

Lazzarato, Maurizio. 2004. *Les Révolutions du capitalism*. Paris: Empêcheurs de penser en rond.

Lazzarato, Maurizio. 2008. *Le Gouvernement des inégalités: critique de l'insécurité néolibérale*. Paris: Amsterdam.

Lazzarato, Maurizio, and Antonio Negri. 1991. "Travail immatériel et subjectivité." *Futur Antérieur* 6 (Summer): 87–99.

Lebsky, Maksim. 2016. "The Economy of Rojava." *Co-operation in Mesopotamia* website, March 17, 2016. https://mesopotamia.coop/the-economy-of-rojava/.

Leverink, Joris. 2015. "Murray Bookchin and the Kurdish Resistance." *Roar Magazine,* August 9, 2015. https://roarmag.org/essays/bookchin-kurdish-struggle-ocalan-rojava/.

Madra, Yahya Mete. 2016. "Democratic Economy Conference: An Introductory Note." *South Atlantic Quarterly*, 115, no. 1: 211–222. https://doi.org/10.1215/00382876-3425035.

Mbembe, Achille. 2003. "Necropolitics." *Public Culture* 15, no. 1 (Winter): 11–40. https://doi.org/10.1215/08992363-15-1-11.

Mesopotamian Ecology Movement. 2016a. [Homepage of the MEM]. http://mezopotamyaekolojihareketi.org/Anasayfa (site discontinued). Accessible via Wayback Machine, March 7, 2021. https://web.archive.org/web/20160312212100/http://mezopotamyaekolojihareketi.org/Anasayfa.

Mesopotamian Ecology Movement. 2016b. "Final Declaration of 1st Ecology Conference (April 23–24, Wan)." Last modified June 8, 2016. English language translation available on *Initiative to Keep Hasankeyf Alive* website. https://www.hasankeyfgirisimi.net/final-declaration-of-1st-ecology-conference-april-23-24-wan/.

Network for an Alternative Quest. 2012. *Challenging Capitalist Modernity, Alternative Concepts and the Kurdish Quest*, Cologne: Internationale Initiative Edition.

Network for an Alternative Quest. 2015. *Challenging Capitalist Modernity II: Dissecting Capitalist Modernity—Building Democratic Confederalism.* Hamburg: Internationale Initiative Edition.

Öcalan, Abdullah. 2004. *Bir Halki Savunmak.* Istanbul: Çetin Yayınları.

Öcalan, Abdullah. 2011a. *Le Confédéralisme démocratique.* Cologne: International Initiative Edition.

Öcalan, Abdullah. 2011b. *Türkiye'de Demokratikleşme Sorunları ve Kürdistan'da Çözüm Modelleri (Yol Haritası).* [s.l.]: Azadi Matbaasi.

Öcalan, Abdullah. 2013a. *La Feuille de route vers les négociations: Carnets de prison.* Internationale Initiative Edition.

Öcalan, Abdullah. 2013b. Demokratik Uygarlik manifestosu 1-2-3-4-5, A. Öcalan Sosyal Bilimler Akademisi Yayinlari.

Öcalan, Abdullah. 2015. *Manifesto for a Democratic Civilization: Civilization: The Age of Masked Gods and Disguised Kings.* Porsgrunn: New Compass Press.

l'OCL. 2015. "Le Mouvement pour une Société Démocratique (TEV-DEM) c'est quoi?" *Ecologie Sociale.ch* website. October 23, 2015. http://www.ecologiesociale.ch/international/index.php/kurdistan/51-le-mouvement-pour-une-societe-democratique-tev-dem-c-est-quoi.

Oke, Naz. 2017. "Jinwar, village de femmes au Rojava." *Kedistan* website. February 27, 2017. http://www.kedistan.net/2017/02/27/jinwar-village-de-femmes-au-rojava/.

Rancière, Jacques. 2009. *Dissensus: On Politics and Aesthetics.* London: Continuum.

Rojava Information Center. 2020. "Explainer: Cooperatives in North and East Syria – Developing a New Economy." *Rojava Information Center* website: https://rojavainformationcenter.com/2020/11/explainer-cooperatives-in-north-and-east-syria-developing-a-new-economy/.

Schaepelynck, Valentin and Engin Sustam. 2018. "Autogestion." *Le Télémaque* 54, no. 2: 27–36.

Shilton, Dor. 2019. "The Radical Eco-Anarchist Experiment Betrayed by the West, and Bludgeoned by Turkey." *Ecologise.in* website, October 27, 2019. https://ecologise.in/2019/10/27/rojava-the-eco-anarchist-experiment/.

Stefani, Daniel, and Edmund Ruge. 2019. "Experiments in Radically Democratic Education: Rojava's Revolutionary Model at Risk Amid the Geopolitics of War."

Periferias (December 20, 2019): http://revistaperiferias.org/en/materia/the-linguistic
-pedagogy-of-co-existence-at-kurdish-schools/.

Sustam, Engin. 2016. "Rojava: A Municipal Autonomy Experiment in Times of
War," interview by Claudio Pulgar and Charlotte Mathivet, in "Unveiling the Right
to the City, Representations, Uses and Instrumentalization of the Right of the City."
Passerelle Collection published by Ritimo, Habitat International Coalition, no. 15
(September): 124–130.

Sustam, Engin. 2020. *Kırılgan Sapmalar- Sokak Mukavemetleri ve Yeni Başkaldırılar.*
Kalkedon: Istanbul.

Tatort, Kurdistan. 2013. *Democratic Autonomy in North Kurdistan, The Council
Movement, Gender Liberation, and Ecology in Practice.* Porsgrunn: New Compass
Press.

Üstündağ, Nazan. 2018. "The Wounds of Afrin, the Promise of Rojava: An Interview
with Nazan Üstündağ." *LeftEast* website. February 14, 2018. http://www.criticatac
.ro/lefteast/nazan-ustundag-interview/.

Yeniay, Özlem. 2017. "Topraksızlar Hareketi ve Kürt Özgürlük Mücadelesinde
Alternatif Yaşam ve Ekoloji: Tartışmalar/Karşılaştırmalar." ["Alternative Economy
and Ecology within the Landless Rural Workers Movement and the Kurdish
Freedom Movement: Discussions/Comparisons"]. *Toplumve Kuram* no. 12. http:/
/zanenstitu.org/uploads/dosyalar/Kitaplar/Toplum%20ve%20Kuram/12sayi/Topra
ks%C4%B1zlar%20Hareketi%20ve%20Ku%CC%88rt%20O%CC%88zgu%CC
%88rlu%CC%88k%20Mu%CC%88cadelesinde%20Alternatif%20Yas%CC%
A7am%20ve%20Ekoloji.pdf.

Yeşil, Gazete. 2020. "Mezopotamya Ekoloji Hareketi: Askeri operasyonlar
ormanlarımızı yok ediyor." *Yeşil Gazete* website. July 27, 2020. https://yesilgazete.
org/mezopotamya-ekoloji-hareketi-askeri-operasyonlar-ormanlarimizi-yok-
ediyor/.

Chapter 4

Radical or Reactionary Tomatoes?

Organizing against the Toxic Legacy of Capital's Environmentalism

Nicholas Hildyard

These days if you eat out in one of Chicago's award-winning restaurants or bars (pandemic restrictions at the time of writing permitting), the chances are that some of the vegetables on your plate will have been grown on one of the city's rooftop "city farms" (DiNardo 2019). Live in the right apartment block in the right neighborhood and you can benefit without even having to leave your home: for a weekly subscription, rooftop grown fruits and vegetables will be delivered to your door. Chicago is not unique: across North America, rooftop farms and "green roofs" are blooming. According to Toronto-based industry association Green Roofs for Healthy Cities, the number of green roofs in North America has increased by 15 percent in the last seven years (DiNardo 2019). Cities such as New York, Seattle, Toronto, San Francisco, and Washington DC have all introduced new building regulations to encourage greater use of rooftops to grow food or gardens (Stern et al. 2019). Outside of North America, Córdoba in Argentina now requires all rooftops—new or existing—of more than 1,300 square feet to be turned into green roofs (DiNardo 2019); Dubai is developing 12,000 hectares of rooftop horticultural projects (Eye of Riyadh 2017); more than 10 percent of houses in Germany (said to be the world leader in green roof technologies) have green roofs (Li and Yeung 2014); and Singapore is even installing green roofs on buses (Hardman and Davies 2019).

The potential environmental benefits are considerable. In addition to increasing the space for food production, green roofs can reduce air-conditioning use in a building by as much as 75 percent and save energy more widely through reducing the tendency of cities to create "heat islands" (Livingroofs undated); mitigate floods by retaining rainwater (Hardman and

Davies 2020); and increase urban biodiversity. Small wonder that many now see them as integral to the greening of future cities.

But green roofs are not just roofs used to grow vegetation: who owns them; how they are financed; who labors on them and under what conditions; who gets to eat what is produced on them and on what terms; who decides how they are managed and for whom; all combine to set in train widely differing social, economic, political, and environmental trajectories. As Mexican social activist Gustavo Esteva remarks: "Some rooftop tomatoes are reactionary but others are radical" (Esteva 2013). They are reactionary when, for example, they are genetically engineered—the gene splicers are already at work to create tomatoes "bunched like grapes" that are better suited to cramped urban spaces than traditional varieties (Unrein 2019)—and grown by labor employed by venture capital-backed entrepreneurs for sale at a profit to exclusive restaurants that can only serve to entrench the exploitative dynamics of capital accumulation that despoil the environment in the first place. They are radical when, say, cultivated through collective endeavor for use rather than profit, prepared and eaten communally, and bearing the potential to weave relationships of comradeship, trust, and mutuality that are the lifeblood of social formations where survival is not the "isolated right of the individual" (Illich 1983) but the collective goal of common behavior.

On the face of it, the Kurdish movement's commitment to democratic confederalism (Öcalan 2017)—with its embrace of inclusive, participatory decision-making practices; its rejection of patriarchy; its opposition to capitalist forms of production and exchange; and its recognition of "ecology" as a pillar of its politics—provides promising ground for radicalizing not just tomatoes but forests, rivers, streams, lakes, mountains, fields, trading arrangements, highways, power plants, factories, homes, offices, and the livelihoods they support. In part, that radicalizing promise rests on the clear rejection of what the anarchist writer and activist Murray Bookchin (whose influence on the Kurdish movement's thinking has been substantial) called "the shopworn Earth Day approach to engineering nature so that we can ravage the Earth with minimal effect on ourselves" (Bookchin 1987)—the rejection of those who "are simply trying to make a rotten society work by dressing it in green leaves and colourful flowers while ignoring the deep-seated roots of our ecological problems" (Bookchin 1987). In part, it rests on the recognition that the ecological movement is not automatically a force for social justice just because it is against environmental despoilation. Far from it: deeply reactionary strands of thinking and doing permeate environmentalism, from the neo-Malthusianism of "deep ecology" (a prime butt of Bookchin's criticisms) to those who employ "nature" to "justify" sexist and racist social orders or who seek to absolve the perpetrators of destruction by obscuring its deep roots in colonialism, imperialism, white supremacy, and patriarchy (Lohmann 2020a).

TRAPPED BY LANGUAGE—WHOSE "ENVIRONMENT"?

Such reactionary strands of environmental thought have left an inheritance of categories, tools, and concepts that it would be dangerous for any would-be cultivators of radical tomatoes/forests/waste disposal systems/roads/irrigation systems or other sinews of society to take at face value.

Just as women—who have no option but to use languages (such as contemporary English) that are replete with patriarchal assumptions—are often left with no vocabulary to express what is essential for them to express (Spender 1980), so too those who seek to radicalize ecology are sometimes trapped by the received language of capital's environmentalism. The need to deconstruct, decolonize, and rework the concepts that are currently used to "manage" the environment is thus urgent. One way forward may be to learn from steps that feminists have taken—the *jineoloji* program of the Kurdish women's movement is an example (Briy and Anonymous 2019)—to explore and analyze the historical roots of patriarchy in order better to subvert the political and economic interests that continue insidiously to shape patriarchy today.

The challenge is huge. Even a simple word like "environment," which is hard to avoid in any discussions on ecological issues, comes equipped with in-built biases and assumptions that occlude or suppress perspectives that do not reflect mainstream economic and political interests—and in doing so favor outcomes that benefit the dominant social order. Environmental impact assessments, for example, start from the premise that the "environment" is a given: it is simply what is around us. But far from being a self-evident, universal, and unproblematic category, the "environment" and the "fields," "forests," "rivers," "streams," and "hills" that, Meccano-like, are taken to be its equally self-evident, universal, and unproblematic components, are always fiercely contested political spaces.

Take forests. For those who rely directly on them for their livelihoods, the trees and undergrowth, mosses, birds, bees, and flowers represent secure water supplies, fodder for animals, housing materials, medicines for friends and family, a home for local deities, and shelter from army patrols, tax collectors or (for playful children) adults (Hildyard et al. 2001). By contrast, for many middle-ranking forest department officials, "forests" are defined instead by the information that passes across their desks: the latest scientific paper on planting regimes; budgets for planting; tenders for logging; catalogues advertising new logging equipment or the latest jeep; curricula vitae; training schemes, and opportunities for promotion. For logging company accountants, forests may be no more than board feet of timber; for many pharmaceutical researchers, they are pools of "biodiversity" from which to extract patented drugs; while for harried executives in polluting industries, they have become "sinks" to be created (or preserved) to offset carbon dioxide emissions (Hildyard et al. 1997). Degradation of forests, therefore, has radically different meanings for different groups of people because of differing consequences—inevitably giving rise to different approaches to tackling environmental degradation. For many in government, business, and international organizations, such degradation—together with the protests it provokes—tends to be viewed as a threat to their political and economic interests. For them, the environment is not what is around their homes but what is around their economies (Hildyard et al. 1995).

The preferred response of many planners, politicians, development practitioners, civil servants, and heads of industry lies in increasingly global forms of management that are instrumental and (inevitably) top-down. The world is split up into fixed ends and available means. Then, in a process that is taken to be synonymous with rationality, the means are matched to the ends. In doing so, nearly everyone and everything is transformed into tools whose effectiveness in "helping us get from A to B" is the prerogative of the managers themselves to decide and measure. Acting on "objective data," managers plan, mobilize, and "clear space for action." Others, whose lack of skills and autonomous ends are either assumed or enforced, are "empowered" only in so far as they can be used instrumentally to carry out the managers' designs. People become "obstacles" to be removed or cajoled into "collaboration"; the physical landscape a terrain to be reordered, zoned, and parceled up according to some preconceived Master Plan (Hildyard et al. 1997). By contrast, for grassroots groups who rely on the forests for their livelihoods, the debate is often not only over such technical questions as how to conserve soil or what species of tree to plant but also over how to create or defend open, democratic community institutions that ensure people's control over their own lives. One central demand made by group after group is for a great deal of authority to be vested in the community—not in the state, local government, the market, or the local landlord, but in those who rely on the local commons for their livelihood. For these groups, the struggle is for more than the mere recognition of rights over the physical commons: critically, it is also a struggle to restore or to defend the checks and balances that limit power within the local community (Hildyard et al. 1997).

Which of these differing interests gets heard and implemented is an outcome of their relative organizing power, itself a reflection of their economic and political position within society, of racism and gender discrimination, and of other inequalities. Those interests and inequalities—and their social, economic, and ecological justice dynamics—should be at the very heart of impact assessments: only then can the drivers of destruction be understood and addressed. Instead, they are currently largely hidden, not just by reactionary language but also by techniques, such as cost-benefit analysis (Lohmann

1997; O'Neill 1993) that are deliberately designed to truncate or exclude open-ended democratic discussion of values, power relations, and the like. Forces of destruction get a free pass. Business-as-usual ecology with all its reactionary tomatoes rolls on largely unscathed.

CLIMATOLOGY: DON'T ASK, DON'T ACT

"Climate" offers another example. As my Corner House colleague Larry Lohmann (2020b) has argued, climate movements have inherited a conception of climate science that rules off-limits most of what is most important to understanding and addressing damaging climate change. As Lohmann records:

> In 2014, Sir John Houghton, founding member of the Intergovernmental Panel on Climate Change, gave an interview explaining that UN climatologists were not permitted to mention the carbon locked up in fossil fuels in their analysis of climate change, but only carbon that had become more mobile in the form of CO_2. To follow what happens when carbon atoms cross one of the internal borders of the earth's geophysical system into the atmosphere is "science," Houghton said. But to analyze their movements *toward* that border "is not a science question." (Lohmann 2020a)

In other words, climatology is not allowed to ask why the climate is changing. It cannot delve into "the politics and history of extraction, enclosure, labour exploitation, colonialism, white supremacy, patriarchy and violation of nature's rights" (Lohmann 2020b)—and consequently leaves the structural drivers of climate chaos unexamined, untouched, and largely untouchable.

Confined within this limited framework, climatology views the problem of climate change merely as a problem of molecules that are in the wrong place. The result, argues Lohmann, is that the climate crisis is treated "in more or less the same way that the far right treats the immigration crisis" (Lohmann 2020a): control the movement of the wayward molecules by turning them away at the border, locking them up (through underground carbon capture and storage schemes or "sequestering" carbon in trees) or discouraging their migration (through carbon taxes to reduce carbon dioxide emissions or Green New Deals to spread new sources of electricity generation that do not emit carbon dioxide).

But attempting to control carbon dioxide molecules without addressing "historically-rooted patterns of capital accumulation, white supremacy, unrelenting imperialism and ruthless patriarchy" (Lohmann 2020a) has simply served to keep the whole destructive juggernaut rolling (Hildyard 2016). Two decades of "policing migrant carbon atoms" has delivered rising, not falling, carbon emissions (Redd Monitor 2019, 2020), threatening to fry all of us. For capital, however, the strategy is a bonanza. Mark Carney, the UN special envoy on Climate Action and former governor of the Bank of England, has described the "net zero" carbon economy as "the greatest commercial opportunity of our time" (Green Horizon Summit 2020). Likewise, mining companies are licking their lips. Jon Samuel, group head of Responsible Business Partnerships at the mining multinational Anglo American, reports: "The transition to green economy, particularly low-carbon energy, is going to be one of the biggest boosts for the mining sector in generations. . . . It's a very big opportunity for us" (Gowling WLG 2020).

While "zero carbon" promises profits for capital, however, it has unleashed new inequalities, new despoliations, and new injustices for the already oppressed. Witness the African children mining cobalt for distinctly reactionary—and racist—electric car batteries (Sanderson 2019); or the hectares of land in the Global South being grabbed by multinationals to grow equally racist and reactionary biofuels for European and North American motorists (Grain 2013); or the plans to cover swathes of North Africa with imperialist and reactionary solar panels to provide European companies with electricity to turn gas into "clean" hydrogen (Corporate Europe Observatory et al. 2020).

THE REVOLUTIONARY PROMISE OF PLURALITY

A further obstacle that the intended radicalism in democratic confederalism's ecology has inherited is the Cartesian dualism that sets Man (yes, "Man") against Nature, an inheritance whose bloodline and trajectory is predictably as steeped in racism, patriarchy, and the demands of capital accumulation as climatology and mainstream environmentalism (Merchant 1980). Even though decades of struggle by environmental justice movements have toned down the talk of Man's "dominance over Nature" in favor of a rhetoric of "working with Nature," the Man-Nature dualism continues to thrive in self-serving assumptions that Nature is "there to serve" Man; in Neo-Malthusian views that ascribe poverty, malnutrition, and hunger to an inherent clash between Man, on the one hand, and the limits set by Nature, on the other (even though the "limits" that matter are generally socially constructed) (Haila 2000); and in the thinking behind efforts to specify precise, enforceable "ecological boundaries" (Rockström et al. 2009) within which humans must live if large-scale ecological damage is to be avoided—a program that (although commendable for taking ecological destruction seriously) threatens to license a profoundly undemocratic response to the ecological crisis facing humanity by legitimizing the policing of both Man (Who? Laborers and poorer people? Or capital and the 1%?) and Nature (Whose "Nature"?) by an unelected elite of ecological managers.

Decolonizing and dismantling the man-nature dualism is central to the Kurdish movement's approach to ecology. Indeed, it insists that respect for plurality (if it is to mean anything) must extend beyond the human realm to embrace the multiplicity of nonhuman forms of life on which all life depends for its collective survival (Hunt 2017). To pay more than lip service to that recognition—to actualize it through practice—would indeed be transformational (Hildyard 2018): but the challenges of doing so are immense.

The gravitational pull of centuries of Cartesian thought and capitalist practice—from which mainstream environmentalism does not seek to break free—is relentlessly anthropocentric. To reject that anthropocentrism is, for many, to enter unfamiliar territory. If the environment is not to be viewed simply as "what is around us" (the "us" being undifferentiated humans), how else might it be conceived? How can nonhumans be accorded meaningful representation in discussions about collective survival? Is "Nature" to be represented through "experts"? Or through daily actions that develop a new conversation with the natural world? Does the very idea of a "Nature" stand in the way of these conversations? Or is "Nature" a concept which "will have to wither away in an 'ecological' state of human society" (Morton 2007, 1), not least because absolutist nature-human boundaries are rendered meaningless once the myriad interactions between humans and nonhumans are contextualized (Haila 2000) and recognition is given to the ways in which these interactions co-produce the world in which we all live.

These are not questions that can be left to the experts. They are intensely political and demand wide public debate. They invite activists for social and environmental justice to broaden our view of what constitutes the political and to take seriously not only the power relations among humans but also the power relations between humans and nonhumans and to be relentless in bringing this expanded notion of the political into our practice and our theorizing. How might our notions of "justice" be changed if we treat the oppression of nonhumans as seriously as the oppression of humans? What new inequalities might we become aware of? How would that change our practices and concepts of what constitutes solidarity?

To ask such questions is inevitably to open up still further areas of inquiry: How might our view of "the collective right of all to survival" change if we acknowledged the agency of nonhumans and their role in the co-production of nature? How might that shift our view of what constitutes labor? How might greater attention to the unpaid labor of nonhumans assist our understanding of how value is created for capital? How might this inform our critiques of contemporary capitalism, particularly as it moves further to enclose the natural world through extracting new forms of rent from environmental services?

And would an expanded notion of labor that took account of non-human labor change how we view class (in the same way, perhaps, as the recognition of the unpaid work of women in social reproduction has expanded our notion of labor beyond factory labor)?

NEW ALLIANCES—BUT WITH WHOM?

Decolonizing environmentalism, climatology, the nature-human divide, and the many other inherited concepts of capital's greenery will not be easy, not least because the concepts are so deeply entrenched: even in this short chapter, I am aware how often I have myself lapsed into mainstream thinking even while trying to question and dismantle it.

But three ways forward suggest themselves.

One is to recognize that "environment," "climate," the "nature-human divide," and "ecology" itself are all political processes in the making: the issue is how activists organize around them and the political choices they make (and with whom) when doing so. Growing radical tomatoes requires more than finding a new word for "environment" or "climate" or redefining these terms in the abstract: it requires building alliances capable of reshaping the practice of "environment" and "climate" through new ways of doing and thinking. The allies that are most likely to assist in this are those whose activism is rooted in the everyday struggles of people the world over to understand their oppression in all its various historical and contemporary manifestations—and who seek to challenge such oppression through defending and experimenting with ways of living rooted in commons-focused resistance to capital accumulation, patriarchy, imperialism, and racism (Hildyard 2016). It is here—within what Raúl Zibechi termed "societies in motion" (Zibechi 2010, 76)—that the most promising vectors for transformative change will be found (Caffentzis 2009; Holloway 2010). The Kurdish movement's practical experimentation in democratic confederalism offers one vibrant example. Others are to be found in Latin America where the resistance of social movements to extractivism and other ecological injustices has spawned a lively debate over the "rights of nature," now enshrined in at least one national constitution. Fruitful alliances are also likely to be found with social movements whose experience of oppression has led them to question the "whiteness" of mainstream climatology (Lohmann 2020a).

A second way forward would be to recognize that lasting, politically effective alliances are unlikely to flourish where activists insist that "their" causes must take precedence over those of others or attempt to rush the slow processes of relationship-building that generate the trust and mutual understanding that make for effective solidarity and co-operation. It is through such patient, inclusive relationship-building that people come to see something of their own struggle in someone else's, and vice versa; where they come to identify with others who may have quite different interests and to whom they may previously have been indifferent or even opposed; and where they are drawn together not so much because they come from or are "embedded in absolute sameness," but because they come to realize that their life courses are being "determined by ultimately similar processes and outcomes" (Palmer 2014, 49). A social justice that is rooted in and shaped by such processes of discovery demands a practice and approach to politics that is very different from the approach adopted, either through inclination or bureaucratic imperative, by many professionalized nongovernmental organizations. This is not an activism that rejects reports or demonstrations or lobbying; but it is an activism that emphasizes mutual learning and unlearning, an understanding of each other's histories and political context, the building of relationships of care, and respectful dialogue; and it places these processes over and above short-term, in-and-out, often opportunistic, priorities of "campaigns." Above all, it is activism that emphasizes patience, listening, solidarity, and comradeship as the basis for collective action in defense of commons; and, beyond that, as seedbeds for cultivating the "disciplined, self-denying, careful, tasteful friendships" that renegade priest and social activist Ivan Illich (1996) identified as the "supreme flower of politics" and the very basis of community. As Illich (1996) remarks, a society will only be "as good as the political result of [its] friendships"; those founded on mutuality and a shared commitment to collective survival—on commoning—will produce a very different society from those based on competition, accumulation, and "capital-ing" (Lohmann 2015), one that is better equipped to resist the temptations of capital and its enclosing web of depredations (Esteva 2014).

A third recognition that offers a solid base for moving forward is that excavating the history and politics of environmentalism—its class roots, its role in legitimizing discrimination and expropriation, the interests that have benefited and the resistances that have been triggered in response—is not a luxury that activists can set to one side. It is essential to alliance-building and to liberating activists from the vocabularies and practices that limit our ability to reconceptualize our relationships with the natural world. To decline the invitation to probe into histories of patriarchy, racism, and capital accumulation, and to decline to accord equal respect to those whose ecology is rooted in such analysis (and even some would-be radical climate activists within Extinction Rebellion are guilty of this [Lohmann 2020a]) is to send the wrong message. It is to signal that the oppressions that are part and parcel of the daily lives of millions are somehow "secondary" or "irrelevant" to "saving the environment." To take such a course is fatally to misunderstand the root causes of the crisis, and also leads to an organizational dead end by cutting campaigners off from the very movements that are most active in seeking to move society beyond what the social philosopher André Gorz (1968, 7) called "reformist reforms" toward system-busting "non-reformist reforms." In doing so, it does not bring change closer, but makes it more distant.

The traps laid by the toxic inheritance of capital's environmentalism are clear. Circumventing them is the challenge. The feminist movement has already pioneered new organizational practices that have taken seriously the struggle of women against the oppressions of patriarchy. Similar radical experimentation will surely be needed to organize against the many oppressions of capital's environmentalism. It is a road that will be fraught with conflict and disappointment: but, if the revolutionary promise of democratic confederalism's ecology is to be realized, it is a road that cannot be avoided.

REFERENCES

Bookchin, Murray. 1987. "Social Ecology versus Deep Ecology: A Challenge for the Ecology Movement." *Green Perspectives: Newsletter of the Green Program Project*, nos. 4–5 (Summer 1987). http://theanarchistlibrary.org/library/murray-bookchin-social-ecology-versus-deep-ecology-a-challenge-for-the-ecology-movement.

Briy, Anya, and Anonymous. 2019. "The Theory and Practice of the Kurdish Women's Movement: An Interview in Diyarbakir." *OpenDemocracy*. January 3, 2019. https://www.opendemocracy.net/en/theory-and-practice-of-kurdish-women-s-movement-interview-in-diyarbakir/.

Caffentzis, George. 2009. "The Future of the 'Commons': Neoliberalism's 'Plan B' or the Original Disaccumulation of Capital?" *New Formations* 69: 23–41. https://newxcommoners.files.wordpress.com/2013/01/caffentzis-planb.pdf.

Corporate Europe Observatory, Food and Water Action Europe, and Re: Common. 2020. "The Hydrogen Hype: Gas Industry Fairy Tale or Climate Horror Story?" Brussels: Corporate Europe Observatory. https://corporateeurope.org/sites/default/files/2020-12/hydrogen-report-web-final_0.pdf.

DiNardo, Kelly. 2019. "The Green Revolution Spreading Across Our Rooftops." *The New York Times*. October 9, 2019, https://www.nytimes.com/2019/10/09/realestate/the-green-roof-revolution.html.

Esteva, Gustavo. 2013. "Aid—No thanks! If anyone wants to do you any good, run away." Presentation to 'Giornata di dialogo tra movimenti,' Florence, April 8, 2013.

Esteva, Gustavo. 2014. "Commoning in the New Society." *Community Development Journal* 49, suppl. 1 (January 2014): i144–i159. http://cdj.oxfordjournals.org/content/49/suppl_1/i144.full.

Eye of Riyadh. 2017. "Building Sustainable Green Infrastructure with Green Walls and Green Roofs." *Eye of Riyadh*. January 13, 2017. https://www.eyeofriyadh.com/news/details/building-sustainable-green-infrastructure-with-green-walls-and-green-roof.

Gorz, André. 1968. *Strategy for Labor: A Radical Proposal*. Boston: Beacon Press.

Gowling WLG. 2020. "Mining Investment: Portfolio Strategies, Working through COVID-19 and ESG Risks and Opportunities." Interview, Gowling WLG, May 21, 2020. https://gowlingwlg.com/en/insights-resources/on-demand-webinars/2020/mining-investment-portfolio-strategies-covid-19/.

Grain. 2013. "Landgrabbing for Biofuels Must Stop." *Grain*. February 21, 2013. https://www.grain.org/article/entries/4653-land-grabbing-for-biofuels-must-stop.

Green Horizon Summit. 2020. *Summit News*. November 9, 2020. https://www.greenhorizonsummit.com/event/a777a3fa-9c35-4901-a077-41d295a85990/websitePage:650f08ac-4ffe-4fdb-a107-0fc39e38d583.

Haila, Yrjö. 2000. "Beyond the Nature-Culture Dualism." *Biology & Philosophy* 15, no. 2 (March):155–175. https://doi.org/10.1023/A:1006625830102.

Hardman, Michael, and Nick Davies. 2019. "Why More Cities Should Adopt Green Roofs." World Economic Forum and The Conversation. https://www.weforum.org/agenda/2019/10/urban-green-roofs-trees-environment/.

Harvey, David. 2008. "David Harvey on the Geography of Capitalism: Understanding Cities as Polities and Shifting Imperialisms." *Theory Talks* no. 20: David Harvey. http://www.theory-talks.org/2008/10/theory-talk-20-david-harvey.html.

Hildyard, Nicholas. 2016. "Energy Transitions: Some Questions from the Netherworld." Lecture on "Energy Transition—Why and for Whom?," Vienna, April 18, 2016. Sturminster Newton, UK: The Corner House. http://www.thecornerhouse.org.uk/sites/thecornerhouse.org.uk/files/QuestionsNetherWorld.pdf.

Hildyard, Nicholas. 2018. "Where's the Revolution in Democratic Confederalism's 'Ecology'?" Presentation to Seminar on Ecological and Gender Dimensions of the Democratic Confederalist Approach in Kurdistan: An Inter-cultural Dialogue, SOAS, London, February 3, 2018. http://www.thecornerhouse.org.uk/sites/thecornerhouse.org.uk/files/Where%27s%20the%20revolution.pdf.

Hildyard, Nicholas, Larry Lohmann, Sarah Sexton, and Simon Fairlie. 1995. "Reclaiming the Commons." Sturminster Newton, UK: The Corner House. http://www.thecornerhouse.org.uk/resource/reclaiming-commons.

Hildyard, Nicholas, Pandurang Hegde, Paul Wolvekamp, and Somasekhare Reddy. 1997. "Same Platform: Different Train - Pluralism, Participation and Power." Sturminster Newton, UK: The Corner House. http://www.thecornerhouse.org.uk/resource/same-platform-different-train.

Hildyard, Nicholas, Pandurang Hegde, Paul Wolvekamp, and Somasekhare Reddy. 2001. "Pluralism, Participation and Power: Joint Forest Management in India." In *Participation: The New Tyranny?*, edited by Bill Cooke and Uma Kothari, 56–71. London: Zed Books.

Holloway, John. 2010. "Forward to the German Edition." In *Dispersing Power: Social Movements as Anti-State Forces*, edited by Raúl Zibechi, xv–xvii. Oakland: AK Press.

Huff, Amber. 2018. "Report from Rojava: Revolution at a Crossroads". Steps Centre, July 3, 2018. https://steps-centre.org/europe-hub/report-from-rojava-revolution-at-a-crossroads/.

Hunt, Stephen. E. 2017. "Prospects for Kurdish Ecology Initiatives in Syria and Turkey: Democratic Confederalism and Social Ecology." *Capitalism Nature Socialism* 30, no. 3 (2019): 7–26. https://doi.org/10.1080/10455752.2017.1413120.

Illich, Ivan. 1983. "Silence is a Commons." *The Coevolution Quarterly*, Winter 1983. http://www.preservenet.com/theory/Illich/Silence.html.

Illich, Ivan. 1996. "Ivan Illich with Jerry Brown: We the People, KPFA—22 March 1996." http://www.wtp.org/archive/transcripts/ivan_illich_jerry.html.

Li, Wai Chin, and Ka Ka Annie Yeung. 2014. "A Comprehensive Study of Green Roof Performance from Environmental Perspective." *International Journal of Sustainable Built Environment* 3, no. 1 (June): 127–134. https://www.sciencedirect.com/science/article/pii/S2212609014000211.

Livingroofs. Undated. "Green Roofs Benefits—Improved Energy Performance." https://livingroofs.org/energy-conservation.

Lohmann, Larry. 1997. "Cost-Benefit Analysis: Whose Interest, Whose Rationality?" Sturminster Newton, UK: The Corner House. http://www.thecornerhouse.org.uk/resource/cost-benefit-analysis.

Lohmann, Larry. 2015. "Reflections on Rimaflow." Sturminster Newton: The Corner House and Rome: Re:Common.

Lohmann, Larry. 2016. "Neoliberalism's Climate." In *Handbook of Neoliberalism*, edited by Simon Springer, Kean Birch, and Julie MacLeavy, 480–492. https://www.researchgate.net/publication/286450117_Neoliberalism's_Climate.

Lohmann, Larry. 2020a. "White Climate, White Energy: A Time for Movement Reflection?" Sturminster Newton, UK: The Corner House. http://www.thecorner house.org.uk/resource/white-climate-white-energy.

Lohmann, Larry. 2020b. "A Different Climate Agenda: The Need Always to Change the Subject away from Carbon." Presentation to Oilwatch, October 2020. Sturminster Newton, UK: The Corner House.

Marsh, Alistair, and Benjamin Robertson. 2020. "Carney Calls Net-Zero Greenhouse Ambition 'Greatest Commercial Opportunity.'" *Bloomberg*. November 9, 2020. https://www.bloomberg.com/news/articles/2020-11-09/carney-calls-net-zero-ambition-greatest-commercial-opportunity.

Merchant, Carolyn. 1980. *The Death of Nature: Women, Ecology and the Scientific Revolution*. New York: Harper & Row.

Morton, Timothy. 2007. *Ecology without Nature: Rethinking Environmental Aesthetics*. Cambridge: Harvard University Press.

Öcalan, Abdullah. 2017. *Democratic Confederalism*. International Initiative Edition in cooperation with Mesopotamian Publishers: Cologne. http://ocalanbooks.com/downloads/EN-brochure_democratic-confederalism_2017.pdf.

O'Neill, John. 1993. *Ecology, Policy and Politics: Human Well-Being and the Natural World*. London: Routledge.

Palmer, Bryan. 2014. "Reconsiderations of Class: Precariousness as Proletarianization." *Socialist Register* 50: 40–60.

Redd Monitor. 2019 "'Nature Cannot be Fooled': Kevin Anderson on Mitigation as if Climate Mattered," September 5, 2019. https://redd-monitor.org/2019/09/05/nature-cannot-be-fooled-kevin-anderson-on-mitigation-as-if-climate-mattered/.

Redd Monitor. 2020. "Interview with Larry Lohmann, The Corner House: 'Carbon Markets do not Need to be 'Fixed.' They Need to be Eliminated," October 22, 2020. https://redd-monitor.org/2020/10/22/interview-with-larry-lohmann-the-corner-house-carbon-markets-do-not-need-to-be-fixed-they-need-to-be-eliminated/.

Rockström, Johan, Will Steffen, Kevin Noone, Åsa Persson, F. Stuart III Chapin, Eric Lambin, Timothy M. Lenton, *et al.* 2009. "Planetary Boundaries: Exploring the Safe Operating Space for Humanity." *Ecology and Society* 14, no. 2: 32. http://www.ecologyandsociety.org/vol14/iss2/art32/.

Sanderson, Henry. 2019. "Congo, Child Labour and your Electric Car," *Financial Times*. July 7, 2019. https://www.ft.com/content/c6909812-9ce4-11e9-9c06-a4640c9feebb.

Spender, Dale. 1980. *Man Made Language*. Pandora: London.

Stern, Maya, Steven W. Peck, and Jeff Joslin. 2019. "Green Roof and Wall Policy in North America: Regulations, Incentives, and Best Practice." Green Roofs for Healthy Cities: Toronto. Available at: https://static1.squarespace.com/static/58e3eecf2994ca997dd56381/t/5d84dfc371cf0822bdf7dc29/1568989140101/Green_Roof_and_Wall_Policy_in_North_America.pdf.

Unrein, John. 2019. "These Tomatoes may Someday Grow in Space." *Supermarket Perimeter*. December 27, 2019. https://www.supermarketperimeter.com/articles/4525-these-tomatoes-may-someday-grow-in-space.

Zibechi, Raúl. 2010. *Dispersing Power: Social Movements as Anti-State Forces*. AK Press: Oakland.

Part II

POSITIVE INITIATIVES FOR ECOLOGICAL CHANGE

Chapter 5

Ecology Structures of the Kurdish Freedom Movement

Ercan Ayboğa

MESOPOTAMIA ECOLOGY MOVEMENT

With the widespread introduction of capitalist economy and relations by the four colonialist states Turkey, Iran, Iraq, and Syria to Kurdistan in the 1950s came a systemic and destructive exploitation of people and nature. In the 1970s, the construction of the first large and damaging investment projects—particularly dams, oil drilling, and mining—was realized and agriculture started to be industrialized (particularly in the area now known as Rojava). All investments, which led to the displacement of hundreds of thousands of people to the fast-growing cities, aimed also at benefiting from a cheap labor force to be exploited and a deepening of the ongoing cultural assimilation. Since the 1980s, society has largely lost its characteristics of solidarity and communality in life and economy, and Kurdistan has become fully part of the "national market" of each of the four states.

Among the many Kurdish political resistance organizations of recent decades, the most revolutionary, emancipative, and broad-based is the Kurdistan Workers' Party (PKK). In the 1980s, the PKK initiated armed (guerrilla) resistance against the Turkish state and gained support from Kurds in the three other parts of Kurdistan. As new organizations emerged, including civil ones, that were affected by the PKK ideologically and politically, a substantial movement developed known as the Kurdish Freedom Movement (KFM).

Figure 5.1   Logo of the Mesopotamia Ecology Movement. *Source*: © Mesopotamia Ecology Movement.

ECOLOGY DISCUSSIONS AND
THEORETICAL APPROACH

In the 1990s, the KFM, and particularly its leader Abdullah Öcalan, started to discuss the ecological question, against the background of the systematic destruction of livelihoods through the Turkish state's war on Kurds. This included the systematic bombing of mountains, the torching of thousands of hectares of natural forest, and the displacement of more than two million civilians from villages (Human Rights Association of Turkey [IHD] 2001) mainly destroyed by the Turkish Army, resulting in growing and brutal urbanization. In the 1990s, the Turkish state with growing financial capacities also started hundreds of dams, agricultural schemes, mining, and other projects of exploitation, causing dramatic physical changes to the landscape and population. Öcalan assessed these changes and how they caused deep social, cultural, ecological, and political impacts. In parallel, he analyzed the emergence of neoliberal capitalism. In his analysis of these developments, he raised the question of the growing alienation of humans from nature. He also included climate change in his thinking, which he considered as an acceleration of ecological destruction by capitalism. These discussions impacted thousands of political activists in the KFM.

Since Öcalan's abduction in 1999, through an international plot, and imprisonment on İmralı Island, a broad discussion has started on restructuring theory and practice in the KFM, based on broad self-criticism. As part of this process, Öcalan also convinced the PKK to end the armed struggle. Strategic priority was given to political-civil struggle, so consequently social movements and thinkers have been discussed intensively, and Marxist-Leninist revolutionary approaches have been criticized strongly, without leading to reformism. When a new political concept called "democratic confederalism" was declared in 2005, the outcome was a critical, inclusive, and systematic review of radical thought, with new perspectives for the Kurds and on relations with other people in the Middle East. An ecological approach to life was stressed as much as radical democracy, which goes beyond parliamentarism, and gender liberation. These three pillars and the whole concept are expressed with the overall aim of an ethical, solidarity-based society in harmony with nature and distinct from capitalist modernity.

To realize such a society, different regions aim to organize themselves as strong autonomous structures within states, according to the principle of "democratic autonomy." Consequently, the objective of a "Kurdish state" has been given up, as it is considered as a legacy of nationalism. Each of the three pillars of democratic confederalism cannot be thoroughly developed without links to the other two. However, the initial starting point is women's liberation. Öcalan states that with patriarchy's overcoming of matricentric society, institutionalized hierarchical structures had emerged and spread among human societies and characterized the upcoming states until the present day. Long before explicit social classes came into being, the first oppressed and exploited class were women. While women are an oppressed gender in all patriarchal societies, they are also importantly regarded as having a stronger relation to nature than men, who are usually seen as more attached to power. This political-ideological formation led also to the historic domination and destruction of nature. The starting point of the so-called ecological genocide (ecocide) we are currently facing, therefore, predates capitalism. Thus, the struggle for an ecological, liberated, and solidarity society also means the struggle against patriarchy and liberation of women.

The KFM views nature as the body of all living beings, including humans. Humans are part of nature and do not stand over it or any species. Nature is regarded as alive and animated, and all living beings are part of one common ecosystem that offers enough opportunities to live for every living being. Nature was and is the source of food, housing, and all other material needs of life. Therefore, throughout history, there was for (almost all) the people always a strong connection with nature in daily life, so it was omnipresent. Based on thorough adherence to ecological principles nature should be treated respectfully and not as a resource for profit.

Under capitalist modernity, humans living in urban centers are usually weakly connected to nature and understand less the relation to nature. Nature had, and has, a multidimensional meaning in life and is essential for the development of culture and identity, as well as spirituality. Due to the alienation between human beings, which contributes significantly to the alienation between nature and human beings, nature is overexploited. Despite everyone experiencing the impacts of grave ecological damage in recent decades, the destruction of nature continues and is even accelerating. Thus, the current approach of human-driven capitalist modernity is a state of betrayal of humans to nature, to their own body.

In this sense, if human beings would meet only their needs, nature would not experience serious destruction and the ecosystems would have the capacity to recover (reproduce) themselves. But what are the real needs of humans today? This question should not be left only to biologists or economists, rather it relates to the question of democracy; that is, whether a society can take decisions under broadly and radical democratic conditions free from imposed exploitative-extractive economy policies and corruption. We believe that in a socially and gender liberated, solidarity-based, radical democratic, and ecological society, there will be no pressure to over-extract "material and elements" from nature. At this point, we connect ecology and democracy with each other.

Nature conservation and even restoration by humans should be goals for strategic action. So, each struggle against ecological destruction is essential and a necessary step to raise ecological awareness and re-establish a better relation to nature for human communities. But it is not enough to defend the contested area of the natural world alongside a human community, because the related investments are caused by the dominant political economic system which will continuously implement exploitative projects. That is why an ecological approach leads the KFM to criticize all processes in the present society, particularly the way of producing and consuming, feeding, housing, transport, and leisure. Only when there is an economy, based on solidarity and communality, can major ecological destruction can be prevented in the long term.

To create such an alternative, "ecological industry" is necessary. Öcalan proposes "ecological industry" in his texts, and Kurdish ecologists continue to discuss the idea. This term appears controversial as industrial activities have played a major part in the destruction and pollution of the natural world, as well as in the current climate crisis, and are seen to concentrate economic and political power. Thus, the question is raised as to how to reorganize industry and technology. In this sense, the definition of eco-technology needs to be discussed. For technology to serve all living beings, the capacity and management of industry must be fundamentally reorganized from an ecological perspective, and the existing concept of economic growth must be broken.

The increasing ecological awareness in the KFM is related also to the long-term existence of the guerrilla in the mountains of Kurdistan. When not fighting, the guerrillas discuss the entire range of social and political issues in their political educational program. The concepts for an ecological life are often practiced in the guerrilla areas on a small scale as a community and then proposed to activists in the "normal" society. Consider that the guerrillas live outside of capitalist modernity; their living conditions are completely communal, based on solidarity. Another crucial aspect is that the guerrillas live in harmony with nature, so their perspectives are relatively free from capitalist and hierarchical frameworks.

PRACTICE IN NORTH KURDISTAN

The year 1999 was also a turning point for two other developments for Turkey's Kurdish region, where around fifteen million people live today. The first ecological movements emerged, and many municipalities were won in elections. In the local elections of 1999, the legal party of the KFM, then called the HADEP (now HDP), won three dozen municipalities. These municipalities— among them the large cities of Amed, Batman, and Wan—became essential elements within the Kurdish freedom struggle. This coincided with a reduction in repressive conditions, mainly because of the halt in the armed struggle. This facilitated more space for the municipalities and other KFM organizations to spread their political ideas and to develop more contact with new and less politically organized parts of society. What has been claimed for years, namely that the KFM has more democratic-social (and also ecological) concepts and is not corrupt, unlike the other parties in the hegemonic system, could be then be implemented at the local level through the municipalities.

Indeed, the KFM-ruled municipalities have been successful in finding solutions to many challenges such as poor-quality drinking water, the lack of facilities for wastewater treatment, garbage mismanagement, the lack of

Figure 5.2 Demonstration at Hasankeyf in 2008. *Source*: © Mesopotamia Ecology
Movement.

green areas in cities, the absence of city planning, poor street conditions and
chaotic traffic, limited sport opportunities, non-existing social services (particularly for women and children), and the conservation of cultural heritage.
Within a few years, the situation improved significantly. If the corruption
decreased, the gentrification (that was experienced in non-KFM-ruled municipalities) and large infrastructure projects that benefited only the few could be
largely stopped. The majority of the population, therefore, was pleased with
the results. Nevertheless, the comprehensive steps necessary to bring about
radical democratic, ecological, and solidarity cities were missing or failed
until the end of the 2000s. For the municipal administration, the notion of
ecological improvement meant mainly issues like more parks, clean drinking

Figure 5.3 Ilisu protest in 2015, a few kilometers away from the site of the dam. *Source*: © Mesopotamia Ecology Movement.

water, and clean streets as they have been developed in a few Turkish cities or in Europe. One reason was a weak political concept and missing experience in regard to local self-governance; another was the low, undeveloped level of ecological awareness. In the 2000s, another factor emerged, in that the state's approach was less directly oppressive toward the KFM than in the 1990s. However, in the challenging framework of neoliberalism, other strategies were used against the KFM-related municipalities. Imposed policies, including limiting subsidies and limited support by state bodies, and administrative centralism, sought to bring the municipalities to a point where they would fail, thus causing them to lose the following local elections and to weaken the KFM's standing among the population in the long term.

With the end of the war in 1999, the first ecological campaigns and movements emerged in Bakûr. Before that date, ecology-related issues were only discussed by individuals in the larger cities. The two most well-known campaigns are those that have targeted destructive dams at the Munzur River in Dersim and the Ilisu Dam on the Tigris River which threatened the 12,000 years old town of Hasankeyf (see Figures 5.2 and 5.3). Local initiatives quickly raised awareness of these issues and gained broad public support. There was substantial support from organizations and media close to the KFM, but also from other parts of society. Thousands of people protested against these dams during the following years. Furthermore, with some partial success in delaying these dam projects, these campaigns generated public debates that have strongly influenced ecological consciousness among Kurdish communities. For the first time, Kurdish society started to talk about the related topics of rivers, dams, energy, cultural and natural heritage, and development on a broader scale, contributing to an increase of critical awareness about these issues. Turkish ecology and leftist activists have also given serious attention to these dam campaigns.

The declaration of democratic confederalism in 2005 further encouraged ecology activists within the KFM. In the following years, there was a steady increase of groups working on issues concerning nature conservation, environmental health, the negative impacts of big infrastructure and energy projects (see Figure 5.9), food production, and social ecology theory. Some of these have been initiated by those with a strong political commitment, while others have been developed by people solely concerned about the grave impacts of governmental policies on their livelihoods and on nature. After Turkey's economic crisis in 2001 and with the new AKP in government, the neoliberal capitalist economy spread to all areas of society in Bakûr. Capitalist modernity unfolded its maximum destructive forces, as the government did everything to enable investments in Turkey, including Bakûr. There was much more capital available, mostly foreign investment, resulting in a construction boom in urban and rural areas. In the 2000s, most KFM-ruled municipalities, and a proportion of KFM activists, were not critical enough in their attitude toward this capitalist development, which not only destroyed and exploited nature, but risked undermining local economic structures and social solidarity.

By the end of the 2000s, however, a more critical perspective on neoliberal development was growing among several parts of the KFM. Around 2010, two developments emerged. First, more people expressed the need to form a coalition of ecological groups and activists in Bakûr. Between 2007 and 2010, the struggle against the Ilisu Dam and the Munzur Dams again became dominant. Second, the KFM structures, organized as an umbrella organization called the Democratic Society Congress (KCD), started to discuss what needed to be done to bring about a "free municipalism," that is, a society based on solidarity, communality, and ecological life.

The introduction of the concept of democratic confederalism initiated two years of discussion, the outcome of which was the KCD, established in 2007 to implement its ideas. Since, the first steps have been taken to set up people's assemblies (councils) at the neighborhood level, in areas where the KFM had a significant basis. The Kurdish women's movement, which had already established a strong self-organization process, has been able to spread its committees and councils. The youth movement also developed in the same way. At higher levels, delegates from these people's assemblies meet social movements, NGOs, parties, municipalities, unions, and other organizations. The first level is the district, the next the province, with the highest being the general assembly of Bakûr. The KCD's general assembly meets regularly—usually every six months—and has 501 members. At these higher levels, usually 60 percent of all delegates come from the neighborhood assemblies and 40 percent from social and political organizations. The inclusion of all willing organizations and actors in the councils above the neighborhood level is a crucial element of the new political project. With this inclusive approach, each constituency, whether social, political, cultural, or gender, has a voice in the discussion process and should be included in the decisions. For the neighborhood assemblies, it is almost obligatory to make decisions by consensus, which is also the aim for the upper assemblies. Unlike parliamentary systems, the principle is that no group should rule over the others.

The KCD's horizontal structures are not enough to properly comprehend the structures of democratic autonomy. It is essential to understand the autonomous social sectors (fields) for the representations of the social movements and mass political organizations, as well. When established, these sectors were mainly representational bodies for women, youth, economy, health, education, diplomacy, justice, municipalities, culture, civil society, and religious and spiritual beliefs. The classification of the sectors has since evolved according to the outcomes of continuous discussions and review. While these autonomous sectors have their own meetings at different levels, they also actively participate in the KCD's assemblies at all levels. For example, each assembly's women's committee is at the same time part of the autonomous women's movement. This connection of horizontal and vertical structures in a direct democracy model is specific to the KFM and, from a global perspective, maybe unique in the way it is implemented.

As discussions proceeded as to the best way to connect deeper gender equality, radical democracy, and ecology within the KCD's local self-governance structures, it was realized that it was necessary to consider how to better confront existing and new challenges. In 2010, many participants took part in the "First Conference on Ecology and Local Administrations," in which social relations and developments were considered as critical. A framework was released which stated that society should be organized based on the four pillars of an "organized society and participative approach," "ecological life," a "gender-liberated approach" and a "participative social economy." Within this approach, one crucial focus was the cities, considering that most people now lived in urban areas. Migration to the cities was ongoing due to the state's economic politics.

In the KCD's early years, ecological policy was positioned within the municipalities sector and not separate. In January 2011, the first Ecology Forum was organized in Amed. Activists and researchers from across Turkey and the Middle East also joined this unique event. For the first time in Kurdistan's history, ecological challenges, struggles, and approaches were debated in a broad and organized way. There were detailed discussions about dams, mining, energy policies, industrial agriculture, urbanization, transport, health, nature conservation, climate crises, and other issues. A group of activists from up to eight ecology-related struggles regularly convened as a consequence of this event. A year later, this group announced the formation of the "Mesopotamia Ecology Movement" (MEM, see Figure 5.1).

At first, the MEM was a network of local groups and interested people rather than a movement. Activists visited each other and discussed how to raise issues within the MEM, and common statements were released. It was a gradual process for the MEM to gain wider social recognition and become part of the KCD with its own delegates. There was no rush to join the KCD directly as long as the common principles were discussed and deepened. An upsurge in the state's intensive repression in 2011–2012 was another factor that delayed the ecology movement's development and integration. At that time, the state not only attacked the guerrillas but also social organizations (9,000 political activists were arrested). A cease-fire between the Turkish state and the PKK in 2013 saw an easing of the political situation, and in this context, two events contributed to the strengthening of social awareness of ecological issues. Students at Amed's Dicle University organized a protest camp at the university site against the destruction of a small forest in the Tigris Valley. With the support of many civil society organizations, they successfully stopped the project. Also, in 2013, the MEM organized a demonstration in which hundreds of people protested against the controversial plans for a huge housing project on a hill, also in the Tigris Valley at Amed. This campaign against a project for which the HDP municipality had given permission, despite broad criticism, was almost unique as the HDP is also part of the KCD. This protest, which could not stop the housing construction, was therefore significant in raising critical objections to aspects of the HDP's city development. Following this and some other conflicts, the HDP began to take notice of the MEM's views and became more cautious in its response to applications from private developers.

From 2013, ecology activists initiated or joined with others in new struggles. Prominent among them were campaigns against the construction of many small hydro dams, with active opposition to those at Amed City, Pasur/Amed, Zilan/Wan, and Dersim. Another important issue was hydraulic fracturing or "fracking," since the Turkish government particularly wanted to expand operations in Bakûr, where it assumed lay the largest reserves, and started test drillings in 2013. Another increasing cause of public concern was the health and water quality of the large Wan Lake. The struggle against a huge coal plant in Sîlopî/Şirnex saw thousands of people take to the streets to protest. For several years, there were also protests against the site of the largest cement factory within Turkey, in Marash Province. The ecology struggles, however, were not limited to the "rural" areas, and many younger people living in urban areas also became motivated to engage with ecological groups and projects. A major contributory factor for this was the fast growth of poorly planned cities, developed with little consideration for social and ecological well-being. There was growing criticism of the new neighborhoods with partly gated communities which, despite the inclusion of parks, largely facilitated the individualization of people. Another cause for discontent was the appearance of large shopping malls, which were destroying the businesses of many small shop owners. While the capital investment behind these malls came from Istanbul, they were developed with the permission of the HDP municipalities. In the face of rising criticism, this ended in 2014, but several of these malls have already been established in all the larger cities. Fortunately, half of the population shares an awareness of the negative impact of their shopping model, which they reject by supporting small shops. Nevertheless, there are many more state-driven projects for large constructions, such as housing schemes and entertainment complexes, or to redevelop existing public areas, which are commercializing cities. The municipalities now usually resist such changes—particularly, given the expression of criticism by the KCD's ecology, economy, women, and youth movements. However, in this centralistic state, the Turkish government often overrides local decisions to implement the contested projects. More positively, the HDP municipalities prevented the destruction of neighborhoods through gentrification, to protect low-income people. Today, therefore many such neighborhoods remain close to city centers or in nice locations. This is a significant difference to cities ruled by other political parties.

In 2015, the MEM restructured itself with the aim to become a broad and more effective social movement. After wide-ranging discussions, which started in 2014, councils in each province of Bakûr were established which offered space to both political activists already working on ecology and in civil society organizations, and also to any interested parties. Within few months, several hundred people joined the assemblies of these councils, set up in some of Bakûr's twenty provinces. Each provincial council formed according to their local needs, with working committees where activists could focus on the issues that are priorities for their provinces. Typical were the committees on energy, water, food and agriculture, biodiversity, urban development and transport, and legal issues; specific and of special interest were committees on animal rights, eco-technology, forests, and ecological economy. For each provincial council, a coordinating committee was created consisting of the co-chairs of each gender on the committees, together with the two co-speakers of each gender, who are elected directly in the assemblies, usually for every six months. There is an aim for many activists to become co-speakers within this system. The provincial coordinators send their delegates to the Bakûr MEM assembly (see Figures 5.4 and 5.5), where a general coordinating committee is elected every six or twelve months. Gender participation and discussions in each MEM structure are a fundamental part of the KCD—it is a requirement that every organization includes this approach on their agenda.

This new form of self-organization made it easy for anyone to join the growing ecology movement, reflecting increasing ecological consciousness.

Indeed, most new activists have never previously been politically active. This has created a new dynamic and enabled discussion on issues, which have not been the focus of attention before. One of the main reasons for this restructuring is that single-issue ecological struggles can lack their intended impact. The MEM, and the broader KFM, has observed this challenge in many Western countries and in the Global South. Indeed, the Kurdish ecological movement's more integrated practice showed how much more effective a coordinated approach can be in its social impact. However, it needs to be noticed that it is easier to bring different ecology activists together in Bakûr, since they have the advantage that most of them are close or sympathetic to the political goals of the KFM, by far the biggest political movement in the region.

The political work of the MEM can be classified into five fields:

i. The struggle against the destructive and exploitative investments and projects of the Turkish state and private companies which causes the biggest ecological destruction. Among them are dams, mining, major roads, fracking, industrial agriculture projects, and urban gentrification.

ii. The Turkish state's war against Kurdish people and the natural world, the main issues being forest fires and the destruction of livelihoods in rural and urban areas. The Turkish Army has again systematically torched forests since the war recommenced; likely 80,000 to 100,000 hectares of forests have been destroyed or damaged (Mesopotamian Ecology Movement 2015). The destruction of livelihoods was extremely intensive in Bakûr in 2015 and 2016, when dozens of neighborhoods were completely or partially destroyed in seven cities, and more than 200,000 people have lost their homes. In recent years, military operations have caused several thousands of people living in villages in mountainous areas to temporarily, or permanently, leave their homes.

iii. Education is another crucial field and a permanent feature of political work, because without self-reflection and criticism, the efforts will be unsuccessful. Education for activists is organized internally and with other interested organizations.

iv. The aim to develop and implement projects for an alternative society and economy, for example, the collection and reproduction of local and organic seeds, the construction of buildings by local and natural materials (see Figure 5.7), and the creation of tree nurseries (see Figure 5.8). This is done by several cooperatives in close collaboration with the economic movement.

v. Work with, and scrutiny of, the KFM-ruled municipalities. Taking over state power at the local level offers advantages, but also leads to risks, such as growing hierarchy, alienation, and corruption. Although since 2016, Turkish state-appointed commissioners and police chiefs have regularly occupied municipalities won in elections by KFM representatives, this aspect remains a crucial issue for the MEM. Without success in the municipalities, the KFM will not enjoy wider success.

Following the restructuring and strengthening, together with its growing actions, projects, and campaigns, the MEM now has stronger representation in the KCD. This means it is better able to present its proposals to the whole KCD. Here it does not act alone, it confers about critical subjects with relevant sectors (such as economy, women, youth, and education) before they are brought to the KCD's provincial or general assemblies. There are still, however, some circles in the KCD which do not care so much about ecological principles, where the MEM has had less success so far. What is most important, however, is that there are continuous conversations about ecological awareness in society. In 2015 and 2016, there were broad discussions in the MEM about what issues could be best brought forward within the framework of the KCD decision-making processes. The MEM organized a program of workshops covering general principles, water, energy, food and agriculture (see Figure 5.6), ecological economy, health, biodiversity and climate, urbanization, and transport. These categories were approved during the first conference in April 2016. The MEM's main goal is, of course, that ecological principles and actions are taken seriously by all organizations and people within the KCD. The KCD should orientate their discussions and policies with ecological principles, in keeping with the approach to gender principles.

From the outset, the MEM also sought to build strategic relations with ecologists in Turkey. As the common "opponent" is the Turkish state and capitalism, it is logical to work closely together to overcome Turkish nationalism. After years of political collaboration, relationships between Kurdish and Turkish ecologists have become much stronger. A critical step in this direction was the creation of the HDK's Ecology Commission with twenty-four groups in 2013. (The HDK is a congress organized across Turkey in collaboration with the KCD.) The next step was the foundation of the umbrella network "Ecology Union" in 2018, which has fifty-two organizations within which almost all ecological organizations are included. In recent years, several common campaigns, calls for action, public meetings, and educational activities have been achieved. However, the Ecology Union has yet to develop sufficient power to influence the political agenda.

The MEM also works with ecological groups in the Middle East, particularly in Kurdish parts of Iraq, Iran, and Rojava. As the rivers connect these regions geographically, the search for just, free, and ecological societies brings activists with different cultural backgrounds together. One example is the First Mesopotamian Water Forum in April 2019 in Silêmanî. The MEM has also built relationships with European, American, and Asian ecologists. There are aspirations to join international networks as is the case with water,

Figure 5.4 Assembly of the Mesopotamia Ecology Movement in 2016. *Source:* © Mesopotamia Ecology Movement.

agriculture, fracking, and energy and to exchange ideas, perspectives, and practice. Considering the growing climate and ecological crisis, this is an urgent necessity. Long-term solutions must be regional and global.

PRACTICE IN ROJAVA

Since its foundation, the authoritarian and racist Ba'ath regime, with its centralistic and capitalistic economy, had limited priorities in regard to Rojava: to exploit its natural resources and labor force with the greatest efficiency, to maximize agricultural production, and to maintain basic public services with minimum input. It scarcely contemplated the ecological consequences of these policies, such as the loss of agricultural and biological diversity. So, with the exception of Afrîn, all forests and wetlands were destroyed in the twentieth century, and almost all local seed varieties have disappeared since the 1970s. Parallel to these developments, Syrian state agricultural policy brought about the intensive application of chemical fertilizers and pesticides to crops in the 1970s. This undoubtedly damaged both soil quality and groundwater. The resulting negative impacts are a grave legacy, yet Rojava today is confronted with even further challenges because of war and embargo.

Until the Revolution of 2011, the level of ecological awareness in Rojava was so low that it was even not comparable with the situation in Bakûr. Only

Figure 5.5 Assembly of the Mesopotamia Ecology Movement in 2016. *Source:* © Mesopotamia Ecology Movement.

within the PYD, the women's movement, and revolutionary structures has ecology been discussed alongside democratic confederalism. While an ecological approach to political practice has spread step by step, it has only been adopted by a small minority of society. Before the Revolution, there were no ecological or environmental groups in Rojava, even of the bourgeois kind. On the other hand, even though discussion about ecological matters hardly existed in pre-Revolution times, Rojava's society was partly based on solidarity, consumerism was relatively undeveloped, and thus the way of living was quite ecological in practice.

After 2011, the Movement for a Democratic Society (TEV-DEM), the broad civil body tasked with developing the Revolution, treated ecology as a part of the municipalities sector, thus, naming this sector "Municipalities and Ecology," forming one of TEV-DEM's nine sectors. With this step early in the liberation process, an emphasis on ecological principles was established from the outset. The decision to treat ecology within the municipalities and not as a separate sector was obvious; there were insufficient political activists with a specific ecology agenda so it would take time to develop the consciousness for an ecological life in wider society.

What have the municipalities done to promote an ecological approach in the first years of the Revolution? One of the first actions was to maintain trees in the parks because of the fuel crisis during the winter of 2012–2013, and then to establish new parks and trees in the cities. The disposal of garbage

Figure 5.6   Agroecology Workshop at Amed in 2016. *Source*: © Mesopotamia Ecology Movement.

was a substantial challenge which was largely solved, and the cities became cleaner. However, the problem with the open disposal sites continues in the present day. A further issue has been the management of drinking water. To address such issues, the communes—self-organization structures with usually up to 150 households—work closely with the municipalities.

These policies by the municipalities were crucial in the beginning and the basis for introducing new ideas and approaches. Such measures however were not nearly enough to bring about an ecological society. So, political activists in TEV-DEM considered how ecological awareness could be further spread within society at large. They concluded that the way forward would be to create successful small projects to show as practical examples that would inspire other people. For instance, vegetables and trees have been planted in the courtyards or around the buildings of all newly created institutions. The activists wanted to use these to demonstrate that a society can achieve real autonomy if it becomes self-sufficient in producing all, or most, of its needs. Such cultivation starts to reclaim the food production that the Syrian state took away from the people of Rojava by imposing industrial agriculture based on monoculture.

The principles of an ecological society have also been discussed intensively within education, alongside radical democracy, communalism, and

Figure 5.7 Traditional House Constructed by the Mesopotamia Ecology Movement at Amed, in 2017. *Source*: © Mesopotamia Ecology Movement.

questions relating to gender. Education is undertaken at the communal level and, also, within the many academies which have been founded in all the cities. Among them, the Academy for Municipalities and Ecology has the strongest ecological approach. Indeed, it became apparent that ecological consciousness grew most through such discussions in the field of education.

In 2015, when the Revolution was stabilized following the defeat of ISIS, a new process started in Rojava to consider ecological awareness and life. All sectors began to explore more thoroughly the measures that would need to be undertaken to fulfill this pillar of democratic confederalism. Within TEV-DEM, it was understood that ecology is not a specific subject that can be implemented by a single sector. However, Kongreya Star, the women's movement, and the economic sectors, in particular, took important steps and decisions which are crucial to developing the ecological principle.

Kongreya Star had started to talk about how to develop ecological life since the beginning of the Revolution. But now it started to orientate its practice more firmly on an ecological basis. The first step was to reorganize and manage buildings and land in an ecological way so that sites maintained by the women's movement have become some of the greenest in the cities and villages. The most well-known example is the women's village, "Jinwar" completed in 2018, which has been constructed in an entirely ecological way.

Figure 5.8   Tree Nursery at Amed in 2018. *Source*: © Mesopotamia Ecology Movement.

Since 2015 the economic sector has systematically discussed how to implement ecological criteria in economic activities. The main goal of the economy
is to democratize economic activities and to prevent monopolization, with
targets for self-sufficiency throughout the liberated region of north and east
Syria. Cooperatives play a critical role in the solidarity economy in this sense.
In 2015 there was an initiative to develop robust principles for the cooperatives, since clear goals and principles were lacking in the first years of the
Revolution. One result was that all cooperatives must now operate according
to ecological principles. This includes a ban on chemical fertilizers, a move
which became possible when domestic production of organic fertilizer was
implemented in 2016, and the use of fewer pesticides. Also, local and organic
seed varieties are increasingly produced and used in Rojava. Nowadays all
200 cooperatives work on these principles. If the long-term objective to reorganize half of the economy through cooperatives is achieved, this would make
a huge difference—currently, they amount to around 5–7 percent of economic

activity. Another significant economic issue is to make the society less dependent on petrol and natural gas in the longer term. This can only be possible with reduced energy consumption and the adoption of new energy sources.

A more immediately achievable improvement for the economy would be to convince farmers to produce vegetables for local use and to plant more trees in their villages. Today, most villages are greener when compared to 2011 when trees were mostly absent. Considering local conditions in this territory, this positive development makes a difference and is even more effective than planting completely new forests. It helps to diversify production and increase biodiversity. Furthermore, it helps to raise the groundwater level which has dropped dramatically in the last four to five decades.

The health sector also started at an early stage to recover traditional treatments which largely depend on plant-based remedies, produced with ecological methods. This movement connects human health to that of the natural environment.

Figure 5.9   Anti-Nuclear Symbol in Kurdish. *Source:* © Mesopotamia Ecology Movement.

Despite these important developments and better recognition of the importance of ecology, a specific ecological structure or movement was still missing. The Internationalist Commune (founded in 2017) launched the project "Make Rojava Green Again" (MRGA) at the beginning of 2018. Many internationalists come to Rojava and discuss with local revolutionaries how they can learn from, and contribute to, the Revolution. This campaign, therefore, provides opportunities for internationalists to contribute technical skills and expertise in specific fields; mainly in connection with ecology as many have been active in their homeland in ecology groups or ecology-related sciences. So, one objective is to propagate and plant native trees in gardens and parks and, in the middle term, for forests. Another goal is to test small wastewater treatments, to limit the contamination of rivers and groundwater, and then to implement successful methods in the communities. When developing such projects, MRGA works closely with different sectors, including the economic sector and the municipalities. This connection is essential if this campaign is to be successful and to have more of an impact on society, especially young people, in north and east Syria.

Today, all Rojava's sectors have become increasingly aware that it is essential to include ecology in their discussion and practice. Only in this way will a liberated, communal, and exploitation-free society, respecting and in balance with the natural world, be possible. A review of the Revolution in Rojava also allows us to see that there have been significant changes in the ecology sector, even though these have been less evident than with the sectors for women or direct democracy. However, while the systematic development of ecological awareness and practice may have started later, it now has the potential to progress rapidly, even exponentially.

REFERENCES

Human Rights Association of Turkey (IHD). 2001. "Joint Press Release About the Issue of Returning to Village." *Human Rights Association* website. May 30, 2001. https://ihd.org.tr/en/joint-press-release-about-the-issue-of-returning-to-village/.

Mesopotamian Ecology Movement. 2015. "Report on the Recent Forest Fires in North Kurdistan (SE Turkey)." Written by the international delegation—organized by the Mesopotamian Ecology Movement. October 12, 2015. https://www.has ankeyfgirisimi.net/wp-content/uploads/2015/10/Forest-Fires-Report_2015-10.pdf.

Chapter 6

From an Interview with Menekşe Kizildere, HDP Ecology Commission Co-Spokesperson

Menekşe Kizildere

WHAT ARE THE HDP ECOLOGY COMMISSION'S OBJECTIVES?

From the beginning, ecology was a main political area of the party. Gender equality—especially equality for women—human rights, democracy, and ecology; these are the things that the party is based on.

The main movement is the Kurdish part of the HDP. It is under a lot of political pressure especially, so it is very hard to set ecological policy under these circumstances. But the party succeeded because its approach was different from the other political parties, the other opposition. They were still very state-based, very development-based, and thinking that the environment is apart from everything, that it is a soft-policy topic that is not related to current life, the economy, the class issue, or to other democratic problems. And they didn't work in a democratic participatory way. But for the HDP, the ideology behind ecology is a very political topic, it is connected to democracy and human rights, and it is connected to all the environmental belonging rights, the other livings' rights, future generations' rights. And it's a crisis, so any crisis cannot be apolitical, it's political! So that's the main ideology behind our ecology policy and that's our motivation.

In our party, the representation is different. For the other parties, when we look at their members of parliament, they are mostly the people who are coming from business. They are the owners of the companies. Or rich, bourgeois people. But our party was trying to set equal representation of women and equal representation of others. And, also, you cannot see any mining company owner, or coal company owner, or any factory owner in our party. So, that's why we are close to all ecology defenders and the local grassroots organizations. They may not like us politically—for instance, they can be very nationalist, or very Kemalist, or they can be from other ideologies—yet they know we are the people that they can reach directly. They don't like to mention the HDP, but they can reach our member of parliament, and they know we are the only people who will stand with them. We are not at the side of capital, we are with the people, we are defending rights, not just human rights but also all of the environmental belonging rights. That is our approach and the starting point.

WHAT ARE SOME OF THE PRACTICAL CHALLENGES THAT YOU ARE FACING?

The Turkish energy sector, for example, is pretty much all run by companies: dissemination, production, and other parts. It's very centralized. The companies are even deciding what their role will be in legislation. The energy ministry, ministry of the environment, and also the treasury, they are too much under the pressure of the companies, especially five companies which are surrounding our president and running everything.

But there is no view of the people. We are trying to set other policies for the people to decide, and we are especially capacity-building within the party and telling the people there are other opportunities. Because, in Kurdistan, energy is a tool that the government is using over its people. For instance, they have a lot of loans, because they cannot pay their very high bills for electricity, especially the farmers who have animals. So, they are cutting their electricity. They cannot go on farming. Their animals are dying because of water scarcity. They are using it for political pressure against the people. Electricity could be a tool or a gun against the people, especially against the Kurdish people. So that is our motivation—let's start talking about the community energy. How can we produce our own energy and get out of this old bullshit system of capital and profit? But it is a very hard job, what we are trying to do. And also trying to raise awareness of the idea of political ecology.

WHAT DO YOU THINK ARE THE COMMISSION'S SUCCESSES?

We are keeping all these things on the agenda, whatever the government do. We are keeping the climate crisis on the agenda of the parliament. So, for instance, the government doesn't ratify the Paris Agreement, and the way that parliament works is that they want to push it under the first lady's attempts to raise the plastic issue, or the waste issue. Climate is a huge problem, but the first lady is dealing with soft issues apart from everything. There is a huge greenwashing because of the impact of capital on these issues. We are trying to keep the country connected to the international agenda. Because climate diplomacy, climate policy is a very international topic, and it is all connected to the other sectors.

Politically, somebody should keep the right discussion on the agenda of the parliament. We keep that role. As a party, we give lots of questions to the parliament about ecology, targeting the climate crisis. So, after this agenda work, the government had to set up a commission. This is ongoing work, and we will see what the outcome of the commission will be.

So, we are trying to keep all these topics under discussion, while the government is trying to shrink democracy and the main area of discussion with the security topic. That is, when they talk about the HDP, it is all about terrorism, the security problem. Whatever they say, we are trying to keep people with the reality. When we started to talk about the climate crisis at first, they kept blaming us on the social media, with all the trolls. "There are forest fires. You are terrorist party. You are killing people, how can you talk about the nature?" But all that is ignorance. But it's happening. People are watching, they are learning, and it's a long process, but we have to be patient.

We [in the HDP Ecology Commission] are working together with the party's economy, labor, and health commissions. We started a campaign a couple of months ago as four commissions, we are visiting actors in the local areas and finding out their problems. For instance, we are then putting Van's problems to the parliament for a week, raising their voices through all the speeches of the members in parliament. And then some of them are solved.

IS THERE INTERNATIONAL POLITICAL SUPPORT FOR WHAT YOU ARE TRYING TO ACHIEVE?

There is no solid solidarity, but we are trying to keep the agendas. For instance, we are trying to keep the new green deal, still, and the climate crisis discussions. But I can't say there is political solidarity on this issue, from the international actors. Sometimes, the European Greens are supporting us. I can't say that there is a wider mechanism for political solidarity.

But the international discussion is going in a new way. People are starting to discuss, and know that climate change is not apart from democracy, it is not apart from rights and that everything is rights-based. I have also been observing COPs [UN climate change conferences], I think for six years. From post-Paris and before Paris, everything was different. All the discussions started to move that way, people are starting to think about it as a class issue, as a democracy issue (Interview by Skype with Stephen E. Hunt, May 15, 2021).

Chapter 7

Greening and Feeding the City

The Difficult Path to the Implementation of Political Ecology in Diyarbakır/Amed, 2015–2017

Clémence Scalbert-Yücel

Since the first election of the pro-Kurdish party HADEP at the head of local governments in many Kurdish towns and cities in 1999, these municipalities have worked to reappropriate and decolonize the cityscape (Genç 2014; Gambetti 2009). As the most important Kurdish city in the country, Diyarbakır/Amed[1] (hereafter Amed) became a "laboratory" (Dorronsoro and Watts 2013, 102) for the implementation of the Kurdish political project. This project, departing from the struggle for a Kurdish nation-state, was reformulated at the beginning of the 2000s as a project of "democratic autonomy," strongly influenced by the thinker of social ecology, Murray Bookchin, embracing local, participatory, and ecological modes of self-government (Jongerden 2019; Akkaya and Jongerden 2012). Inspired by Bookchin's political thought, ecology is at the heart of this political project which aims at the transformation and improvement of the sociopolitical and organization of life in Kurdistan, with the suppression of all relations of domination—including economic and gender domination—and exploitative relationships to nature.

The aim of this chapter is to investigate how democratic autonomy and political ecology have been embraced in Amed, in an effort to reclaim the city and the urban space. Parks and gardens were an ideal place to start examining this question. Green spaces have been essential parts of contemporary cities, developed as places of exercise and health, recreation and leisure, and education but also of the control of urban dwellers. As such they are sites of the (re)production of social orders and urban norms. Yet, with the introduction of new forms of urban agriculture and urban gardening practices, these spaces can turn into sites of contestation and conflict over the use of space and the political and economic power relationships (Glatron and Granchamp 2018, 2). Examining the gardens, their conceptions, implementations, and works enables us to understand how the project of democratic autonomy, with its ecological and economic components, is put in action in the city, and the challenges such practices encounter.

This work is based on different periods of fieldwork conducted between June 2015 and April 2017, when I met with diverse actors involved in the (trans)formation of the green spaces and their uses in the city. Indeed, pro-Kurdish municipalities are one of the interconnected actors forming the Kurdish movement network (Drechselová 2018, 126; Watts 2010, 21). Besides (ex-)municipal workers and employees, I met members of the Democratic Society Congress (DTK), and of the Diyarbakır Ecology Association (DEA), as well as several persons who cultivated the gardens. The time in which the fieldwork was conducted was marked by the dramatic rise of authoritarianism in Turkey and a turn toward a securitarian and military approach to the resolution of the Kurdish question by the state. Most of the fieldwork was conducted in the aftermath of the Kurdish peace process and the violent destruction of Suriçi, the old city of Amed during the autumn and winter of 2015–2016. That period was also marked by the "dismantlement of the Kurdish municipal system" (Drechselová 2018) in the wake of the July 15, 2016, military coup attempt, with the appointment by the central government of a trustee at the head of Amed municipality in November 2016[2] and the dismissal of many municipal employees and public service workers through a series of ordinance laws (KHK) (İHD Diyarbakır).[3] The work thus provides a recording of a time passed—the time of the pro-Kurdish municipalities—and offers insights into the experiences of the actors involved in the gardens. It provides an opportunity to observe the work of the Kurdish movement and individuals for democratic autonomy and the practices of reclaiming the city in the face of dispossession, as it was ongoing. For some, while it felt like time was on hold during this uncertain and tense period, it was also the time of necessary dreams. These dreams (and deep commitments) kept inspiring local people's actions.

In the first section of this chapter, I present the way in which urban parks and gardens have been a key element of the reconstruction of a modern Kurdish city from the 2000s. I discuss how urban parks created and/or managed by the municipality have progressively embraced the new ecological consideration of the Kurdish movement mainly on a symbolic level. In the second section, I analyze the emergence of new forms of urban agriculture through the development of municipal *bostans* the primary aims of which were to achieve food sovereignty. This enables me to analyze political ecology, in practice, and the difficulty of implementations of projects and thoughts. In the last section, I analyze the *bostans* set up as communal gardens through the DEA which, though short-lived, continued their work even after the shrinking of the public and urban spaces.

GREENING THE CITYSCAPE: PARKS AND GARDENS AS SITES OF A MODERN KURDISHNESS

Pro-Kurdish municipalities developed "symbolic politics" that "helped routinize explicitly Kurdish norms and practices, re-marked the cultural and physical landscape as Kurdish" (Watts 2010, 143) and have revalorized Kurdish identity as a modern identity. Doing so, they worked at "reconstructing Diyarbakır and its inhabitants" as "modern" in order to "refute social stereotypes of Kurds as dirty and primitive" (Watts 2010, 145). Watts quotes Osman Baydemir, the Mayor in 2005: "Our fundamental vision is to make Diyarbakır as a city that lives up to the European vision: a city dedicated to protecting the environment and natural resources, its people, and its heritage" (Watts 2010, 145). This modern city also marked a rupture with the informal habitat characteristic of the 1990s marked by violence, migration, and war (Genç 2014, 283). Progressively too, the ecological vision of the movement is blended into the vision and work of decolonization and reappropriation of the municipality. In this section, I examine how the city parks designed and run by the municipality integrated these three interlinked dimensions: Kurdishness, modernity, and ecology. I focus more specifically on the way ecology and ecological narrative are progressively integrated into the design of the parks, showing that they are also a site of conflict among various actors of the Kurdish movement and, in particular, the municipality and the DEA.

The creation of parks in the city developed extensively under the rule of these parties and was planned to continue at the end of the past decade.[4] As noted by Genç, "building city parks on every scale alongside wide arterial roads has become a *sine qua non* feature of good and successful municipal governance in the eyes of both residents and administrators" and a sign of a city "reborn from its ashes" (Genç 2014, 250). They were associated with the creation of a modern city to which people had to adapt the norms. This is highlighted for instance by the Mayor of Bağlar, Cabbar Leygara, who claimed the villagers "always want to sit on the ground [. . .] but we need to teach them how we live in the city" (quoted by Watts 2010, 146).[5] The landscaped parks are part of a modern city with its norms. They are part of a Kurdish modernity as shown by their names, the sculptures, memorials, and events they host (Bozarslan 2015; Jongerden 2009). Progressively, the municipal green spaces, their conceptions, management, and use evolve to integrate the movement's "vision" of the city, encompassing ecology and the ideals of democratic modernity as stated in the Strategic Plan (Diyarbakır Büyükşehir Belediyesi 2015).

Discussions with the ex-director for the city's green spaces, from 2009 to the date of his dismissal in February 2017, highlight the way that the parks were conceived as sites of a modern Kurdishness and reflect the ecological approach of the Kurdish movement. On the municipal land, a variety of ideas and projects had been taking shape before they were halted with the appointment of the trustee. They included the creation of a botanical garden, with aromatic and medicinal plants collected from Dersim to Hewler; a *gulistan* (rose garden) which would showcase the twenty-four varieties of roses local to Amed which had been identified in the Hevsel Gardens by researchers; the revival of the Hevsel violet (*menekşe*), which had been rediscovered in a private garden and propagated in the municipal nurseries. The municipality was further working on local seeds and varieties and aimed at establishing a seed bank (Interview, Amed, April 2017). All these projects had an underlying heritage dimension. Such gardens and conservation projects aimed at keeping and making the character of the city. Concern with natural and horticultural heritage can be understood in the context of the heritage policies in Amed around the Hevsel Garden and the fortress inscribed on the UNESCO World Heritage list as "cultural landscape" in 2015, highlighting the harmony between human activity and its natural environment (Boucly 2019, 319–321), as well as in regard of the ecological turn of the Kurdish movement, and the revalorization of peasant heritage.

Most of these projects had not fully seen the light at the time of our last meeting. I was able to visit Mardin Kapı Park with the ex-director for the city's green spaces in April 2017. He presented it to me as a "model" park, designed and created in a "Kurdish mindset," reflecting the ambitions and ideology of the city council at that time. We walked together to the park located just outside the Mardin gate by the old city wall, overhanging the Hevsel Gardens and the Tigris Valley, a location that makes the connection between the wall and its natural environment highly visible. He told me that the relief has been left intact to keep the "naturality" of the site, in harmony with the wall. The paving stones have been chosen with care: the black stone, used in most of old Amed's buildings, symbolizes the urban, while the slate (*teht*), coming from the Lice area, symbolizes the rural (Personal conversation, Amed, April 2017).

The park reflects this "Kurdish mindset" that integrates the rurality, a rurality connected to the city, as this had been developed in the writing of Abdullah Öcalan, and incorporated by local Kurdish activists. Öcalan had written: "I have no doubt that what secures an ideal life for humanity is not the structure of the cancerous city of all modern civilizations but the ecological village. Only the city in complete harmony with ecological village can be (a place) tolerated" (Öcalan 2013, 88). Impossible here not to think of Murray

Bookchin's thought. Since *Our Synthetic Environment* (1962) and in *Crisis in our Cities* (1965), Bookchin had underlined the need to redevelop a sense of place, in small communities, in a familiar environment. He had promoted a "complete way of life" with access to both countryside and town and to their specific resources (Bookchin 1962, chapter 7). Both Öcalan and Bookchin are widely read by Kurdish activists and have permeated their discourse and action. Democratic modernity as defined by Öcalan as an alternative to capitalist modernity is not only a project for the present and the future but also has deep roots in the Kurdish past (Öcalan 2013, 426): the Kurdish society which has broken its links with rurality and agriculture must rebuild them. This need was stressed by many in the city.

The discussion and walks across the city with the dismissed director of the city's green space testified the embrace of the ecological project and the Kurdish identity project by the municipality and in its works. The fieldwork and different encounters, however, show that this adoption has not been straightforward and unilinear, but rather was still partial and the subject of critical debate (for general discussion, see Ayboğa and Pale 2019). The role of activists and groups of reflection seemed to have been important in the municipal team's gradual adoption of the political vision. The DTK or the DEA aimed at presenting policy propositions to the municipality and it seemed that several projects mentioned above were inspired, if not launched by DEA members, for instance, the local seed bank. Yet a DEA activist stressed the fact that, though they share the same ideology with the municipality, the municipality does not grasp its full content (Interview, Amed, April 2017). The disagreement around the use and function of the city parks is an example of some differences of approach. In particular, the issue of planting fruit trees in the city has been an element of conflict between the actors: for the municipal workers, fruits falling from the trees would spoil the pavements, whereas for the ecological activists, they would offer free food to the inhabitants of the city.

Today, Mardin Kapı Park stands as a testament to the ambitions and visions of the past municipality. As for the other projects and dreams, some have been picked up by the trustee (Diyarbakır Büyükşehir Belediyesi 2017), but the spirit behind them will probably have been distorted as many encountered during the fieldwork worried. The ex-director of the city's parks and gardens deplored the prospect that the newly appointed municipal team would create a "new landscape," disconnected from the city's soul (Interview, Amed, April 2017). In the aftermath of the destruction and depopulation of Suriçi, and the takeover of the municipal government by the central state, the competition between different hegemonic projects (Genç 2014, 2016) is more than ever visible and violent, making the internal dissensions within the Kurdish movement obsolete. In the second section, I will return to some of these dissensions, more specifically regarding food-growing projects in the bostans and food sovereignty.

FEEDING THE CITY:
THE MUNICIPAL BOSTANS AT WORK
OR THE DIFFICULT EMBRACE
OF DEMOCRATIC ECONOMY

The bostans, small-scale polycultural agricultural production with a commercial purpose (equivalent of market gardens), were common features of cities in Turkey. In Amed, commercial agriculture is still being practiced just outside the old city wall, in the gardens of Hevsel (Gisclard and Raymond 2018). In the past decade, a few contemporary bostans had seen the light in different neighborhoods, embracing communal and democratic economy—as part of the democratic autonomy political project. Democratic economy proposes an "alternative to the paradigm of neoliberal developmentalism" (Akbulut 2017, 232 and 231) as discussed during the 2014 Democratic Economy Conference (Yeniay 2017, 48–49). In Amed, the bostans aimed at the reappropriation and (trans)formation of the city by reclaiming the space, some means of production, and food sovereignty. The bostans, as an experiment in democratic economy, are a good place to observe and analyze the complexity and challenge of its practice.

BOSTANS FOR FOOD SOVEREIGNTY AND
RECONSTRUCTING AN AGRICULTURAL KURDISTAN

The bostans were set up by the municipality in 2015, with the aim of providing for the city's impoverished population, and more particularly for those working as seasonal migrant workers in agriculture. The attention to seasonal agricultural workers is to be understood in the context of the reformulation of the PKK ideology and its views on the resolution of the Kurdish question, on the one hand, and the framework of the Peace Process, on the other. With the redefinition of the primary aim of the PKK in 2005 to create an "Ecological Democratic Confederalist Society" and democratic autonomy, the issue of landlessness, unemployment, and proletarization of the rural migrants uprooted from their villages during the 1990s war has been identified as a key issue. "Ecologist-rural communes" were designated as the basic economic entities (Yarkın 2015, 28–29; Öcalan 2010, 334) and perceived as the solution to address this issue. "A new agricultural society would be recreated through ecological communes that have communal values and 'food sovereignty' as its goals" (Yarkın 2015, 37; see also Öcalan 2013, 266).

Such goals became more tangible, and started being put in practice, at the time of the discussions about the peace process (Koç 2013). To this effect, the DTK's Commission for Labor, Migration, and Poverty organized a Symposium of Mesopotamia's Agricultural Seasonal Workers in April 2013.

The issue of seasonal agricultural workers was identified as arising from the destruction of the socioeconomic fabric of the rural society (Koç 2013). The symposium drew an action plan to address this issue, first, by restoring the conditions of a dignifying life in Kurdistan's villages and cities, facilitating the return to villages in which people can make a living, and, second, by improving seasonal workers' working conditions and life ("Mevsimlik Tarım İşçileri Kurultayı Sonuç Bildirgesi" 2013). The bostans project in Amed endeavored to address this issue. The dismissed director of the Municipal Department for the Local Economy explained the bostans' work and their underlying principles (Interview, April 2017): they aimed at redeveloping agricultural livelihoods and breaking the idea that agriculture cannot exist in cities. The ambition was also to recreate the link between the people and their land. Highlighting the ambition of building food sovereignty, the bostan project was called "I want to feed myself on my land."

The first garden was set up in the village of Ambar (Bismil district) in 2015. It was organized as a commune run by four families (thirteen working people). It produced lentils sent to Rojava on a 100 *dönüm* or 10 hectares (1 *dönüm* is a decare, or a 1,000 square meters), and vegetables on 50 dönüm of municipal land. In the city, three dönüm were allocated (Yenişehir, Mahabat bulvarı, or seventy-five neighborhood) to be turned into a bostan in 2015 and 2016. In 2016, another bostan was created on 300 dönüm in the outskirts of the city, on Sılvan Road (Interview with the dismissed director of the Municipal Department for the Local Economy, April 2017). Though the project was initially aimed at seasonal workers, it was quickly adapted for the needs of the inhabitants of Suriçi who had been evicted, had lost their homes and, very often, their livelihoods altogether, during the war. At the time of the central government's takeover of the municipality, more projects were to be implemented and more communal gardens to be set up on municipal land on the outskirts. All this came to an end.

THE MUNICIPAL BOSTANS IN PRACTICE

Putting the projects into practice and cultivating the bostans was not straightforward. Several issues can be mentioned. First, the people enrolled in the project did not necessarily have an experience of growing. Some may have been seasonal agricultural workers, but their agricultural experience was limited to harvesting. Some had no experience at all in the farming or agricultural sectors: contrary to what the municipal employees thought and declared, people do not necessarily know how to grow food!

Second, the municipality had planned for the provision of material (machinery, fertilizers, seedlings, and seeds) and technical support (such as agricultural engineers' advice). Yet according to the growers I met in December 2016 and April 2017, the support seems to have been inadequate (for instance, too much fertilizer was used which negatively affected the crops), or too late, which led some of the growers to do the job by themselves with their own financial resources.

Third, following the views of the Kurdish movement, the bostan work had to be communal. Yet such organization has encountered many obstacles. In the seventy-five neighborhood bostans, because of the difficulty to recruit growers, the work was mostly done by an extended family. In the Sılvan Road bostan, the DTK wanted the work to be done collectively as a commune, but the municipality gave parcels to families. As a result of this disagreement, the DTK ceased to be involved with the site.[6]

Finally, the financial gains from the bostans are limited. Some of the produces were sold on the piazza at the foot of the metropolitan municipality building; some others were sold directly to the neighborhood's residents. Yet, due to the lack of experience, late support, and inadequate advice, the harvest and its economic benefits were limited and unreliable. Real gains from this experience though, were the moments of joy and peace in the gardens mentioned by some of the growers I met, contrasting with the violence of the war and the dismissals.

These projects seemed to have been designed and implemented in haste, to put into practice some elements of the action plans, with the limited resources of a pro-Kurdish municipality, and altered to respond quickly to swiftly changing conditions. Uncertainty added up to over-hasty designs. As an experiment, there was no guarantee that growers in the bostans would be able to continue after a year. This context made long-term planning impossible. It also made work in the bostans difficult. In the autumn of 2015, curfews prevented the growers from going to the fields for eight days, which led to the loss of the harvest. At the time of the fieldwork, some mentioned the need to occupy more municipal land to grow food and to develop longer-term projects and growing plans. Yet with the dismantlement of the pro-Kurdish municipal system, most action stopped. Nevertheless, as we shall see in the following section, not all dreams collapsed.

CREATING A BOSTANS' CULTURE: THE WORK OF DIYARBAKIR ECOLOGY ASSOCIATION

Shaping the Bostans and Society

During these years, a few bostans were also set up through the DEA, specifically by its Commission for Agriculture, Seeds, and Food (*Tarım, Tohum, Gıda Komisyonu*, or TTGK). The DEA was funded as part of the

Mesopotamia Ecology Movement, an umbrella organization that coordinates local ecological associations across Kurdistan. The MEM was founded in 2011, following the social forums organized in Amed in 2009 and 2011 (Casier 2011) and played an active role in shaping the ecological policy of the Kurdish movement (Yeniay 2017). The TTGK's concerns with food sovereignty, "natural agriculture,"[7] local seeds, and the anti-GMO struggle reflect its sympathy with the international peasant movement and its relationships with La Via Campesina and its Turkish member, Çiftçi-Sen.

The bostans emerged progressively. This was in the context of a striking experience in the refugee camp, set up to shelter Yazidis from Şengal who had escaped the Islamic State's invasion in October 2014, just outside Amed. A key TTGK activist, who worked in the camp as a municipal employee, stressed how the experience had been both shocking as formative. In March 2015, refugees started to grow vegetables; eighty-five family gardens were planted. Refugees sowed seeds that activists had collected previously in the local villages around Amed. The first seed bank—described as a "seed commune"—was created in 2016 in the camp; there the TTGK grew many seedlings to be distributed for free outside (Interview with a DEA activist, November 2016). This experience seems to be as much important—if not more—than the influence of national and international struggles, together with the intersection of the association in the Kurdish movement, and in the network of actors administering the city.

Following this experience in the camp, two bostans were set up through the TTGK on municipal land in the Bağcılar and Mezopotamya neighborhoods and one was set up in Toplu Konut neighborhood (Yenişehir), where it was run by a resident in agreement with the local Administration for Collective Housing (TOKI). In total, they covered four dönüm. The first two bostans were run by the neighborhood assembly, and the cultivation of the garden, according to another TTGK member, depended on the existence of a neighborhood assembly, which could facilitate the enrolment, and mobilization of local residents in the gardens: the TTGK approached the neighborhood assembly to discuss the establishment of the bostans in their vicinity. The two bostans in Bağcılar and Mezopotamya were located adjacent to the assembly buildings. A few other bostans were established in the Kurdish language nurseries, called Zarokistan, set up and run at that time by the city council (Interview with a DEA activist, November 2016).

In practice, the role of TTGK was to launch the gardens through the neighborhood assemblies. TTGK also provided the seedlings to the assemblies, participated in the plantation, in informing the local people, and explaining how to run and cultivate as a commune. Indeed, all bostans had to be cultivated as such—based on solidarity and collective growing rather than division into individual plots. Once this was done, local inhabitants were left to cultivate the land. Run by the local community, and in the vicinity of the neighborhood assemblies, the gardens were well looked after, I was told.

The bostans' aim, according to the members of the TTGK, was to "create a dialogue and organise people; to create communes; develop self-production" (Interview with two members of the TTGK, November 2016). As such, they shared some of the aims of the municipal bostans but did not pretend to provide a living to the growers. According to the members of the TTGK, this work contributed to "create a bostan culture in Kurdistan" (Interview with two members of the TTGK, November 2016). Although the Gezi occupation was not mentioned openly, one cannot help but think of the experience of the bostan set up in Gezi Park during the 2013 occupation, and the ones created in the aftermath of its destruction, which carried anti-government and ecologist values (Fautras 2016), values shared by the activists of the Kurdish ecological movement. In Amed, the work of the contemporary bostans is a work of resistance against the neoliberal enclosure of the urban space and the means of production aiming at the (trans)formation of place and society.

CLOSURE OF THE PUBLIC SPACE, PRESERVATION OF THE *BOSTANS'* WORK

In our meetings, the activists of the DEA and the TTGK mentioned many projects but also that all initiatives had stopped in the uncertain context of the first appointment of the trustee in the fall of 2016, and the following normalization of this situation, leading to the loss of many jobs. The future of the bostans was threatened. Yet the context, marked by uncertainty and the loss of many of the activists' jobs, also reinforced the idea of the importance of self-sufficiency and autonomy. The end of the pro-Kurdish municipal system, and the political and economic crisis bringing hardship into many people's lives, highlighted in a tangible fashion the importance of the ecological and alternative economy project of the Kurdish movement. After the dismantlement of the pro-Kurdish municipal system, activists have had to find alternative places to conduct their work. Leaving the city though does not necessarily mean the abandonment of political vision and activism.

Roughly fifteen years ago, a few individuals, including members of the TTGK had bought a field of approximately four hectares (forty dönüms) on the city's outskirts. It is on three dönüms of that field that the commission started moving and organizing the work of the bostan in 2017 (Interview with a member of the TTGK, April 2017). On this private land, the communal work of the bostan reorganized itself as a political space. Today the bostans continue their work as communal gardens in that privately owned space. Such a relocation had been possible with a handful of passionate activists. Though the experience could only be accessible to some persons with economic resources and with mean of transportation to this land outside the city, this experience of the bostan as a commune on a private land enables the work and the struggle to continue.

CONCLUSION

Through an analysis of city parks and gardens in Amed, this chapter has shown how the actors of the Kurdish movement have attempted to reclaim the urban space and means of production and subsistence in line with their vision of a dignifying life for the individual and wider Kurdish society, at a time of renewed state violence and dispossession. It has highlighted some divergences in visions and practices, and some difficulties in shifting from traditional institutional ways of doing things to new practices informed by democratic autonomy thought. Yet, the goal and dream of reclaiming a destroyed relationship to the village, agricultural practices, and henceforth alternative mode of living and subsistence had started to be orchestrated in action, enrolling municipal workers, activists, and inhabitants of deprived neighborhoods. Actions have produced more dreams and ideas, and all did not collapse after the end of the pro-Kurdish-municipal system. In the context of the shrinking of the urban and political space, these dreams have continued, moving the impetus in other spatial contexts. This context, which makes the implementation of the Kurdish project of democratic autonomy extremely difficult, also highlights its very importance and potential for rebuilding a dignifying life. Yet, as shown by the example of the communal bostan on private land, this remains mostly possible for those with the necessary capital and resources to adapt, and first and foremost to own, one's land.

NOTES

1. Majority Kurdish city known as "Amed" in Kurdish and called "Diyarbakır" in Turkish.
2. The Ordinance Law 674 of September 1, 2016, enabled the modification of the municipal law and the appointment of trustees at the head of the municipalities whose mayors or members of the council are facing legal charges (many were accused of supporting terrorism). On September 11, trustees were nominated by the central government at the head of twenty-eight municipalities. On October 25, 2016, the co-mayors of Amed were arrested and a trustee appointed on November 1, 2016. At the time of writing, they remain in jail and the city is still administered by a trustee.
3. During the state of exception that lasted two years after the coup attempt of July 15, 2016, thirty-one ordinance laws were published, which led to the dismissal of at least 130, 000 civil servant workers (Yaşam Hakları Derneği 2020).

4. In Amed, the municipality aimed to increase the amount of green space area per person from 3.4 to at least 3.8 square meters (Diyarbakır Büyükşehir Belediyesi 2015, 81).

5. The reference to modernity is present in the *Strategic Plan 2015-2019*, stating that old parks had to be turned into "places of modern life" (Diyarbakır Büyükşehir Belediyesi 2015, 81).

6. Interview with a municipal worker, ex-member of the DTK's economic commission and specialist on the issue of agricultural seasonal workers, December 2016, and interview with a grower, April 2017.

7. Activists support "natural agriculture," which is respectful of the environment, yet not certified organic since the certification process can be costly for producers.

REFERENCES

Akbulut, Bengi. 2017. "Commons against the Tide: The Project of Democratic Economy." In *Neoliberal Turkey and its Discontents: Economic Policy and the Environment in the Justice and Development Party Era*, edited by Fikret Adaman, Bengi Akbulut, and Murat Arsel, 231–45. London: I.B. Tauris.

Akkaya, Ahmet Hamdi, and Joost Jongerden. 2012. "Reassembling the Political: The PKK and the Project of Radical Democracy." *European Journal of Turkish Studies* 14. http://ejts.revues.org/4615.

Amnesty International. 2016. *Turkey: Displaced and Dispossessed: Sur Residents' Right to Return Home.* December 6, 2016, Index number: EUR 44/5213/2016. https://www.amnesty.org/en/documents/eur44/5213/2016/en/.

Ayboğa, Ercan, and Egit Pale. 2019. "The Democratization of Cities in North Kurdistan." In *The Right to the City and Social Ecology: Towards Democratic and Ecological Cities*, edited by Federico Venturini, Emet Değirmenci, and Inés Morales, 110–17. Montréal: Black Rose Books.

Bookchin, Murray. 1962. Our Synthetic Environment. New York: Knopf. http://dwardmac.pitzer.edu/Anarchist_Archives/bookchin/syntheticenviron/osetoc.html.

Bookchin, Murray. 1992. Urbanisation without Cities. The Rise and Decline of Citizenship. Montréal: Black Rose Books.

Boucly, Julien. 2019. La fabrique nationale du patrimoine mondial. Une étude politique de l'action publique patrimoniale en Turquie et à Diyarbakır. These de Doctorat, EHESS. Paris.

Bozarlslan, Mahmut. 2015. "Diyarbakır'ın isimsiz parkları." AlJazeera Turk. January 31, 2015. http://www.aljazeera.com.tr/al-jazeera-ozel/Diyarbakırin-isimsiz-parklari.

Casier, Marlies. 2011. "Beyond Kurdistan?: The Mesopotamia Social Forum and the Appropriation and Re-imagination of Mesopotamia by the Kurdish Movement." *Journal of Balkan and Near Eastern Studies* 13, no. 4, 417–32.

Diyarbakır Büyükşehir Belediyesi. 2015. *Stratejik Plan 2015–2019.* Diyarbakır: Diyarbakır Büyükşehir Belediyesi.

Diyarbakır Büyükşehir Belediyesi. 2017. *Stratejik Plan 2017–2021*. Diyarbakır: Diyarbakır Büyükşehir Belediyesi.

Dorronsoro, Gilles, and Nicole F. Watts. 2013. "The Collective Production of Challenge: Civil Society, Parties, and Pro-Kurdish Politics in Diyarbakır." In *Negotiating Political Power in Turkey: Breaking up the Party*, edited by Elise Massicard and Nicole F. Watts, 99–117. London: Routledge.

Dreschselová, Lucie. 2018. "Le démantèlement du système municipal kurde et ses retombées genrées dans le sud-est de la Turquie." *Confluences Méditerranée* 107: 125–36.

Fautras, Agathe. 2016. "Les nouveaux bostan d'Istanbul: quelle pérennisation pour les jardins de la contestation?" *European Journal of Turkish Studies* 23. https://doi.org/10.4000/ejts.5400.

Gambetti, Zeynep. 2009. "Decolonizing Diyarbakır: Culture, Identity and the Struggle to Appropriate Urban Space." In *Comparing Cities: The Middle East and South Asia*, edited by Kamran Asdar Ali and Martina Rieker, 97–129. Karachi: Oxford University Press.

Genç, Fırat. 2014. *Politics in Concrete: Social Production of Space in Diyarbakır, 1999–2014*, PhD dissertation, Atatürk Institute for Modern Turkish History, Boğaziçi University, Istanbul.

Genç, Fırat. 2016. "Suriçi in Destruction-Regeneration Dialectic." Heinrich Boll Stiftung, April 15, 2016. https://tr.boell.org/en/2016/04/15/surici-destruction-regeneration-dialectic.

Gisclard, Marie and Raymond, Richard. 2018. "L'agriculture urbaine comme opportunité de construction politique: Réflexions à partir des jardins de l'Hevsel à Diyarbakır." In *Les jardins de l'Hevsel, paradis intranquilles*, edited by Martine Assénat. Istanbul: Institut francais d'etudes anatoliennes. https://doi.org/10.4000/books.ifeagd.2289.

Glatron, Sandrine, and Laurence Granchamp. 2018. "Places and People of Urban Gardens. Elements for an Introduction." In *The Urban Garden City. Cities and Nature*, edited by Sandrine Glatron and Laurence Granchamp, 1–14. Cham: Springer. https://doi.org/10.1007/978-3-319-72733-2_1.

İnsan Halkları Derneği. Diyarbakır Şubesi. 2019. Seçme ve Seçilme Hakkına Yönelik İhlaller Araştırma Raporu. Belediyeler yönelik kayyım atamaları belediye eşbaşkanları ve meclis üyelerinin gözaltına alınması-tutklanması. Diyarbakır: İHD.

Jongerden, Joost. 2009. "Crafting Space, Making People: The Spatial Design of Nation in Modern Turkey." *European Journal of Turkish Studies* 10. https://doi.org/10.4000/ejts.4014

Jongerden, Joost. 2019. "Learning from Defeat: Development and Contestation of the 'new paradigm' within Kurdistan Workers' Party (PKK)." *Kurdish Studies* 7, no. 1: 72–92.

Koç, Ferda. 2013. "Kürt işçisi de çözüm istiyor." Sendika.org. March 26, 2013. https://sendika.org/2013/03/kurt-iscisi-de-cozum-istiyor-ferda-koc-99752/.

"Mevsimlik Tarım İşçileri Kurultayı Sonuç Bildirgesi." April 10, 2013. https://sendika.org/2013/04/mevsimlik-tarim-iscileri-kurultayi-sonuc-bildirgesi-102257/.

Öcalan, Abdullah. 2010. *Demokratik Uygarlık Manifestosu. I. Kitap. Uygarlık. Maskeli Tanrılar ve Örtük Krallar Çağı.* Abdullah Öcalan Sosyal Bilimler Akademisi Yayınları. https://archive.org/stream/DemokratikUygarlkManifestosu/Demokratik%20Uygarlık%20Manifestosu_djvu.txt.

Öcalan, Abdullah. 2013. *Demokratik Uygarlık Manifestosu. VI. Kitap. Ortadoğu'da Uygarlık Krizi ve Demokratik Uygarlık Çözümü.* Abdullah Öcalan Sosyal Bilimler Akademisi Yayınları. https://pndk.org/sites/default/files/files/content/reberti/Dördüncü%20Kitap%20-%20Ortadoğu%27da%20Uygarlık%20Krizi.pdf.

PKK. 2005. PKK Yeniden İnşa Bildirgesi. http://archive.ph/zFTt4#selection-247.0-247.27.

Watts, F. Nicole. 2010. *Activists in Office. Kurdish Politics and Protest in Turkey.* Seattle: University of Washington Press.

Yarkın, Güllistan. 2015. "The Ideological Transformation of the PKK Regarding the Political Economy of the Kurdish Region in Turkey." *Kurdish Studies* 3, no 1: 26–46.

Yaşam Hakları Derneği. 2020. *Kanun Hükmünde İhlaller. KHK Mağdurlarına Yönelik Çalışma Hakkı İhlalleri ve Ayrımcılık Araştıması.* Istanbul: Yaşam Hakları Derneği. https://bianet.org/system/uploads/1/files/attachments/000/003/249/original/KHK_Raporu_-_Çalışma_Hakkı_İhlalleri.pdf?1606816532.

Yeniay, Özlem. 2017. Topraksızlar Hareketi ve Kürt Özgürlük Mücadelesinde Alternatif Yaşam ve Ekoloji: Tartışmalar/ Karşılaştırmalar. *Toplum ve Kuram* 12, 33–56.

Chapter 8

Regenerating Kurdish Ecologies through Food Sovereignty, Agroecology, and Economies of Care

Michel P. Pimbert

People's initiative to promote ecological approaches to agriculture and the re-localization of decentralized food systems[1] are remarkable in war-torn Kurdistan. Efforts to generate Kurdish ecologies through agroecology, food sovereignty, and economies of care are uniquely based on traditions of social ecology, a rejection of patriarchal relations, and democratic confederalism.

A BRIEF HISTORY OF CHANGING AGRICULTURAL ECOLOGIES IN KURDISTAN

Historical studies of agriculture in Syrian and Turkish Kurdistan have described highly sophisticated and locally adapted traditional land use practices. For several centuries, agriculture was based on a highly diverse combination of plants and animals, adapted to summer drought and a winter growing season. Four complementary strategies reduced risks: outfield cultivation of grains and legumes; a diversity of garden vegetables, condiments, medicinal plants, and herbs; orchard crops providing wine and fresh or dried fruits and, where possible, olive oil. Last, several options helped to integrate animal husbandry with agriculture as a source of manure and alternative proteins and fats. In Butzer's words:

> The strong integration of horticulture and arboriculture distinguished this Mediterranean-Near Eastern agro-system. . . . Characterized by a diversified yet distinctive cuisine, this lifeway balanced solutions to risk with equally deep-seated cultural values. (Butzer 1994, 19)

Animal husbandry has also been an important part of agrarian life. Kurdish shepherds settled in the area north of the ancient Fertile Crescent. This region witnessed the domestication of sheep about 8,000 years ago which then resulted in the selection of many locally adapted breeds by Kurdish shepherd communities. Thirty years ago, 97 percent of the sheep population in Turkish Kurdistan was composed of local breeds (Askin et al. 1989). Pastoralism in Kurdistan traditionally has two forms: village-based sedentary pastoralism and pastoralism with vertical or horizontal movements. The latter has many variations, including local transhumance as well as regional or interregional trips by nomadic or semi-nomadic Kurdish communities and tribes (Thevenin 2011).

Historically, Kurdistan has been a smallholder, food-producing region that covered the needs of its population. Wheat and barley production mostly took place in the Syrian Kurdistan region, while most of the vegetables were produced along rivers and areas where irrigation was available. Fruit and date orchards were well suited to Kurdistan's temperate hillsides and to more arid regions where irrigation water is available. Crop and tree farming as well as pastoralism have deeply influenced the ecological features and the biodiversity of landscapes. Over time, local food providers worked with nature to co-create myriads of humanized landscapes in Kurdistan and the wider region. Biodiversity-rich forests, wetlands, and grasslands have coexisted with farming and pastoralism as part of a complex land use mosaic adaptively managed by local communities.

However, this ecological diversity and complexity was largely suppressed with the imposition of colonial resource extraction and industrial agriculture based on genetically uniform monocultures. As part of this historical shift, Syrian and Turkish Kurdistan have become major importers of food over the last decades.[2] Like the rest of the region, the capacity of Kurdish agriculture to feed its population and its role in the economy has been severely affected by poorly conceived "modernist" policies, colonial designs, violent conflict and war, and cheap imports of foodstuffs.

In the 1960s and 1970s, the Ba'athist regime began to create a centralist planned economy in which northeastern Syria became the nation's breadbasket. The communal, traditional mode of farming was replaced by huge, industrialized wheat monocultures managed by state companies that relied heavily on chemical fertilizer and pesticides. After the 1990s, the Ba'athist regime introduced some neoliberal reforms in which foreign companies partly replaced the state-owned businesses. However, overall, the ecologically destructive mode of production remained the same. Modernizing agricultural and environmental policies also undermined nomadic or semi-nomadic practices in the region (Walliser 2011), forcing many pastoral communities to become sedentary.

Today, farms in northeastern Syria are ecologically damaged by the monoculture of wheat in the region of Cizîrê and of olives in the region of Afrîn. Cizîrê's holly oak forests and its wetlands have been replaced by extensive wheat monocultures that have destroyed habitats for wildlife (such as gazelles, birds, fish, and the Caucasian squirrel). The expansion of the agricultural frontier for barley, cotton, grapes, and olive monocultures has also led to dramatic losses in the biodiversity of ancient forests, grasslands, and wetlands in Syrian Kurdistan. Deforestation and impoverishment of the soil are particularly severe in the region of Afrîn. There has also been a significant loss of biological diversity as locally adapted crop varieties and livestock breeds have been displaced by modern hybrids and genetically uniform varieties, promoted by the Syrian government and seed corporations (FAO 2019).

The model of industrial and export-oriented agriculture embraced by Syria's government is a form of capitalist accumulation based on the relentless dispossession of Kurdish communities and the wealth of ecosystems they depend on for their food, agriculture, and livelihoods. This extractive process is exacerbated by neighboring Turkey. For example, the aridity of the Kurdish environment and availability of water is increasingly made worse by Turkey's extractive siphoning of underground water in Kurdish Syria and by holding back water in Turkish dams for crop irrigation and other purposes. These externally driven ecological dislocations in Syrian and Turkish Kurdistan are deeply undermining community and socio-ecological resilience to shocks and stresses, such as climate change and market volatility.

KURDISH PEOPLES' RESPONSE TO DESTRUCTIVE INDUSTRIAL AGRICULTURE

Over the last decade, many Kurdish communities have responded to the destruction caused by externally imposed industrial agriculture and capitalist extractivism by advancing alternatives grounded in the social ecology of Murray Bookchin and the democratic confederalism of Abdullah Öcalan. Reflecting on the ecological dimension of democratic confederalism, Bookchin said:

The views advanced by anarchists were deliberately called *social* ecology to emphasise that major ecological problems have their roots in social problems—problems that go back to the very beginnings of patricentric culture itself. The rise of capitalism, with a law of life based on competition, capital accumulation, and limitless growth, brought these problems—ecological and social—to an acute point; [. . .] Capitalist society, by recycling the organic world into an increasingly inanimate, inorganic assemblage of commodities, was destined to simplify the biosphere, thereby cutting across the grain of natural evolution with its ages-long thrust towards differentiation and diversity.

Bookchin went on to argue that to reverse this trend:

Capitalism had to be replaced by an ecological society based on non-hierarchical relationships, decentralised communities, eco-technologies like solar power, organic agriculture, and humanly scaled industries—in short, by face-to-face democratic forms of settlement economically and structurally tailored to the ecosystems in which they were located. (1998, 154–55)

Bookchin's social ecology and libertarian municipalism, which inspired Öcalan's views on democratic confederalism (Öcalan 2011; 2017), is based on a model of anarchism in which autonomous communes federate at multiple scales. However, Bookchin has enriched this classical model of anarchism by emphasizing the central importance which ecology must have, because humankind has no future if we do not protect nature (Bookchin 2005). This was also the view of several social anarchists at the end of the nineteenth century in Europe. There are indeed similarities between Bookchin's vision of an ecological society, and the libertarian communist future which Peter Kropotkin described in *The Conquest of Bread* (Kropotkin 2015) and *Fields, Factories and Workshops* (Kropotkin 1898), including an agrarian-industrial mutualism in which most economic activities are re-localized in mixed agricultural/industrial villages and where production is controlled by those directly engaged in it.

Bookchin's proposals for a regenerative ecological agriculture and a network of local food systems are informed by his social ecology perspective (Bookchin 1976). His and Abdullah Öcalan's proposals for a self-governing communalist society also converge with the food sovereignty paradigm developed by *La Via Campesina* and other global social movements.

The term food sovereignty was first brought to international attention at the World Food Summit organized by the United Nation's Food and Agriculture Organization in 1996. It was put forward by La Vía Campesina,[3] an international movement which coordinates organizations of small and medium-sized producers, agricultural workers, rural women, and indigenous communities from Asia, America, and Europe.

Food sovereignty aims to recreate the realm of democracy and freedom by regenerating a diversity of autonomous food systems in both rural and urban areas (Pimbert 2008). It is thus grounded in the idea that farmers[4] and other citizens—men and women—can and should govern themselves by engaging in the practice of direct democracy. The Declaration of the 2007 Nyéléni

Forum on Food Sovereignty affirms the centrality and primacy of "peoples"[5] in framing policies and practices for food, agriculture, environment, and human well-being:

> Food sovereignty is the right of peoples to healthy and culturally appropriate food produced through ecologically sound and sustainable methods, and their right to define their own food and agriculture systems. It puts those who produce, distribute and consume food at the heart of food systems and policies rather than the demands of markets and corporations. It defends the interests and inclusion of the next generation. It offers a strategy to resist and dismantle the current corporate trade and food regime, and directions for food, farming, pastoral and fisheries systems determined by local producers. Food sovereignty prioritizes local and national economies and markets and empowers peasant and family farmer-driven agriculture, artisanal fishing, pastoralist-led grazing, and food production, distribution and consumption based on environmental, social and economic sustainability. Food sovereignty promotes transparent trade that guarantees just incomes to all peoples as well as the rights of consumers to control their food and nutrition. It ensures that the rights to use and manage lands, territories, waters, seeds, livestock and biodiversity are in the hands of those of us who produce food. Food sovereignty implies new social relations free of oppression and inequality between men and women, peoples, racial groups, social and economic classes and generations. (Nyéléni 2007)

Kurdish communities—but by no means all—have understood the need to regenerate the ecology and economy of local food systems within a food sovereignty framework rooted in libertarian municipalism and democratic confederalism. From July 2012 onward, the forces of the Ba'athist regime moved out of Rojava, the companies of the Syrian state were expropriated, and the councils took control of the agricultural land. The new Kurdish administration was based on newly created communes that mostly consisted of a village and the surrounding hamlets. These communes distributed the land among their families according to need and ability to manage it. Some pieces of land remained in the hands of the higher councils and were earmarked for the first cooperatives.

Early on, Kurdish people spoke of the need to diversify agriculture for the sake of self-sufficiency. Some communities also envisioned re-localizing food and agriculture to restore local ecologies, regenerate biodiversity, and reduce greenhouse gas emissions in Kurdish-liberated territories.

However, these are huge challenges for a war-torn society defending itself against Islamic fundamentalism and the state-sponsored violence of Turkey and Syria. In this profoundly disabling context, Kurdish initiatives for ecology in food and farming point to a direction of travel rather than to fully formed working alternatives based on food sovereignty and democratic confederalism. Three mutually reinforcing pathways for regenerating Kurdish ecologies are critically important in this regard, as discussed below.

AGROECOLOGY FOR FOOD SOVEREIGNTY

From its initial emphasis on ecology for the design of sustainable agriculture, agroecology now emphasizes the study of the ecology of food systems (Gliessman 2015). A transformative agroecology for food sovereignty seeks to reduce dependence on corporate suppliers of external inputs and distant global commodity markets.

Agroecological approaches are based on the following:

(a) Re-embedding agriculture in nature, relying on functional biodiversity and internal resources for production of food, fiber, and other benefits. Resilient agroecological systems mimic the structure and function of natural ecosystems: biodiversity-rich fruit orchards and agroforestry systems, intercropping, genetic mixtures, mixed farming, agro-sylvo-pastoral production systems. The principles on which these agroecological systems are based originate in the knowledge-rich practices of indigenous and peasant communities. They include the following practices:

(i) adapting to the local environment and its diverse micro-environments; (ii) creating favorable soil conditions for plant growth and recycling nutrients, particularly by managing organic matter and encouraging soil biological activity; (iii) minimizing losses of energy, water, nutrients, and genetic resources by enhancing the conservation and regeneration of soil, water, and agro-biodiversity on the farm and in the neighboring landscape; (iv) diversifying species, crop varieties, and livestock breeds in the agroecosystem over time and space—including integrating crops, trees, and livestock at the field and wider landscape levels; (v) strengthening the "immune system" of agricultural systems through the enhancement of functional biodiversity—natural enemies of pests, allelopathy, and antagonists, for instance, by creating appropriate habitats and through local adaptive management; (vi) enhancing beneficial biological interactions and synergies throughout the system and among the components of agricultural biodiversity, thereby promoting key ecological processes for sustainable production and resilience to stresses and shocks.

Examples of these regenerative practices exist in different parts of liberated Kurdistan. The Internationalist Commune has planted 2,000 trees in 2018 and has generated some 50,000 design plans for the reforestation of areas belonging to the commune, as well as other zones in the Cizîrê region. For example, the Committee for Natural Conservation in the Reforestation of the Hayaka

Natural Reserve, close to Dêrik, plans to grow 50,000 trees on the banks of Lake Sefan within five years, thus contributing to the regeneration of biodiversity and ecological functions, such as pollination and carbon sequestration. In 2015, agricultural cooperatives planted over 50,000 fruit trees in Cizîrê, as part of their attempt to diversify production and move away from genetically uniform wheat monocultures dependent on the use of toxic pesticides and expensive chemical fertilizers (Bance 2019).

(b) Farmers distance themselves from markets supplying inputs (such as seeds and seedlings, fertilizers, and pesticides). Reduced dependence on commodity markets for inputs enhances farmers' autonomy and control over the means of production. For example, agricultural cooperatives have set up plant nurseries to provide farmers, gardeners, and municipalities with a large number of diverse fruit trees (olive trees, pomegranates, peach trees, and grapevines), forest plants, or decorative plants, such as rose bushes, as well as ornamental plants.

(c) A rediscovery of forgotten resources: organic manure and the soil's capacity to improve the yields and nutritional quality of foods; renewable energies (solar, wind, and biogas) and their decentralized and distributed micro-generation in towns and cities. The tightening of the embargo in 2015 led to a fourfold increase in the price of fertilizers and a reduction in agricultural production, especially in Cizîrê. In response, an international crowd-funding campaign[6] was organized, and decentralized facilities have started to be established in several districts of Rojava to produce organic compost and fertilizers by recycling waste from cities and farms. As part of these efforts for greater self-reliance, local knowledge on the nutritional and medicinal values of plants for people and livestock is being slowly brought back into food and farming in Rojava. Indeed, much of agroecology draws on the experiential knowledge of peasant farmers and pastoralists which it combines with the science of ecology.

(d) Farmers diversifying outputs and market outlets, often with the help of citizens. A greater reliance on alternative food networks that reduce the distance between producers and consumers while ensuring that more wealth and jobs are created and retained within local economies: short food chains and local food webs in which decentralized small-scale cooperatives engage in food processing (e.g., bakeries and cheese making) as part of the social economy.

Overall, agroecological initiatives in Kurdish territories show a mixed picture. There are different levels of change toward agroecology practices in rural, peri-urban, and urban areas (figure 8.1). Several agroecological initiatives aim for more efficient use of external inputs, such as fertilizers, or have replaced toxic agrichemicals with less harmful substances, such as organic composts or botanical insecticides, in largely unchanged

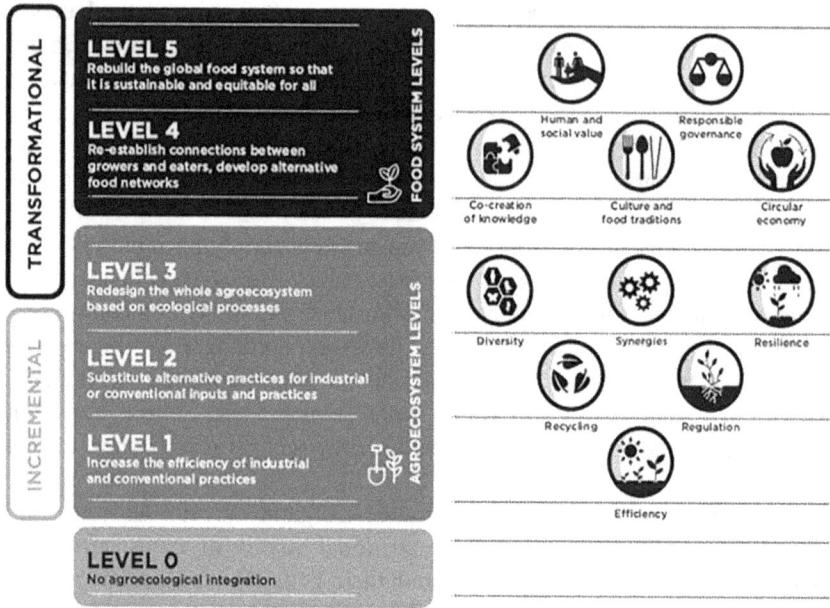

Figure 8.1   Five Levels of Food System Change and FAO's Elements of Agroecology. Adapted from Gliessman 2015 and FAO 2018. *Source*: © Biovision.

monocultures. Transitions toward agroecology practices tend to be incremental here.

In contrast, more transformative agroecological changes seek to redesign farms and the surrounding landscape to imitate the structure and function of natural ecosystems. For example, the ecology of homogenous and degraded landscapes is revived by planting intercrops and polycultures, mixed fruit tree orchards, bio-diverse urban/peri-urban gardens, and the revival of agroforestry systems. This diversification of agro-ecosystems allows farms to generate their own internal biological pest control and nutrient cycling, thereby reducing dependency on external suppliers of industrial inputs. These transformative agroecological practices enhance the autonomy of farmers and food sovereignty.

Last, there are also examples of food system transformation that link biodiversity-rich agroecological production with local markets and food networks in Kurdish territories—Level 4 in figure 8.1.

This coexistence of different levels of agroecological change (figure 8.1) is common in many countries today where several disabling factors prevent society-wide agroecological transformations—Level 5 in figure 8.1 (Anderson et al. 2021; HLPE 2019). However, large- scale shifts toward a transformative agroecology are even more constrained in war-torn Kurdish territories where significant amounts of resources are allocated to defense against violent external aggressors.

International solidarity that facilitates mutual learning on agroecological practices would help diversify and restore Kurdish food and farming ecologies. For example, evolutionary and participatory plant breeding can significantly increase genetic diversity in crop monocultures (Ceccarelli and Grando 2020). In Iran, evolutionary plant breeding (EPB) in barley and wheat allowed farmers to access and develop a great amount of crop biodiversity in a relatively short time. In EPB, farmers start by planting a large mixture of hundreds or even thousands of different varieties of a crop. These genetically diverse varieties are left to cross freely between each other. Seed harvested is replanted the next year for another crop cycle in the same environment. The plant populations thus evolve under different types of agronomic management and adapt to diseases, insect pests, drought, extreme temperatures, and salinity. As a dynamic and inexpensive strategy for agroecological diversification, EPB rapidly enhances the adaptation of farmers' crops to climate change. It also creates living gene banks under the control of farmers (Rahmanian et al. 2016). Mutual sharing of farmer knowledge on this agroecological practice could help Kurdish farmers re-diversify genetically uniform wheat monocultures, thereby enhancing adaptation to climate change and improve nutrition.

AN ECONOMICS OF CARE AND SOLIDARITY

Democratic confederalism envisions a communal economy in which alternative economic institutions are necessary not only for self-government (Öcalan 2011). They are also needed to enable ecological sustainability and keep human activities within safe planetary limits (Steffen et al. 2015). New Kurdish economic institutions must also help reverse enduring patriarchal relations and the structural violence against women. Ecology, gender justice, and democracy thus need to be at the heart of an economics of care and solidarity that can sustain food, agriculture, and the well-being of people and nature.

Building such a communal social economy is a huge challenge in Kurdish-liberated territories. In practice, this has involved creating cooperatives, communes, and communal economic networks. Cantons have funded the setting up of new cooperatives, most of which are agricultural cooperatives. It is noteworthy that the number and size of women's collectively self-managed cooperatives have markedly increased since 2012. Many of them focus on food processing, including, for example, a woman's cheese-making cooperative in Dêrik, women's bakeries in Derna village and the city of Serekaniyé, and a women's agricultural cooperative working to diversify crops grown near Amudé. In a strongly patriarchal context, women's cooperatives are an important way for young and older women to become less and less dependent on their fathers or husbands.

To date, the growth in number and impact of these cooperatives and economic communes are constrained by a lack of funds, access to reliable energy sources, and appropriate facilities and machinery for running community food processing units and small-scale industries. However, these self-managed local organizations signal how the economic infrastructure of agri-food systems could be re-built on a large scale in Kurdish territories. Each link in the agri-food chain offers potential economic niches for many more people organized in cooperatives—as millers, bakers, butchers, carpenters, ironworkers, and mechanics, local milk processors, and cheese makers. The fluid network of decentralized economic communes and cooperatives involved in food production, processing, distribution, and waste recycling demonstrates that a significantly greater number of livelihoods and ecologies could potentially be sustained by scaling out agroecology-based food systems.

The communalization of the economy has greatly facilitated ecological regeneration on farms, pastoral lands, watersheds, and municipal gardens. This is largely because the communalization of the economy by the Kurdish freedom movement enables more equitable access to water, land, seeds, and energy. These are common goods available for use by the whole of society. As Cemil Bayik said:

> As long as society is the communal proprietor of these goods, no individual can exploit them. Moreover, such a society cannot become subject to economic domination . . . Least of all should water, land and energy belong to a State. A State that claims to control them is despotic and fascistic. (Cemil Batik quoted in Knapp et al 2016, 207)

The transformative potential of agroecological practices is thus further enhanced by gender-inclusive forms of economic organization, in which the means of production are socialized and communally shared. Economic exchanges mediated by the solidarity-based cooperatives bring into closer proximity food producers and food eaters in re-territorialized systems that reduce dependency on commodity markets and global value chains.

However, regenerating the local ecology of autonomous food systems also partly depends on a solidarity economy rooted in a deep sense of care, gift relations, and an intimate entanglement with nature. Talking about how women and plants mutually care for each other in nature, Azime Efrîn says:

> There's especially a lot of plants in Mesopotamia, because of the nature, the rich water, the mountain environment, so there's all these various types that grow here.

Truly, nature, it's a part of life. Because without nature, there can be no life. And women and nature should not be divided.

And nature's defence is women.

And nature is inside every woman. So every woman needs to see herself responsible for nature. Because nature is living. For us to live, we have to see nature as a part of ourselves. (Jineolojî 2020)

This intimate ecological relationship opens possibilities to cultivate mutualities of care between people and nature by re-embedding food provisioning and production in diverse local ecologies and economies. Regenerating ecologies, autonomy, fertility, health, and mutual entanglements with nature depends on a cyclic economy based on care and solidarity. Such a care economy values women's productive and reproductive labor, as well as the labor of nature. In this life-affirmative dynamic, women cooperatives and their communities engage in a caring and permanent regeneration of the humanity-nature metabolism (Salleh 2017). This caring subsistence perspective (Mies and Bennholdt-Thomsen 1999) is reflected in the science of woman (*Jineoloji*) espoused by many Kurdish women who are now struggling for freedom and equality by opposing patriarchy's violence against women and nature (Öcalan 2013; Strangers in a Tangled Wilderness 2015).

As an ethics of care and solidarity that rejects hierarchies and patriarchal domination, jineolojî enables life to unfold by supporting nature's strivings for ever more diversity, complexity, and adaptation to local context. As such, Kurdish women and men who embrace jineolojî are particularly well placed to develop bold agroecological transformations (figure 8.1) that can heal the relationship between people and nature. Kurdish ecologies are indeed slowly being diversified and restored through agroecological approaches, ecological and permaculture design, widespread recycling and reuse, a focus on "doing more with less," and the re-localization of production, supply chains, and consumption as part of a new agrarian-industrial mutualism between towns and countryside.

However, further delinking Kurdish local economies from global commodity capitalism requires developing and promoting circular systems that mimic natural ecosystems at different scales—from individual farm plots to entire cities. The building blocks for that do exist, and include enhancing functional biodiversity, the ecological clustering of industries, recycling, and localized production and consumption in specific territories dedicated to sustainable living (Jones et al. 2012). Circular systems that combine food and energy production with water and waste management can reduce carbon and ecological footprints while maintaining a good quality of life through controlled processes of de-growth in consumption and production (Latouche 2011, 2019; Pimbert 2012). This process of decolonial de-growth would be designed for local control by communities and aim to strengthen collective tenure, conviviality, gender justice, autonomy, and direct democracy in territories. International solidarity for mutual learning and knowledge exchange could help strengthen this radical shift toward food sovereignty, agroecology, and a care economy in Kurdish territories.

A global network of food sovereignty and de-growth movements could help Kurdish communities enhance ecological sustainability and decolonize economics through:

i) popular education and training on how to develop processes of self-managed research and grassroots innovations needed to scale out food sovereignty, agroecology, economies of care, and bio-cultural diversity to more people and places;

ii) farmer-to-farmer training for transformative agroecology on ecological diversification of agricultural landscapes as well as solidarity-based economic exchanges and short food webs between food producers and food eaters;

iii) knowledge-sharing on the development of ecologically sustainable agrarian-industrial mutualisms based on nested circular systems and economies of care that enhance diversity, decentralization, gender justice, and direct democracy.

In turn, global food sovereignty and transformative agroecology movements have much to learn from the Kurdish freedom movement, including changing patriarchal relations and masculinist identities through education of boys/men and the creation of safe empowering spaces for women. Like the Kurdish women who espouse jineolojî, many members of La Vía Campesina recognize that "if we do not eradicate violence towards women within the movement, we will not advance in our struggles, and if we do not create new gender relations, we will not be able to build a new society" (La Vía Campesina 2008).

International farmer exchanges for mutual learning, intercultural dialogues, and power-equalizing research with activist scholars can all contribute to the momentum needed to strengthen people's alternatives to the destructive logics of colonialism, heteropatriarchy, white supremacy, and capitalist exploitation (Pimbert, 2018).

DEMOCRATIC CONFEDERALISM

One of the clearest demands of the food sovereignty movement is for people to exercise their fundamental human right to decide their own food and farming policies (Nyéléni 2007). Democratizing the governance of rural and municipal food systems means enabling small-scale producers and other citizens—both men and women—to directly participate in the choice and design of policies and institutions, decide on strategic research priorities and investments, and assess the risks of new technologies.

Similarly, decisions on how to adaptively manage ecosystems need to be directly made by people who live in, or close to, these ecosystems and natural resources (land, water, seeds, and trees) they use for food and agriculture. According to the Convention on Biological Diversity, "the objectives of management of land, water and living resources are a matter of societal choices" and "management should be decentralized to the lowest appropriate level." For forests, grasslands, wetlands, and agricultural landscapes "the closer management is to the ecosystem, the greater the responsibility, ownership, accountability, participation, and use of local knowledge" (Secretariat of the Convention on Biological Diversity 2007).

Putting peoples' participation at the center of ecology can best be done through an expansion of direct democracy in decision-making in order to replace models of representative democracy. According to Öcalan, this radical democratization of society is possible because

> in contrast to centralised administration and bureaucratic exercise of power, confederalism proposes political self-administration, in which all groups of the society and all cultural identities express themselves in local meetings, general conventions, and councils. . . . Politics becomes part of everyday life. (Öcalan 2011, 26)

The autonomy of Kurdish society depends on institutions being self-organized and self-administering through small self-governing decentralized units that can confederate into larger structures for coordinated action over large areas. Based on diversity and gender equality, this anti-centrist and bottom-up approach is key for the regeneration of Kurdish ecologies associated with crop farming, pastoralism, and other food provisioning activities.[7]

Indeed, democratic autonomy and self-administration are particularly good at empowering local organizations to carry out their roles in sustaining food systems and the wider ecologies they are part of. The knowledge co-created in local organizations (such as farmer groups and cooperatives) is both practical and political. For example, local organizations of peasant farmers and pastoralists develop knowledge, institutional rules, and policies (informal or formal) that enable coordinated and collective action for ecosystem management and agroecological practices. This knowledge is generated as part of the day-to-day activities mediated by local organizations that have been set up for different purposes within communities (Pimbert 2009, 2018), such as

i) sustaining the ecological basis of agri-food systems—including producing knowledge and joint actions for the local adaptive management of
land and the development of reliable bio-physical indicators to track and
respond to change, including climate change;

ii) coordinating human skills, knowledge and labor to generate both use
values and exchange values in the economy of the agri-food system, as
well as organize economic exchanges over large geographical areas;

iii) governing agri-food systems—including decisions about people's access
to food and natural resources (such as land, water, and seeds) as well as
collectively generating the political knowledge needed to shape policies
and institutions.

In addition, new Kurdish organizations were especially created to coordinate local processes of social learning based on critical education and
self-governance. Remarkably, the project of democratic confederalism has
generated a web of new organizations and institutions for direct decision-
making by Kurdish people: commune, neighborhood or village community
people's council, district people's council, and the peoples council. At each
of these four levels in which power flows bottom-up—from the level of the
commune—there are eight commissions responsible for different areas (such
as defense or justice). Each commission has two spokespersons—one man
and one woman. Democratic confederalism involves a network of citizen-
based (as opposed to government) bodies or councils with members or delegates elected from popular face-to-face democratic assemblies.

Several Kurdish organizations with different functions, powers, and
responsibilities are thus usually involved in facilitating the adaptive management and governance of agri-food systems and the ecosystems they are
embedded in. Such "nested organizations" and their polycentric networks
operate at different scales and act in complementary ways (Ostrom 2010).
These interlinked organizations provide the institutional landscape that is
needed to manage the social and ecological realms in which agri-food systems are nested. They provide the organizational fabric that enables coordinated and collective agency by networks of peasant farmers, pastoralists,
wild food and medicinal plant collectors, peri-urban and municipal gardeners.
Decentralized and distributed decision-making usually ensures that there are
no standard "one-size-fits-all" policies and agreements. Each decision tends
to be tailored to its specific context through deliberations and negotiations
within and between organizations—from the level of communes upward and
between Kurdish territories.

This governance architecture greatly enables democratic autonomy and
collective action to manage ecosystems and organize economic exchanges
over large geographical areas in Kurdish territories. Such polycentric

networks are also potentially well suited to tackle major ecological crises caused by agriculture (such as habitat destruction, chemical pollution, and climate change) because it creates a confederated web of city neighborhoods, municipalities, villages, and rural areas that can coordinate Kurdish peoples' actions at multiple scales.

As such, confederated networks of democratic organizations are vital for regenerating Kurdish ecologies through food sovereignty, agroecology, and economies of care.

CONCLUSION

Kurdish ecological initiatives in food, agriculture, and land use are uniquely rooted in democratic confederalism, libertarian municipalism, and the rejection of patriarchal relations. This Kurdish version of food sovereignty affirms the right of men and women to democratically govern themselves and co-produce local food systems, ecologies, economies, knowledge, and culture. Decentralized, distributed, and democratic decision-making is crucial for reviving and sustaining diverse ecologies associated with food, farming, pastoralism, and other forms of land use. Moreover, Kurdish agroecological practices that re-localize food systems help reduce carbon and ecological footprints while enhancing food and nutritional security.

However, this is an unfinished journey and emergent process. As Ozlem Tanrikulu, representative of the Committee for the Reconstruction of Kobane and president of UIKI (Kurdistan Information Office in Italy), said: "We have not yet succeeded in developing a system like the one we are planning, but we have been able to lay the foundations for understanding the principles and models of production and adapting them to our project" (ANF News 2020).

Kurdish ecological initiatives that regenerate the diversity of nature and culture in food and agriculture offer important insights for global social movements working for food sovereignty, agroecology, and economies of care based on de-growth and gender justice. In turn, these global solidarity movements can help Kurdish communities strengthen their struggle to develop life-affirmative and just alternatives to the ever-expanding process of commodification of nature and social relations.

NOTES

1. A food system gathers all the elements (such as environment, people, inputs, processes, infrastructures, and institutions) and activities that relate to the production, processing, distribution, preparation, and consumption of food, and also the outputs of these activities, including socio-economic and environmental outcomes (HLPE 2019).

2. At the beginning of the uprising in March 2011, Rojava's main agricultural products were wheat, cotton, and to a lesser extent, meat, olives, pistachios, eggs, and dairy products.

3. La Vía Campesina (LVC) is an international movement that brings together peasant organizations of small- and medium-sized producers, agricultural workers, landless people, women farmers, migrants, and indigenous communities from Africa, Asia, the Americas, and Europe. It is an autonomous, pluralistic movement, independent of all political, economic, or other denominations. LVC comprises about 164 local and national organizations in 73 countries and represents about 200 million farmers altogether. See: https://viacampesina.org/en.

4. "Farmers" refers here to smallholder peasant/family crop and livestock farmers, herders/pastoralists, artisanal fisherfolk, landless farmers/rural workers, gardeners, forest dwellers, indigenous peoples, hunters and gatherers, and any other small-scale users of natural resources for food production. The majority of the world's food producers are small family farmers.

5. People is a group of persons who belong to the same culture, ethnicity, race, or nation. When more than one of such groups is referred to, "people" becomes "peoples."

6. See https://mesopotamia.coop/feed-the-revolution/.

7. In Article 57 of its "Social Contract," the Democratic Federation of Northern Syria declares that democracy is the "way of achieving the balance between economics and ecology."

REFERENCES

Altieri, Miguel A. 1995. *Agroecology. The Science of Sustainable Agriculture*, 2nd ed. Boulder, CO: Westview Press.

Anderson, Colin A., Janneke Bruil, Jahi M. Chappell, Csilla Kiss, and Michel P. Pimbert. 2021. *Agroecology Now! Transformations Towards More Just and Sustainable Food Systems*. London: Palgrave.

ANF News. 2020. "Rojava: An Identity Built Through Agriculture." *ANF News*. September 24, 2018. https://anfenglishmobile.com/features/rojava-an-identity-built-through-agriculture-29796.

Askin, Yücel, Firat Cengiz, Ayhan Elicin, Mehmet Ertugrul, and Reçit Sonmez, 1989. "La production animale." *Agricultures Méditerranéennes: la Turquie*. Serie B (1): 79–88. Montpellier: Centre International de Hautes Études Agronomique Méditerranéennes.

Bance, Pierre. 2019. "Reverdir le Rojava." *Autre Futur*. June 11, 2019. Accessed November 28, 2020. http://www.autrefutur.net/Reverdir-le-Rojava.

Bookchin, Murray. 1976. "Radical Agriculture." In *Radical Agriculture*, edited by Richard Merrill, 3–13. New York: New York University Press.

Bookchin, Murray. 1998. *Remaking Society. Pathways to a Green Future*. Montréal: Black Rose Books.

Bookchin, Murray. 2005. *The Ecology of Freedom: The Emergence and Dissolution of Hierarchy.* Edinburgh: AK Press.

Butzer, Karl. W. 1994. "The Islamic Traditions of Agroecology: Cross-cultural Experience, Ideas and Innovations." *Ecumene* 1, no. 1: 7–50.

Ceccarelli, Salvatore, and Stefania Grando. 2020. "Participatory Plant Breeding: Who Did It, Who Does It and Where?" *Experimental Agriculture* 56, no. 1: 1–11. https ://doi.org/10.1017/S0014479719000127.

FAO [Food and Agriculture Organization of the United Nations]. 2018. *The 10 Elements of Agroecology: Guiding the Transition to Sustainable Food and Agricultural Systems.* Rome: FAO.

FAO. 2019. *The State of the World's Biodiversity for Food and Agriculture.* FAO Commission on Genetic Resources for Food and Agriculture Assessments. Rome: FAO.

Gliessman, Steve. R. 2015. *Agroecology: The Ecology of Sustainable Food Systems,* Boca Raton, FL: CRC Press.

HLPE [High Level Panel of Experts]. 2019. *Agroecological and Other Innovative Approaches for Sustainable Agriculture and Food Systems That Enhance Food Security and Nutrition.* Rome: Committee on World Food Security.

Jineolojî. 2020. "Women, Plants and Nature." *Jineolojî* website. December 6, 2020. Video. http://jineoloji.org/en/category/fields-of-jineoloji/ekoloji/.

Jones, Andy, Michel P. Pimbert, and Janice Jiggins. 2012. *Virtuous Circles: Values, Systems, Sustainability.* London: IIED and IUCN CEESP.

Knapp, Michael, Anja Flach, and Ercan Ayboğa. 2016. *Revolution in Rojava: Democratic Autonomy and Women's Liberation in Syrian Kurdistan.* London: Pluto Press.

Kropotkin, Peter A. (1898) 1913. *Fields, Factories, and Workshops, or, Industry Combined with Agricultures and Brain Work with Manual Work.* New York: Benjamin Blom.

Kropotkin, Peter A. (1913) 2015. *The Conquest of Bread.* London: Penguin Classics.

Latouche, Serge. 2011. *Vers une société d'abondance frugale: Contresens et contro-verses sur la décroissance,* Paris: Fayard—Mille et une nuits.

Latouche, Serge. 2019. *La décroissance.* Paris: Humensis-Que sais-je?

La Via Campesina. 2008. "The Maputo Declaration." Accessed November 28, 2020. https://viacampesina.org/en/declaration-of-maputo-v-international-conference-of-la-via-campesina/.

Mies, Maria, and Veronica Bennholdt-Thomsen. 1999. *The Subsistence Perspective: Beyond the Globalised Economy.* London: Zed Books.

Nyéléni. 2007. *Declaration of Nyéléni Forum for Food Sovereignty.* Sélingué, Mali: Nyéléni Accessed December 23, 2015 at http://nyeleni.org/spip.php?article290.

Öcalan, Abdullah. 2011. *Democratic Confederalism.* International Initiative Edition. Oakland: PM Press.

Öcalan, Abdullah. 2013. *Liberating Life: Women's Revolution.* International Initiative Edition. Oakland: PM Press.

Öcalan, Abdullah. 2017. *The Political Thought of Abdullah Öcalan.* London: Pluto Press.

Ostrom, Elinor. 2010. "Polycentric Systems for Coping with Collective Action and Global Environmental Change." *Global Environmental Change* 20, no. 4: 550–57.

Pimbert, Michel P. 2008. *Towards Food Sovereignty: Reclaiming Autonomous Food Systems.* London: Rachel Carson Centre and International Institute for Environment and Development. http://www.environmentandsociety.org/mml/towards-food-sovereignty-reclaiming-autonomous-food-systems.

Pimbert, Michel P. 2009. "Local Organizations at the Heart of Food Sovereignty." Chap. 4 in *Towards Food Sovereignty: Reclaiming Autonomous Food Systems.* London: Rachel Carson Centre and IIED. http://www.environmentandsociety.org/mml/book-chapter-pimbert-michel-local-organizations-heart-food-sovereignty.

Pimbert, Michel P. 2012. "Fair and Sustainable Food Systems: From Vicious Cycles to Virtuous Circles." *IIED Policy Brief*, London: IIED.

Pimbert, Michel P. 2018. *Food Sovereignty, Agroecology, and Bio-cultural Diversity: Constructing and Contesting Knowledge.* London: Routledge.

Rahmanian, Maryam, Maede Salimi, Khadija Razavi, Reza Haghparast, Salvatore Ceccarelli and Ali Razmkhah. 2016. "Evolutionary Populations: Living Gene Banks in Farmers' Fields in Iran." *Farming Matters,* April 2016. Accessed November 28, 2020. https://www.ileia.org/2016/04/16/evolutionary-populations-living-gene-banks-farmers-fields-iran/.

Salleh, Ariel. 2017. *Ecofeminism and Politics: Nature, Marx and the Postmodern,* 2nd ed. London: Zed Books.

Secretariat of the Convention on Biological Diversity. 2007. "Principles of the Ecosystem Management Approach." February 7, 2007. https://www.cbd.int/ecosystem/principles.shtml.

Steffen, Will, Katherine Richardson, Johan Rockström, Sarah E. Cornell, Ingo Fetzer, Elena M. Bennett, Reinette Biggs, et al. 2015. "Planetary Boundaries: Guiding Human Development on a Changing Planet." *Science* 347, no. 6223: 1–15.

Strangers in a Tangled Wilderness. 2015. *A Small Key Can Open a Large Door. The Rojava Revolution.* San Bernardino, CA: Combustion Books.

Tatort Kurdistan. 2013. *Democratic Autonomy in North Kurdistan, The Council Movement, Gender Liberation, and Ecology—In Practice.* Translated by Janet Biehl. Porsgrunn: New Compass Press.

Thevenin, Michael. 2011. "Kurdish Transhumance: Pastoral Practices in South-east Turkey." *Pastoralism* 1, no. 23. https://doi.org/10.1186/2041-7136-1-23.

Walliser, Yannes. 2010. "L'agriculture du Kurdistan irakien." *Études rurales* 186: 133–148.

Chapter 9

Free Life Together

Jinwar, the Women's Eco-Village[1]

Fabiana Cioni and Domenico Patassini

WINTER

In Paris, at the heart of the city in a building of the 10th Arrondissement, the lifeless bodies of three women are lying on the ground with their heads in a pool of blood: a cold-blooded execution on the night of January 9, 2013. French investigators reviewing the evidence concluded that the Turkish secret services (MIT) were "implicated" (Seelow 2015; France 24 2019) in the triple murder in the heart of Europe, an action without noise, without a break-in, inside the Kurdish Information Office. Ömer Güney, the suspect charged with the assassinations, died mysteriously shortly before the start of a legal trial against him. The victims, Sara (Sakine Cansiz), Rojbîn (Fidan Doğan), and Ronahî (Leyla Şaylemez), were all activists of the Kurdish women's liberation movement.[2]

Sara had obtained political asylum in France and led the movement in Europe: a life dedicated to the struggle for the sake of freedom that began with the meeting of Abdullah Öcalan and the founding of the PKK. Arrested at twenty-two years old, she spent twelve years in prison in Turkey for organizing women against patriarchal oppression. Subjected to torture, she never gave in. On regaining her freedom, she was responsible for the PKK's finances and political strategy. Together with Sara, the women's movement was able to organize itself autonomously. In everyday life in the mountains, they developed a community approach based on eco-systemic solidarity. Their imaginations were advanced, even envisioning an ecological and state-free society. The guerrillas helped to develop a dialectical practice of assessment and criticism of patriarchal society: they proposed and practiced alternatives to existing hierarchies and power.

Inside the women's camp, the common experience became a progressive liberation from the patriarchal and hierarchical mentality that permeated Kurdish society. To enter the camp, any man had to be willing to metaphorically kill his "toxic masculinity." It was necessary for him to be willing to become an actor in the process of collective analysis: in particular, to challenge patriarchal hierarchical structures, male privilege, the use of language as an expression of power and competition.

The camp presented itself as an academy to design an ecological society, developing a holistic view of social and environmental issues. The stimulus of the mountain and its rhythms was vital. The camp was subjected to fierce bombing by the Turkish air force. Yet, *jineolojî* has spread, and the dialectical critique of power from women's perspectives has entered Kurdish society, enabling radical changes.

Öcalan first used the neologism "*jineolojî*" in 2008, in *The Sociology of Freedom*, recently published in English (Öcalan 2020). Here he states that women can break up the structures of domination, putting in place a social revolution capable of placing humanity and the biosphere at the core of life. His ideas regarding jineolojî form part of his wider prison writings, which are a dialectical critique of the power system from the beginnings of human history to "capitalist modernity."

In January 2014, the predominantly Kurdish but multiethnic population of northeastern Syria proclaimed the democratic autonomy of the cantons of Cizîrê , Kobanî, and Afrîn, commonly known as Rojava, and the "Social Contract" designed to ensure the coexistence of cultures, religions, and languages in the region. The population includes Muslim Kurds and Yazidis, Arabs mainly Sunni, Assyrian, Aramaic, Christians, Armenians, Chaldeans. There are also communities of Turkmens, Chechens, and Circassians. In the "Social Contract," ecology, women's liberation, and democratic confederalism are the founding pillars of society. Women meet independently, make decisions about community welfare, start cooperatives, and design plans for the future. They are coordinated through Kongreya Star,[3] an umbrella association that brings together all-female formations. In the territories of the autonomous administration, dialogue is the basis for policies on reconstruction, self-defense, and life-long learning.

A WOMEN'S ECO-VILLAGE

The place chosen to found the first women's village, Jinwar: literally the "place of women," (from the *Kurmanji* words, "*Jin*," meaning "woman," and "*war*," meaning "place") is outside the city of Dirbesîyê, just before a *tell*.[4] The first settlement has been developed on this land, previously nationalized by the Ba'athist regime and exploited as a monoculture,[5] now a source of an ecological approach to architecture, economy, and society. The women's associations: *Kongreya Star, Mala Jin* (House of Women), Congress of Free Women in Rojava (*Kongreya Jinen Azadi*), *Jineolojî Committee*, Cooperative of the Families of the Fighters, and the Committee for Diplomacy of the Women of Rojava are responsible for this project. Together, these organizations form the Women's Village Committee, *Gunde Jine*, which has initiated the design of a self-sufficient, self-built ecological farming village.

In 2017, the women started the self-build construction project, using ancient traditional raw earth technology. Beyond its intrinsic efficiency in terms of energy and climate, this technology creates a real social ecosystem; it helps women to regain self-confidence and, in collective practice, the sense of freedom. They chose materials and social technology with a high symbolic and aesthetic value: clay bricks, natural binders, and wood. With these materials, they are reviving a deep cultural heritage, expanding their manual skills, cementing their knowledge, and also stimulating collaboration. "Through the bodies," as Heval Rosa (pseudonym), a leading organizer at Jinwar says, "we learn and communicate," underscoring that "we share a collective intent to carry out a project that goes beyond abstract urban schemes, which needs creativity to be realized and asks for a particular land management system": community management, in which the land is owned by the Committee (Fabiana Cioni, personal conversation with Rosa during my stay at Jinwar in May 2019). The village council administers and manages the settlement. During the work, the community acquires skills that make it possible to adapt the initial project according to their disclosed needs. The dimensions of the apartments and the volumes are modified; in the most recent modules, adobe "pseudo-domes," dome shells arranged in concentric circles, are added to flat roofs, improving comfort and thermal efficiency.

The raw material is locally accessible. It is a common good from the earth, much as the women who first developed the art of clay modeling took their materials as an offering from the "Cosmic Goddess." Adobe is a locally sourced material, developed according to the spatial context. It guarantees the best performance in terms of environmental comfort during hot summers and cold winters. Thermal regulation and energy saving allow the management of humidity variations in the internal environments, also absorbing electromagnetic radiation. Using raw earth, the community creates multidimensional connections between ecological, cultural, and social components. These connections take the form of a spiral and form a circular process. The energy needs of collective buildings, such as the Women's Academy, are met by photovoltaic panels, with the aim of extending them to private homes as well. The region depends on the outside for this technology, with the difficulties that importation entails in the context of war and embargo. For the most part, the use of energy-saving solar energy is combined with processes that incorporate local natural resources to achieve the safeguarding of the ecosystem. This also achieves economic benefits, due to the reduction of transport costs and the circumvention of a hard embargo. In this process, cultural traditions are recovered, collective history is strengthened by interpreting in an original way the relationship between past and present. In collective work, all these elements vibrate in a nonlinear way, they create unexpected synergies with recursive effects on the social and ecological level. Self-construction not only favors the democratization of society and develops autonomy but also enhances creativity and innovation.

The housing typology is uniform: single-story and single-family house with four modules different in size and internal distribution that rework the features of traditional homes. The entrances are protected by loggias; the bathroom is separate from the spacious living room whose size changes with the number of women living there. The living room is a flexible place, used to receive guests and as a collective bedroom. Typological innovations include also the tiled kitchen, with its large worktop for efficient and hygienic food preparation, flanked by accessory rooms, proportional to the household size, and spaces designed for craft activities. The two loggias rework the Middle Eastern custom of living in the contiguous open space as if it were integrated into the home. It is a place endowed with seasonal flexibility, enabling an appropriate relationship with the outside world (shoes are stored here because you must be strictly barefoot to enter the houses). In the summer, it hosts the *çay* (tea) for breakfast or dinner and, in the hottest periods, it is a breezy place to rest at night, see figures 9.1, 9.2, and 9.3.

The area in front of the main loggia (the entrance faces the central square) is cultivated as a private garden with fruit trees and flowers: a green space that evokes the hortus conclusus of the villages or the historic urban centers. Figures 9.4 and 9.6 show women who take care of their garden and plant fruit trees that were donated to the village by the Agriculture Committee of Dêrik. In Jinwar, the public space delimited by the houses looks like a large garden, whose features are creatively managed by the resident women. At the heart of the settlement are the circular square and the collective kitchen with storehouse and pantry. The residents gain use of the house for the duration of their stay in the village, enriching it with decorations and paintings, and managing it according to their needs. The adobe house requires constant maintenance, for which women share their practical knowledge. Since rain and wind tend to wash away the protective layer of earth and straw plaster, women need to repair the eroded parts each spring.

A self-built village helps to create social relations, thus avoiding the alienation of city life and enabling women to experience a practical alternative that inspires awareness of humans' ecological role. Construction and architecture

Figure 9.1   Main Lodge at Jinwar, Used for Eating Together During the Summer Season, May 2019. *Source*: © Fabiana Cioni.

Figure 9.2   Night Setting at Main Loggia, May 2019. *Source*: © Fabiana Cioni.

Figure 9.3 **Woman Cleaning the Main Lodge, May 2019.** *Source*: © Fabiana Cioni.

Figure 9.4 **Women Taking Care of their Garden and Planting Fruit Trees Donated to the Village by the Agriculture Committee of Dêrik, May 2019.** *Source*: © Fabiana Cioni.

will never be simply rubble or inert building blocks, precisely because they are made of organic materials deemed to be mother earth.

This shift in social and ecological relations is a crucial issue in the reflections and proposals of Öcalan and Murray Bookchin. Öcalan writes:

> The new society can only be achieved if material and ideological culture are balanced and consistent. The synthesis of society's internally balanced and harmonious material and ideological culture with that of nature will result in free nature (or as Murray Bookchin puts it in *The Ecology of Freedom*, "third nature"). This will serve to overcome the contradiction of civilized society's imbalance between nature and society. (Öcalan 2015, 163)

FREE LIFE TOGETHER

The eco-village was inaugurated in November 2018. All women who wish to be welcomed can access it without distinction of religion, language, or culture. In Jinwar, there are women living with their children, widows of martyrs for liberation, or those separated from violent husbands. Others are women who wish to train in organic agriculture, study, acquire holistic knowledge, and participate in collective experiences to be reproduced elsewhere.

Jinwar's participatory planning was carried out according to a collective decision-making process set up for empathic listening. With dialogue, shared solutions have been reached, in accordance with the practice of direct democracy in democratic autonomy. This type of participation requires an "organic" time, the time necessary to create opportunities for expression during the stages of the assembly. The design proposal was developed by simplifying the urban template, which started from a composite geometry, to arrive at a primary shape: the triangle. A bird's eye view would show the quasi-final design solution in which the residential core of thirty houses forms a triangle. The collective space with a central plan is located in the vegetable garden toward which all the housing units open. It is an elementary geometry with a strong symbolic value: triangle, circle, or spiral are among the oldest forms related to the Mother Goddess. The triangle, simple or superimposed with the signs either "V" or "W," becomes part of the ritual writing developed by women during the Neolithic Revolution and has been used since the Lower Paleolithic.[6] The triangle is one of the symbols associated with the Goddess in the personification of "Life-giving Nature."[7] Spirituality was for millennia found in nature and revered through the body of woman.[8] Thus, human communities developed "ecological" societies with care for social interaction, well-being, and mutual support at their heart.

Women in Jinwar seek in the distant past the roots of an ecological relationship with the natural world. They are aware of preserving reminiscences from the Mesopotamian Neolithic civilization within popular customs handed down from mother to daughter. Their sense of continuity with the remote past, for instance, is experienced during the baking of bread, using processes and a type of oven thought to have been used since prehistoric times. Such traditions, therefore, are not yet completely eradicated by "capitalist modernity," the term Öcalan uses to define the period in which we are presently living. This process of knowledge takes place in jineolojî practices. Jineolojî was born after more than forty years of the Kurdish women's movement's struggle, as a practice for an ecological society able to manage common goods, economy, health, training, and self-defense as an integrated community strategy.

At one of the triangle's vertices is the academy, the fulcrum of collective life where all meetings are held. Training on various aspects of jineolojî also takes place here. When COVID-19 spread, specific health training initiatives were promoted, in which, in addition to the measures to protect against contagion, the causes of its diffusion were assessed, together with the role played by the globalized neoliberal economy and the unlimited exploitation of nature. The academy is a single building with two floors, with a large room on the ground. Upstairs on the terrace, solar panels have been installed. For the floor and the support pillars, reinforced concrete was used, since the academy could become a target of an armed attack, potentially by air.

The school's classrooms are opposite the academy and are designed to converge toward the central elliptical garden. The spaces are of various sizes, all with a circular plan so as to create contexts of direct and equal physical relationship. The circle fosters dialogue and a radically different teaching which, in Rojava, is experienced at all levels of training. The pedagogical method is similar to the libertarian approach and to that promoted by Paulo Freire in *The Pedagogy of the Oppressed* (2005), and Augusto Boal, in *The Theatre of the Oppressed* (Boal 1995; Howe, Boal, and Soeiro 2019).

To take responsibility for the care and healing of the body is to reclaim and defend prime social territory from patriarchal institutions and practices, a principle that finds expression in the *Sifa Jin*, the Centre for Natural Medicine. The center is located near the entrance, alongside the communal vegetable plot, from which it emerges with the four pseudo-domes that mark the vertices of the base rectangle, see figure 9.5. With *Sifa Jin* women regain their holistic knowledge of the body, approaching an ancestral epistemology that took care of human health and the environment (a perspective reflected in ancestral people's affirmation that "the plant is us, we are the plant"). At the same time, women become more and more aware of the destiny of exploitation that unites them with Nature.[9] This commonality enables the circle of

Figure 9.5 **View of Sifa Jin's Building, May 2019.** *Source*: © Fabiana Cioni.

sharing practical knowledge of the past and reconciliation with nature and its cycles. This is fundamental to recover the ancestral wisdom of herb care; to know when to harvest and how to produce healing oils and herbal teas. The residents also use allopathic medicine.

Women who decide to stay in Jinwar agree to live communally,[10] sharing everything. The community meets whenever necessary, in a circle, deciding how to respond to material and psychological needs, how to organize the shifts of daily activities: taking care of animals and plants, preparing collective lunch, managing the shop and stocking it, making bread and other foodstuffs. In summer, harvesting in the open fields is carried out during the cooler hours. Everyone can participate, but no one is obliged to toil. In practice, after the first day's initiation, everyone, including the guests, works together, having fun and discovering unexpected perspectives of humanity. The wages of external seasonal work, only undertaken by some, are divided in a circle.

The social ecosystem is built on the community's needs and operates outside the logic of the competitive market and planned obsolescence. Food self-sufficiency is achieved with a circular model of which the human community is an integral part:[11] food waste, animal excrement, by-products, inedible and non-usable parts of cultivated plants become organic food and energy sources. The non-compostable and nonreusable waste is mainly composed of food packaging that tends to decrease as new agricultural practices become fully operational and that is collected by the municipal service. The women's growing attention toward littering positively affects the behavior of girls and boys: open and collective spaces are cleaned regularly as if they were part of homes.[12]

Energy needs are met with the integrated use of photovoltaic panels until self-sufficiency is achieved; water, a common good, is managed by limiting waste and making use of rainwater collection tanks.

Tekmîl contributes to the continuous updating and adaptation of this strategy. While loosely translated as "relationship," a more conceptual meaning, as applied by the Kurdish freedom movement, would be "revolutionary constructive criticism." The tekmîl is an instrument of the revolutionary process, a fundamental element of the practice of radical social change. This practice is rooted in the ability to question oneself. In the assembly, the women expose precise and detailed criticisms of the behavior of the others. If misunderstandings or conflicts emerge, self-criticism is required in the same spirit. They try to identify the problem, propose solutions, and when self-criticism is needed, self-assessment of the subjective being in relation to others and to the group becomes praxis.

From an ecological perspective, the tekmîl can be considered as a tool for stabilizing the social ecosystem, to allow the human component to develop an ethics aimed at mutualism and complementarity. A holistic approach helps build an ethics that will lead the group to make rational choices for the good of the whole community. The dialectical relationship between criticism and self-criticism builds trust and increases awareness of collective potential. As criticisms are respectfully expressed and reworked so as to reinforce awareness of individual skills and respect for diversity, it tends to increase the sense of sharing. Tekmîl is an environment of trust in which the participants can open themselves to their companions with their fragility, learning to overcome their own limits; it is the place where the group discovers the limits of personalities forged by patriarchal and capitalist mindsets. Assembly meetings always end in a proactive manner.

Tekmîl is based on the ethical principle of *Hevaltî*[13] according to which the people of a community work together with the aim of taking care of all the components of the biosphere. Human beings not in competition, they share practical and psychological issues. Hevaltî is a moral cornerstone of the "free life together" founded on trust, a key concept of the ecological society shared by the Kurdish freedom movement. Tekmîl is practiced at all levels of the administration, either in military or civil domains, and indicates that society shares the modern anthropopoiesis[14] concept: "making humanity."

Food self-sufficiency strategies are inspired by this principle in the community. Women are wondering how to progressively recover land depleted by monoculture, before starting to cultivate it following principles of permaculture and organic farming, excluding private property: food sovereignty and land justice are the key messages.[15]

To date, the available fertile soil almost ensures the entire food needs of the community. What to produce, and how to produce it is decided by the collective within the framework of organic farming, with crop rotation without using chemicals or synthetic components, having limited machinery, and collecting rainwater or groundwater for irrigation. Cattle raising is integrated with agricultural production. A problem for agricultural cooperatives is represented by the unavailability of fertile seeds, because the market supplies only modified seeds made in Israel. However, for seed-bank access, the strong international link with the "Sem Terra" movement (MST) is helpful. When it occurs, the agricultural surplus is not brought to the market, since the growers prefer the perspective of relational economy and "vital exchange" with the contiguous populations.

An important activity, not only from an economic point of view, is the production of carpets. Women make everything they need with their hands and imagination: carpet making, in particular, is a cultural institution with ancient roots.[16] For this reason, the women of Jinwar have taken the loom as a symbol of perseverance and aesthetic behavior. Referring to the traditional looms, they created simple but functional prototypes using the poor materials available (wood, nails, threads). As Mahatma Gandhi previously, the collective reinvents its practices and puts itself to the test in front of the loom.

In conclusion, Jinwar is a groundbreaking experience that contains many elements of a free and ecological society, together with the contradictions and difficulties encountered on the path of communalism in the circumstances of ongoing conflict and an embargo. Here, the social ecology project is doubly strategic as it helps to create a new society and a model of self-defense. The issue of self-defense catalyzes most of the efforts, including economic ones, of the autonomous administration. The parasitic dynamics backed by the Syrian regime and the conflict initiated by Turkey, now occupying a border strip, have created hundreds of thousands of displaced people, a situation that shows no sign of easing. Jihadist forces and the Turkish military disrupt the supply of drinking water: the Turkish state exercises this form of blackmail during the summer season by blocking the stream that runs alongside the Jinwar settlement, as reported by Jinwar's Council. Systematic crop fires seriously jeopardize the possibilities of feeding the embargoed population, undermining the economy of the entire region. Defense requires the construction of air-raid shelters, supervision of a continuous collective mobilization that would not be possible without the ecological village project and the organizational model of the community as discussed above. Both are essential factors to ensure continuous self-(re)construction. The experience of self-construction and management of the Women's Council is a precious asset,

Figure 9.6   Tree Planting In Front of Sahmaran Mural, June 2019. *Source*: © Fabiana Cioni.

not only to carry out other "gender" projects in the region, but also to favor the return of the population to the villages by recovering agricultural land and pasture and reactivating traditional building practices.

NOTES

1. The chapter is related to PhD research undertaken at the Università Iuav di Venezia, conducted in the field by Fabiana Cioni with the supervision of Professor Domenico Patassini. The authors would like to acknowledge the collaboration of Necibe Qaradaxi (from the Bruxelles Jineolojî Center) and Jinwar's Council.

2. Sara (1958, Dersim) in charge of the Kurdish Movement in Europe, Rojbîn (1982, Elbistan) representative of KNK (National Congress of Kurdistan) with important diplomatic functions, and Leyla (1989, Mersin) in charge of the youth movement.

3. Kongreya Star derives from *Yekitiya Star* (Union of the Star), a movement founded in 2005. The Ba'athist regime has brutally repressed their activists with torture and imprisonment.

4. Artificial hill or mound, very common in Mesopotamia, which often conceals settlements of archaeological interest.

5. Eighty percent of the arable land had been previously nationalized. The Autonomous Administration returned most of the land to the communities. One part was made available to agricultural cooperatives, another entrusted to families who requested it. Some large estates are still owned by Arabs who obtained them

during the "Arab belt" policy implemented by the Ba'ath regime after 1970 (Rojava Information Center 2019).

6. These abstract symbols were used since the Acheulean and were inherited by the Neolithic. Such nonfigurative signs are considered by Gimbutas as a writing system used for more than 15,000 years (Gimbutas 1996, 79–93).

7. Symbol of the female pubis which, together with the vulva, buttocks, and breasts, is one of the parts of the woman's body to which the generative and parthenogenesis power of the Goddess is associated.

8. The ovarian cycle is approximately twenty-eight days which corresponds to the lunar cycle. In indigenous cultures, the cosmic forces are regarded as female.

9. The philosopher Maria Mies proposes the notion of woman's body as the first colony, thus directly linking violence against women to the limitless exploitation of nature (Mies 2014, 74–110). Such a topic also entered Öcalan's thinking through *Caliban and the Witch: Women, the Body and Primitive Accumulation*, where Silvia Federici (2004) assesses the role witch hunts and slavery played in original accumulation processes. Against Marx's claim that primitive accumulation was a vital precondition for capitalism and a liberatory defeat of feudalism, Federici states that primitive accumulation is a critical feature of capitalism itself, a necessary expropriation. In this perspective, the expropriation of women's unpaid labor connected to reproduction should be considered as a precondition of the rise of capitalism based upon wage labor.

10. Women's situation in Jinwar is not homogeneous: it varies by age, culture, and language, with a deep respect for these differences: diversity is wealth and creativity. There is no general rule on how long they stay in the village; some remain as long as they need to feel ready to go elsewhere as advocates, having understood how society's power mechanisms can be changed. There are a few women who leave the village after some weeks because they cannot adapt to life together.

11. Eliseé Reclus (1830–1905), anarchist and founder of social geography, wrote in the introduction to the colossal geography of the planet: "l'homme est la nature prenant conscience d'elle meme" (quoted in Clark 1995, 5–8).

12. Mesopotamian culture is characterized by fluidity between external and internal environments. One of the semantic foundations of civic space is as a common good in rural and urban contexts: such interaction is still present in some historical centers, such as Amed (Diyarbakır). Here the women cook together in the open air, showering the alleys with intense perfumes. In Jinwar, in the summer, people sleep on high beds outside, under the loggias or in front of the house.

13. The word comes from *heval* which means friend, companion. *Hevaltî* can be translated as sisterhood or "companionship."

14. The term was first used in the early 1990s in the field of anthropology to indicate a theory developed around the concepts of creativity and complexity. In *Making Humanity. The Dramas of the Anthropopoiesi*, the Italian anthropologist Francesco Remotti "highlights the lack of responsibility of modern Western society with respect to the social role played in the formation of the human being" (2013, 19–77). This fact is instead assumed with ritual aspects in indigenous societies as evidenced, for instance, by Mircea Eliade and Clifford Geertz. The KFM feels this responsibility and tries to develop humanity to be consciously ecological.

15. In social ecology, the food system is considered as a vehicle for change. See practical and theoretical work since the 1980s by Grace Gershuny (2020), member of the Institute of Social Ecology and in the 1990s consultant of the National Organic Program in the United States, about food justice, food agricultural issues, and land justice.

16. In the geometric decorations of the kilim, knots are used as variations of the triangle. Ancient and complex, the knots of the fabric have not yet been fully understood. The loom and weaving objects were objects housed in Neolithic temples (as in Çatalhöyük). In the urban fabric, the weaving activity re-proposed the kitchen environment.

REFERENCES

Boal, Augusto. 1985. *Theatre of the Oppressed.* New York: Theatre Communications Group. (ed. or. 1974, *Teatro de oprimido*).

Bookchin, Murray. 2015. *The Next Revolution. Popular Assemblies and the Promise of Direct Democracy.* New York: Verso.

Bookchin, Murray. 1992. *Urbanization Without Cities. The Rise and Decline of Citizenship.* Montréal: Black Rose Books.

Cansız, Sakine. 2018. *Sara: My Whole Life Was a Struggle.* London: Pluto Press.

Cansız, Sakine. 2019. *Sara: Prison Memoir of a Kurdish Revolutionary.* London: Pluto Press.

Clark, John. 1995. *The Dialectical Social Geography of Elisée Reclus.* https://theanarchistlibrary.org/library/john-clark-the-dialectical-social-geography-of-elisee-reclus.

Federici, Silvia. 2004. *Caliban and the Witch, The Body and Primitive Accumulation.* New York: Autonomedia.

France 24. 2019. "France Reopens Probe into Killing of 3 Kurdish Activists." *France 24*, May 15, 2019. https://www.france24.com/en/20190515-france-reopens-probe-killing-3-kurdish-activists.

Freire, Paulo. 2005. *Pedagogy of the Oppressed.* New York: Continuum (30th-anniversary edition, ed. or. 1970, *Pedagogia del oprimido*).

Gimbutas, Marija. 1996. *The Living Goddesses: Religion in Pre-Patriarchal Europe.* Los Angeles: University of California Press.

Gimbutas, Marija. 2001. *The Language of the Goddess.* New York: Thames and Hudson.

Gershuny, Grace. 2020. *The Organic Revolutionary: A Memoir from the Movement for Real Food, Planetary Healing, and Human Liberation.* 3rd ed. Montréal: Black Rose Books.

Howe, Kelly, Julian Boal, and José Soeiro. 2019. *The Routledge Companion to the Theatre of the Oppressed.* London: Routledge.

Mies, Maria. 2014. *Patriarchy and Accumulation on a World Scale in the International Division of Labour.* London: Zed Books.

Mies, Maria, and Vandana Shiva, 2014. *Ecofeminism.* London: Zed Books.

Öcalan, Abdullah. 2013. *Liberating Life: Woman's Revolution.* Cologne: International Initiative Edition.

Öcalan, Abdullah. 2015. *The Civilization. The Age of Masked Gods and Disguised Kings. Manifesto for a Democratic Civilization,* Volume I. Translated by Havin Guneser. Porsgrunn: New Compass Press.

Öcalan, Abdullah. 2020. *The Sociology of Freedom: Manifesto of the Democratic Civilization, Volume III.* Translated by Havin Guneser. Oakland: PM Press.

Remotti, Francesco. 2013. *Fare Umanità. I drammi dell'Antropo-poiesi.* Bari: Laterza & figli.

Rojava Information Center. 2019. *Beyond the Frontlines. The Building of the Democratic System in North and East Syria.* Accessed January 17, 2020. https:// rojavainformationcenter.com/storage/2019/12/Beyond-the-frontlines-The-building -of-the-democratic-system-in-North-and-East-Syria-Report-Rojava-Information -Center-December-2019-V4.pdf.

Seelow, Soren. "Assassinat de Militantes Kurdes à Paris: La Justice Souligne l'Implication des Services Secrets Turcs." *Le Monde,* July 22, 2015. https:// www.lemonde.fr/police-justice/article/2015/07/23/assassinat-de-militantes-kurdes -a-paris-la-justice-pointe-l-implication-des-services-secrets-turcs_4694801_ 1653578.html.

Chapter 10

Women's Subjectivity and the Ecological and Communal Economy

Azize Aslan

(translated from Spanish by Karen Tiedtke)

Abdullah Öcalan[1] states that patriarchy is "the basis and main womb of hierarchical and class society" (Öcalan 2007, 25). For this reason, the Kurdish movement believes that if a revolution aims to destroy class and power relations, it must also eliminate patriarchy; otherwise, power and the classes will be re-institutionalized on the patriarchal terrain. Anti-patriarchal autonomy, therefore, is a fundamental organizational perspective for the Kurdish struggle; a perspective that constitutes the heart of the ecological aspect of democratic confederalism.

Today, all women who are mobilized within the Kurdish women's movement participate in two organizational areas to coordinate and advance the Kurdish movement. First, they participate in what they call "general organization" where they work together with men. Second, they also join women-only organizations on the basis that they call "autonomous organization" (Kurdish: *Registina Xweser*). However, this self-organization is not just about creating spaces and institutions where only women have their place but also refers to the fundamental organization, or essence, of society. Because as women, oppressed by capitalism and patriarchy, re-organize themselves in their own way, they also revive the original relationship they have established with nature. All anti-patriarchal relationships that women establish with men would mean the rejection of society's dominant relationship with nature. This unified, but at the same time separate organization, which takes place in the civil, political, and military sphere, is based on the practice of the rupture that began with the organization of women at the beginning of the year 1992.

The women's movement later defined this practice as the "theory of rupture" in 1996.

The theory of rupture, in essence, refers to the break with patriarchy that Kurdish women understand to be both a physical and mental rupture (Academía Jineoloji 2016, 11). Mental rupture refers to women assuming total domination and argues that, for gender liberation, women must reflect on their mentality, shaped by patriarchy throughout history. In common with the experience of other women in left organizations (Gutiérrez Aguilar 2014), the early years of women's experience in the PKK show that all mixed spaces have indeed been patriarchal, even if they are socialist. With a series of determinations and analyses ranging from questioning gender relations to questioning power, the denial of the state, the definition of patriarchy as the oldest system of domination and hierarchy, and the fact that society cannot be liberated if women are not truly emancipated in a struggle for freedom; all of this has an important presence in women's critical reflections and organizations. Therefore, Kurdish women today understand and feel the need to acknowledge a crisis in the functioning of mixed spaces and at the same time to create their own spaces to undo patriarchy. Thereby, they become a source of collective strength to recreate and redesign mixed spaces. So, one can understand physical rupture as being a step toward reflexive and mental rupture, but, at the same time, it is the creation of autonomous organization and of feminine spaces where women can discover their views and their way of communication and can make feminine decisions. It is necessary to emphasize that there is no rupture in the separatist sense embraced in the current of separatist feminism (Frye 1983); the autonomous organization of Kurdish women has always implied a return to mixed spaces, but with feminine consciousness and strength. This organizational perspective is the most important characteristic that distinguishes the Kurdish women's movement from other feminist movements.

Another important characteristic that distinguishes Kurdish thought is that the struggle for women's liberation is expressed not as a struggle for equality (equality of public rights for women and men), but as a struggle for gender emancipation. In other words, women consider the idea of revolution to be a phenomenon that aims to liberate not only women but also all the sexes and, therefore, society, because they believe that the enslavement of men follows the enslavement of women (Öcalan 2013, 9). Gender emancipation, through a struggle that occurs simultaneously, separately and in common, would ensure the emancipation of society.

The women who have been discussing revolution and social transformation in the context of the women's social contract and women's confederalism since the 2000s are turning the Rojava Revolution into a women's revolution, that is, an anti-patriarchal revolution. The women of the Kongreya (North Syrian Women's Congress 2016), organize themselves in committees, build communes, assemblies, cooperatives, academies, self-defense units (YPJ-AsayîşaJIN), and women's houses (MalaJIN) of their own, are not symmetrical in any sense, but an antagonistic organization, in this sense an antagonistic autonomy that determines and constantly remembers the principles of mixed autonomy. In this way, they create their own autonomous spaces, decision-making mechanisms, modes of collective work, and political formation in *jineolojî*. The women of Rojava today assert shared voices, co-representation, co-presidency, and co-participation (50 percent) within all areas of autonomy. The presence of women is not only about physical participation, but, with the right of veto in all mixed spaces, ensures that each collective decision taken in Rojava becomes first a decision of women, and then a decision of society as a whole. Kongreya Star, following the principles of democratic nationhood and plurality to include women of all ethnicities and religions, creates autonomy for women living in the area (Kongra-Star 2016, 5). The Women's Congress plays an active role in the organization of women in every area of life: education, culture and art, economy, self-defense, social affairs, conflict resolution and justice, local government, press and media, ecology, and diplomacy (Kongra-Star 2016, 3).

Each of these sectors has an autonomous organization and demonstrates a confederal relationship under the Kongreya Star. This chapter focuses on the experience of women's economy (AborîyaJIN) but, as Kongreya Star spokesperson Evin Swed says, the presence of women in the economy shows a dialectical link with their organization in other sectors of women's confederalism:

> MalaJIN is where the women's justice commissions that deal with the problems of women are located; right now, of the problems that come from there, we see that women are being abused because they don't have economies; but we would be wrong if we said that if women worked then everything would be fine; in a society where justice is in the hands of men, even if women had money, they would be oppressed; that is, everything must be reorganized together. This is why we are organized in Kongra-Star, not just in one area, but in all areas related to women's lives. (Swed 2019)

SOCIAL ECONOMY

Near the end of her life, Hevrîn Khalaf, co-spokesperson for the Economic Council declared: *Si hay pan y agua, ¡la lucha continúa!*[2] One of the most important areas in which women are organizing themselves within the framework of revolution and autonomy in Rojava is the economy. The purpose of the economic organization called *AborîyaJIN* (women's economy) is to discover the original forms of women's economy, to create an economic process of their own so that women can once again be the subjects and determinants of life and ensure the organization of a social economy in Rojava. In this sense, AborîyaJIN aims to organize a democratic, ecological, and libertarian economy of women based on the perspective of anti-capitalist economics, which the Kurdish movement considers as a communal-democratic economy and which is defined in Rojava as social economy (*Aborîya Ciwakî*).

Social economy is defined as an autonomous, self-sufficient economy. The aim is for the economy to be at the service of the people, where the purpose is for social resources to be public and for the people to decide how to use them. In other words, that social wealth is under democratic management and control. Communes and people's assemblies organize and plan work that is undertaken in cooperatives, so that people can meet their social needs; eventually, property and social classes will disappear. However, the whole process must be accompanied by anti-patriarchal practices so that the social economy that is created can generate social emancipation for all of society.

In this sense, the most fundamental principle that defines social economy is the liberation of women. The transformation of anti-patriarchal relations and gender roles starting from the economy and the presence of women in the economy to reclaim the support systems necessary to sustain life is one of the fundamental pillars of social transformation as defined by the women's movement.

The women's movement, supported by feminist thinking, understands that women become invisible within the capitalist economy because their jobs are unpaid or unrecognized. The main reason for this is that, even though women have paid jobs like men, within the capitalist economy, they are not included in the decision-making processes (Aslan 2016). Therefore, the struggle for the remuneration of women's work is where socialist thought complements feminist thought. The creation of the (female) working class makes no more sense than the commodification of women's labor. The socialization of women's work through collective and community economies, therefore, is advocated in social economy (Academía Jineoloji 2016, 218).

Two core principles underpin ecological production within the social economy. First, the conditions that a woman's outlook is included and that the natural world must be taken as the basis for economic well-being. These conditions are prerequisites for the second principle: that the economy and organization of social life is geared toward collective needs and is based on use value.

Maria Mies attributes women's relationship with the natural world and understanding of the subjectivity of nature as an active agent to factors such as a woman's fertility and her experience of her own reproductive body as nature (Mies 2018, s. 116). The woman who gives birth to a child and whose body produces milk for the child's nourishment is a producer without tools or equipment, so a woman's fertility is the first social labor that can be defined. Therefore, the sanctity ancient societies attributed to motherhood is not related to the phenomenon of God and religion but is due to association with this first social labor. By contrast, the productivity of men follows the development of tools, exploiting social knowledge. In other words, men's productivity is not possible without the social accumulation reproduced by women. The long, repeated, and continuous time that women spent working with nature and the process of learning to care for nature as part of life enabled the agricultural revolution in Mesopotamia, known as the Neolithic Revolution, brought to fruition by the hands of women. The Neolithic Revolution was not the conquest and domestication of nature but the collaboration of the productive and creative work of women with the essential power of nature. Women's work protects, cares for, and defends nature, thus aiding its reproduction. Although nature was appropriated, this appropriation did not constitute a relationship of domination or ownership. Women did not own their bodies or the land, but cooperated with their bodies and the land to "let grow and make grow" (Mies 2018, s. 122). This approach, in allowing nature to flourish, is the most important part of the subsistence economy. The reason why the capitalist system of production and consumption is so destructive to society and nature is that the logic of capitalism does not allow nature to reproduce and renew itself.

The shaping of society on the basis of the separation created between worker-consumer and use value (need)-exchange value (surplus value) is not a phenomenon seen in community life and economy where women are the decision-makers. In communities that produce for their own collective needs, producers are also consumers, so alienation is avoided. In capitalism, concrete labor has become abstract labor for wages and money. The monetary relation created between production and consumption—that is, the exchange value capital regularizes independently of the phenomenon of "necessity" —destroyed the community's subsistence economy and created the capitalist economy and society in which productive activity is carried out with the main ambition to maximize profit.

In this sense, organizing a social economy means letting nature develop and grow, as women have done by creating the subsistence economy. While the subsistence economy is the heart of the social economy based on use value and ecology, the autonomous practice of cooperatives and cooperatives together with the commune, assembly, and academies corresponds to a social organization that transcends this.

ABORÎYAJIN: THE EXPERIENCE OF
WOMEN IN THE SOCIAL ECONOMY

In 2020, women at Jinwar declared: "one of the main tasks of our struggle is to create bread and roses." Today, women's confederal autonomy with this understanding of "organizing life" has reached a significant degree of organization within the realm of the economy (Delal 2018). AborîyaJIN (women's economy) is organized by committees in sectors such as agriculture, industry, the market, and cooperatives under the Kongreya Star. This creates new relationships and communal spaces and organizes strategies to resist and eliminate capitalism. AborîyaJIN's coordination mobilizes civilian women through communes and village assemblies in cities and cantons, giving them the responsibility to develop and maintain the women's economy and making it a fundamental aspect of these organizations' agenda for self-determination. These women have formed the Women's Economic Assembly (Meclîsa AborîyaJIN). In addition to the people's assemblies in the political sphere created by the system of delegation, one of the most important autonomous organizations mobilizing society in Rojava, there is also the practice of gathering in everyday life to form new assemblies in all sectors.

In Rojava, the practice of forming assemblies has gained momentum, embedding discussions about aspects of direct democracy, such as representation and scale. Women both collectivize decisions in the areas they organize through their own assemblies and ensure that the democratization of all decisions taken within mixed assemblies is encouraged on the basis of gender identity. Women's cooperative houses (*Mala Kooperatîfên JIN*) and women's cooperative units (*Yekîtîya Kooperatîfên JIN*) play an active role in establishing women's cooperatives in all sectors of the women's economy (AborîyaJIN). Many economic institutions arise to meet particular needs during the process of organizing the social economy; importantly, however, the assemblies have administrative primacy in coordinating, managing, and mobilizing economic relations. Therefore, cooperatives are not only established as a production unit or institution. Cooperatives develop as a new form of communal relations. For example, women's cooperatives cultivate one-third of the land that has been communalized by the revolution; however, these cooperatives are not institutionally established cooperatives with permanent members. The AborîyaJIN cooperative committee coordinates with the communes for cooperative cultivation on communal lands. According to this process, the different communes designate women who are members of a cooperative to their commune on a two-year rotation principle, to the benefit of the communal lands and the cooperative. As in many parts of the world, women's ownership of land is almost out of the question in the Middle East. In this regard, apart from communal land, women do not own land to establish cooperatives. The aim of the women's cooperative committee, therefore, is to enable large numbers of women to participate in this process and to ensure that female perspectives inform the practices of collective work, cooperatives, and self-management for the whole of Rojava. The most advanced experience so far in women's agricultural cooperatives has been in areas such as organic farming practices, organic fertilizer production, irrigation for farming, and the diversification of production. For decades, when the Syrian state dominated the vast Mesopotamian lands that now form Rojava, there was a system of industrial agriculture that produced only wheat and barley. The region's needs for fruit and vegetables were supplied from neighboring areas, from around Damascus, or from other parts of northern and eastern Syria. Today, however, although this situation continues to a large extent, the women's economy puts the diversity of food production before the status of a self-sufficient economy and has established vegetable and herb gardens under the leadership of women's communes. Women have also increased the production of products previously grown in small quantities, such as sesame, lentils, chickpeas, and soybeans, and planted thousands of fruit trees. In Rojava, where previously no tomatoes had been grown (the Syrian state banned the cultivation of crops other than wheat and barley), women now produce tomato paste in the workshop they have set up, using tomatoes and peppers grown in their own greenhouses. Aware that food derived from uniform industrial products is one of the most significant developments threatening human health, the women's cooperatives, under the umbrella of the women's cooperative union (*Yekîtîya Kooperatifên JIN*), together plan to diversify production and create a complementary network for the exchange of produce among women's cooperatives. They diversify production according to local conditions, both in the field and at the processing stage. The women in the cooperatives plan every product so that it can be processed to satisfy a particular social need. For this purpose, many small canning factories were established. On the one hand, preserving means transforming the product, on the other hand, it is based on the collective historic memory of Mesopotamia's geography.

One of the traditions of the subsistence economy, maintained by women in Mesopotamia over centuries, is the preservation of vegetables and fruits harvested in the summer months. These are then dried in the sun, buried underground, stored in caves, or preserved by various other methods so that they can be consumed during the winter, when they would otherwise be unavailable due to climatic conditions. Such traditional conservation is fundamental to subsistence economies and reminds us of the importance of the natural cycle of summer heat and cold winters in the Mesopotamian region. The production of necessities according to the cycle of nature and local affordances is non-industrial and follows non-capitalist logic. Harvest festivals and peasant festivals to swap seeds, local and cashless markets, organized by the ecological and economic committees, were the practices of the women of Bakur that had a very positive effect on the entire organization of the communal economy. In Rojava, young women have also been planting fruit trees and tying a cloth to each tree to encourage it to grow and bear fruit, being feminine wishing rites, important and fundamental practices for women to reconnect with the land and nature.

By contrast, the industrial system adopts modes of production which are contrary to the natural cycle, illustrating the exploitative character of capitalism and patriarchy which dominate nature. There is a poignant refusal of this domination of nature in knowing that cherries can be eaten in early summer, and raising children who know this; in not demanding cherries from nature in winter, and instead of consuming summer cherries, making them into jams to be consumed in winter. Or while also attaining a new taste, drying tomatoes in the sun in the summer, so as not to eat fresh and expensive tomatoes in winter, is a way of relating to complementary and mutual nature that humanity maintained for thousands of years. However, the market forces that respond to this demand, and make it permanent, rely upon industrialist methods of production, injecting chemicals into the soil and seeds which threatens human health, resulting in both ecological and social destruction. In this sense, the food preservation workshops established by AborîyaJIN, which are based on traditional methods, are a reminder that alternative, healthier modes of production and consumption are possible according to the seasons of nature. The women of the women's cooperative Demsal, established by six women in Heseke in March 2019, described the seasonal way that they process food according to the harvest time in the villages, for example, sometimes making fruit jam, at other times making pickles, and making cheese in the spring (Demsal, 2019). AborîyaJIN's core activities are producing milk, yogurt, cheese, and eggs, via dairy cooperatives, and producing bread in bakery cooperatives. The spokesperson of AborîyaJIN explains the main reason for creating the bakery cooperatives as follows:

We believe that those who continue to give bread in a society also determine the form of justice in society. In Syria, this has been the case for decades. Villages worked and grew wheat on state land and then the state would come and take all that was produced. People were not even able to keep flour for their own consumption. The state was turning the flour processed in their own mills into only one type of bread in their own bakeries. You may have seen that there is still a state bakery here with people queuing in front. Because it is cheap and until now there was no other place to buy bread. Now there are our bakeries, but most people still go to the state bakery. We were surprised to see this situation. In other words, it is a social habit in Kurdistan that mothers make their own bread, and not just one kind, that is, each region has its traditional bread that they make in different ways: tandoori bread, tortillas, lavash, pita, etc. The women make it all. But when we arrived in Rojava, we saw that there was such economic slaughter that women didn't even know how to make bread. Think about it, there are bread sellers who distribute bread in the border villages or people go to buy their bread in the nearest town. Seeing this, we decided to start with women's most basic activity, so we opened the bread cooperatives. The Adar cooperatives are the first cooperatives we established. Now we have bread cooperatives that we have established in all the cities. Women are learning to bake bread. In the Adar bakery you saw today, three types of bread are baked. In this war environment, villages often cannot get bread; therefore, our goal is for each commune or some nearby communes to come together and establish their own bread cooperative. (Delal 2018)

Among the economic alternatives that do not recreate an intermediate activity, women's markets or communal markets established by the Women's Economic Committee (AborîyaJIN) in Dirbesîyê, Dêrik, and other provinces stand out. The trade sector is one of the areas where the patriarchal pact is strongest; women produce, but then become invisible in the sales process. Goods produced by women become commodities and the target of competition through trade in the hands of men. In other words, women's labor is enslaved through markets controlled by patriarchy. Moreover, it is much more difficult for women to enter this sector, because in most Islamic societies in the Middle East it is frowned upon for women to be saleswomen. For this reason, AborîyaJIN has created alternative women's markets and spaces where women can take their products to sell directly to the consumer, without intermediaries, demonstrating that women can unveil another kind of exchange mentality. Delal, the spokeswoman for AborîyaJIN, stated that products were exchanged, especially among women. Although this relationship often takes place using money, the most powerful signifier of capitalism, it also helps to develop autonomous capacity and to establish other types of commercial and market relationships (Interview with Delal, spokesperson for the Women's Economic Committee, 2018).

One of the greatest difficulties encountered in the process of organizing women's economy is that women have lost their knowledge, skills, and forms of production, because capitalist domination has destroyed and stripped their economy for hundreds of years. What we have lost under the patriarchal capitalist system is not only our freedom but also the collective memory we have created through the reproduction of life. In the patriarchal sphere of work, women have been excluded from full participation in economic life, except within the principal domain of care work; the knowledge and skill of production have become a "professional" phenomenon and, therefore, the relationship between the economy and the organization of life has been broken. Today, women in cooperatives try to organize an alternative life by collectivizing their little knowledge for any work that is not domestic. Women of my generation, in particular, do not know how to live with the natural world, with the land, and we are not familiar with herbs and animals; we are living in the gap between our own realities and the essential reality of nature and life. For this reason, the recovery of knowledge and productive skills, on the basis of an autonomous (women's) organization manifests itself as the most essential need for an anti-capitalist economic reorganization.

When looking at established women's cooperatives, we see that women do not move away from the sexist division of labor, or that the AborîyaJIN organization has so far not been able to change this social dynamic, however, the spokesperson of AborîyaJIN expresses that while they share this concern, they are also reconsidering the definition of the sexist division of labor:

> Yes, this critique has its merits, but we believe that the definition of a sexist division of labor needs to be revised. The sexist view which considers the economy and men's capacity of physical labor more valuable, and undervalues the productive skills and emotional labor of women influences this critique. Also linked to this is seeing the industrialized economy as the significant economy. Women's economic abilities are shaped by social needs, but men's economic style is structured by profit and consumption. What kind of work and production is valuable, necessary, and meaningful to society and women? Essentially, this should be discussed. Will we make women work according to capitalist needs? The sexist division of labor is a problem that arises in the realm of industrialism. Let us not think this should be considered as significant as economics. You ask why we are active in these areas of work, because women are more successful in the sectors of agriculture, food, and clothing industry. In our opinion, improving women's economy by making women''s work and production valuable and meaningful is a better method; because competition with men's physical strength yields no results except preparing women for capital. (Dicle 2020)

That the anti-capitalist economy is constructed in an anti-patriarchal way means that, beyond the satisfaction of material needs, it also leads to a far-reaching reconstruction of meanings in society, ultimately redefining the idea of economy. Women define new meanings and interpret the revolutionary process in Rojava by reflecting upon their own practices which they discover through their own subjectivity within the struggle. Women have shown that revolution not only means working collectively but also rethinking and reorganizing all spheres of life communally, discussing all production processes in the commune's assemblies, and taking decisions collectively. Working and deciding together establishes a symbiosis and reconstructs the meaning of revolution with feminine wisdom. As the women of Jinwar say:

> Revolution is not great acts of violence and destruction. Revolution is communities coming together to work the land. Revolution is soil between fingers that no one can take away. Revolution is sharing food and taking care of each other. Our promise to the martyrs is not blood and fire: it is that someday, the children of this land can go back to being farmers instead of soldiers. (The Women's Village, JINWAR, 2020)

Social ecology begins to deteriorate as relations in society move away from the terrain of democratization and toward the terrain of domination. For this reason, creating women's own spaces and trying to transform mixed spaces and decision-making processes with women's full participation is important work that allows us to move toward the conditions of social ecology. The democratization of gender relations in the economy is a fundamental part of the democratization of the economy. One of the situations observed in traditional collective structures, where men and women are together, is that ultimately a decision is made under the domination of men. The percentage of women expressing opinions and participating in the discussion is lower, even though they are equal in number with men. As a result of patriarchal relations, it is very difficult for women to voice their opinions and express themselves in front of the community. The combination of strong (loud) speech of men, the personalization of discussions, and failure to care for the participation of their peers increases this reserve and fear in women. Therefore, for women to be able to express their views and thoughts, the mixed cooperative must make a special effort, working toward the democratization of gender relations without relying on the assumption that "women have the right to participate freely." In other words, everyone in the cooperative must ensure that they have equal influence in the decision-making process. Therefore, having agency in making policies, rather than being merely subjected to their outcomes, is important for women and for all members of the cooperative. This positions cooperatives as a structure in which not only productive relations but also social relations are restored. As in all spaces of the Kurdish movement, the participation of women in mixed cooperatives is 50 percent. This does not mean that women are quantitatively half everywhere, but it does mean that their influence on decisions is equal to half, and that the system of shared voices is applied in all cooperatives.

In addition to the co-representation and shared platform of women within cooperatives, two other practices carried out in the cooperatives in Bakur were important as attempts to achieve self-management based on gender equality. One of the practices was that men's speech was not accepted before a woman took the floor and started speaking in the cooperative assembly, as was done in all mixed spaces in the Kurdish movement. After the co-moderator opens the assembly, a woman is expected to take the floor and start the meeting. It is expected that a woman will volunteer to start the meeting, otherwise, the meeting waits until a woman takes the floor. Sometimes the wait is long, other times the women speak immediately, but this practice has two consequences. One being that after the first speaker, other women will have the courage to take the floor and so enable by opening the way for ongoing women's participation in the discussion. The second is that there is a clear difference in the topics discussed in the assemblies when women take the floor first. Men prefer to talk about profit, while women mostly choose topics that focus on life relationships. Thus, in cooperative assemblies when women have the first word and start the discussion, instead of production, the topics focus on reproduction and nature.

All these practices, in which women transform the organization of the economy, give rise to a favorable atmosphere for the formation of community, democracy, solidarity, as well as mutual and complementary social relations. Furthermore, a non-capitalist economy means not only satisfying our individual and collective needs but may also lead to the end of wars and the disarmament of men. A communal society, based on use values, has a mutually beneficial relationship with nature, undertaking labor that cares for both the natural world and society, based on democratic decisions regarding social needs and production, thus resolving the material causes of man-made wars. Perhaps, as Mies and Shiva argue, if men acquire the caring and nurturing qualities that have hitherto been considered the domain of women, and if, in an economy based on self-sufficiency, mutuality, and self-provision, not only women but also men engage in subsistence production, the latter will have neither the time nor the inclination to pursue their destructive war games (Mies and Shiva 2014, s. 321–22).

NOTES

1. Abdullah Öcalan is the leader of the PKK and one of the leaders of the Kurdish peoples. He has been serving a life sentence (often in isolation) in İmralı Island Prison, Turkey since 1999. Öcalan has written numerous books and articles that inspired the struggle for the freedom of the Kurds. See: http://ocalanbooks.com/#/.

2. "If there is bread and water, the struggle continues!" Hevrîn Khalaf was the co-spokesperson of the Economic Council when I met her in 2018. On October 12, 2019, three days after Turkey launched its military offensive in northern Syria, she was brutally murdered. According to reports from the Syrian Observatory for Human Rights, she was taken out of her car and shot in cold blood on the road. Her autopsy reveals that she was shot, beaten with heavy objects, and dragged by her hair until the skin was separated from her scalp.

REFERENCES

Academia de Ciencias Sociales Abdullah Öcalan. 2012. *Demokratik-Komünal Ekonomi.* Qandil: Azadi Matbası.

Academía Jineoloji. 2016. *Jineolojiye Giriş* (Introducción de Jineoloji). Qandil: Jineoloji Akademisi Yayınları Azadi Maatbası.

Aslan, Azize. (18 de 11 de 2014). Demokratik Ekonomi Konferansı: Bireyci Kapitalist Bir Ekonomiden, Toplumsal Komünal Bir Ekonomiye. Obtenido de Toplum ve Kuram/Zan Enstitüsü: http://zanenstitu.org/demokratik-ekonomi-konferansi-bireyci-kapitalist-bir-ekonomiden-toplumsal-komunal-bir-ekonomiye-azize-aslan/.

Aslan, Azize. 2016. Demokratik özerklikte ekonomik özyönetim: Bakur örneği (Autogobierno económico en la autonomía democrática. El ejemplo de Bakur). Birikim (325), 93–98.

Bookchin, Murray. 2014. *Kentsiz Kentleşme: Yurttaşlığın Yükselişi ve Çöküşü.* Istanbul: Sümer Yayıncılık.

Comisión de Preparación de la Conferencia de Economía Democrática. 2014. Ekonomi Konferansı Bileşenlerine (A los Componentes de la Conferencia de la Economía Democrática). Demokratik Modernite, 71–78.

Coordinadora Andina de Organizaciones Indígenas—CAOI. 2010. Buen Vivir / Vivir Bien Filosofía, políticas, estrategias y experiencias regionales andinas. Lima: CAOI.

D'Alisa, G. 2020. *Küçülme: Yeni Bir Çağ İçin Kavram Dağarcığı.* İstanbul: Metis Yayınları.

Frye, Marilyn. 1983. *The Politics of Reality: Essays in Feminist Theory.* Trumansburg, NY: The Crossing Press.

Gutiérrez, R., & Navarro Trujillo, M. L. 2019. Producir Lo Común Para Sostener y Transformar La Vida. Confluéncias, 298–324.

Gutiérrez Aguilar, Raquel. 2014. *A Desordenar! Por una historia abierta de la lucha social.* México: Paz En El Árbol.

KJA (El Movimiento de Mujeres Democraticas). 2016. El reporte final del taller de "Trabajo de Las Mujeres." Mardin: KJA.

Kongra-Star. 2016. *About the Work and Ideas of Kongreya Star, the Women's Movement in Rojava.* Qamishlo: The Committee of Diplomacy of Kongreya Star.

Mies, Maria. 2018. *Patriarcado y Acumulación a Escala Mundial.* Madrid: Traficantes de Sueños.

Mies, Maria, and Shiva, Vandana. 2014. *Ecofeminism.* London: Zed Books.

Öcalan, Abdullah. 2007. Devlet. Abdullah Öcalan Sosyal Bilimler Akademisi.

Öcalan, Abdullah. 2009. Demokratik Toplum Manifestosu: Özgürlük Sosyolojisi (Vol. III). WEŞANÊN SERXWEBÛN 149.

Öcalan, Abdullah. 2011. Ortadoğu'da Uygarlık Krizi ve Demokratik Uygarlık Çözümü, Hawar yayınları. Hawar Yayınları.

Öcalan, Abdullah. 2012. *Kürdistan Devrim Manifestosu: Kürt Sorunu ve Demokratik Ulus Çözümü.* Diyarbakır: Ararat.

Öcalan, Abdullah. 2013. *Liberar La Vida: La Revolución de Las Mujeres Kurdas*. Colonia: International Initiative Editon.

Öcalan, Abdullah. 2014. Endüstriyalizm (Kapitalizm) ve Ekoloji. Demokratik Modernite (11), 7–24.

Stravrides, S. 2018. Müşterek Mekan: Müşterekler Olarak Şehir. İstanbul: Sel Yayıncılık.

Yusuf, A. 2015. Rojava Deneyimi Bağlamında Sosyal Ekonomiyi Düşünmek: Temeller ve İlkeler. En J. Jongerden, A. H. Akkaya, & B. Şimşek, İsyandan İnşaya Kürdistan Özgürlük Hareketi (págs. 261–290). Ankara: Dipnot.

INTERVIEWS

Delal, H. (11 de Marzo de 2018). La vocera del comité economía de mujeres (AboriyaJIN). (A. Aslan, Entrevistador). Las contradicciones de la revolución en la lucha kurda y la economía anticapitalista de Rojava. https://repositorioinstitucion al.buap.mx/handle/20.500.12371/10300.

Demsal, C. (15 de Diciembre de 2019). La vocera de La Demsal. (A. Aslan, Entrevistador). Las contradicciones de la revolución en la lucha kurda y la economía anticapitalista de Rojava. https://repositorioinstitucional.buap.mx/hand le/20.500.12371/10300.

Dicle. (Octubre de 2020). La vocera del comité economía de mujeres. (A. Aslan, Entrevistador). Las contradicciones de la revolución en la lucha kurda y la economía anticapitalista de Rojava. https://repositorioinstitucional.buap.mx/ handle/20.500.12371/10300.

Swed, E. (25 de 12 de 2019). La vocera de Kongra-Star. (A. Aslan, Entrevistador). Las contradicciones de la revolución en la lucha kurda y la economía anticapitalista de Rojava. https://repositorioinstitucional.buap.mx/handle/20.500.12371/10300.

Part III

SOCIAL MOVEMENTS AND ENVIRONMENTAL ACTIVISM

Chapter 11

Environmental Activism in Rojhelat

Emergence and Objectives

Allan Hassaniyan

This study argues that environmental activism in Rojhelat (Iranian/Eastern Kurdistan) is a platform for protesting the Iranian government's sociopolitical and economic policies toward its Kurdish population. For this reason, Rojhelat has quickly turned into a securitized area, with severe consequences for environmentalists. Popularizing a platform of "culturalization of environmentalism" (CoE), and attempting to alter the approach of state institutions to the natural environment and natural resources in Kurdistan, are among the key objectives of environmental activism in Rojhelat. While environmental education and campaigns are the tools used in persuading Kurdish society to adopt environmentally friendly behavior, the movement's approach to the state has taken a more confrontational path. The diversified focus of environmental activism in Rojhelat allows its conceptualization through a theoretical framework of "environmental humanities," a category which encompasses "a wide ranging response to the environmental challenges of our time, [and] engages with fundamental questions of meaning, value, responsibility, and purpose in a time of rapid, and escalating, change (Rose et al. 2012, 1)."

This study's main sources of primary data are written materials available online. For instance, to analyze the discourse, development, and activities of the NGO Chya Green Association (*Anjomen-e Sabz-e Chya/Çiya-i Sawzi Mariwan*, hereafter Chya), the over seventy biweekly and twenty-seven monthly editions of the magazine *Chya*, published by the NGO from 2008 to 2018, have been invaluable for this study.[1] In addition, twenty journalists, environmental activists, and individuals with links to the environmental movement, from within and outside Rojhelat, have been interviewed.

Environmental activism has been generally defined as "organized participation in environmental issues, comprising an example of environmentally friendly behavior rooted in the political realm," being "expressed in specific activities reflecting a commitment to the environment channeled in formal settings and realized through institutional structures" (Marquart-Pyatt 2012, 684). Some conceptualizations of environmentalism and environmental activism have argued that this kind of activism is as much a social as a political movement, existing to change people's outlook on the world, their beliefs, and behavior. For instance, Timothy O'Riordan writes that environmentalism "interacts with the social, economic and political conditions in which it finds itself, changing current paradigms of thought and action and at the same time resonating to its own successes and failures" (O'Riordan 1981, 3).

ENVIRONMENTAL ACTIVISM IN ROJHELAT

The emergence of environmental activism in Rojhelat dates to the late 1980s. From the beginning of the movement, it was inextricably linked to Kurdish identity. A group of environmental activists, mainly university students, held a memorial for the victims of Saddam Hussain's chemical attack on the Iraqi Kurdish city of Halabje in 1988, during the Iran-Iraq War (1980–1988); they highlighted the destructive impact of chemical weapons on the environment of Kurdistan. Across all the nation-states which occupy Kurdish land in the region, both natural and human environments in Kurdistan have suffered from severe attacks from governments. Not only Halabje but also Sardasht, a city in Iran's West Azerbaijan Province in Rojhelat, suffered from Saddam's chemical attacks during the Iran-Iraq War. The literature on the links between civil war and destruction of the natural environment suggests that the counterinsurgency tactics adopted by Iran, Turkey, Iraq, and Syria against their Kurdish oppositions have attempted to undermine rebels' ability to operate through the deliberate destruction of the natural world, particularly the systematic deforestation of Kurdistan (Gurses 2012, 255; Hassaniyan 2020a, 1–24).

Environmental activism became more organized in the form of NGOs and associations in the late 1990s, when it was given a certain breathing space by the Iranian authorities, who granted permission for the formation of NGOs. The late 1990s and early 2000s were characterized by the establishment of thousands of associations and NGOs all over Iran, among them over 500 environmental NGOs (Doyle and Simpson 2006, 260). Therefore, the period of the first stage of environmental activism should be studied in the context of the relative easing of the sociopolitical situation in Iran, during the presidency of Mohammad Khatami (1997–2005) (Afrasiabi 2003, 432).

Characterized as "environmentalism from below," these environmental NGOs, with their different approaches to environmental sustainability and sustainable development, manifested a challenge to the state-centric approach to economic growth (Afrasiabi 2003, 433). Simin Fadaee has conceptualized environmental activism in Iran within a theoretical framework of social movements and argues that the semi-authoritarian context in which the movement developed "at the same time enables and limits the activities of the environmental groups. The environmentalists challenge the existing norms and structures of the society. Their main aim is to change the people's lifestyle and governmental policies towards environmental issues" (Fadaee 2011, 85).

Despite many similarities in the approaches and values shared by environmental activism in the Iranian center (that is around the large central cities and Persian-speaking regions) and in Rojhelat, it can be argued that while the center-based environmental groups and NGOs focus on "purely" environmental issues, environmental activism in Rojhelat is a diversified form of activism that, in addition to focusing on the environment, actively cherishes Kurdish identity, culture, and language.

In Rojhelat, in the period from 1997 to 2005, dozens of local environmental NGOs emerged, aiming at protecting Kurdistan's natural environment and spreading the message of CoE. This era can thus be referred to as the era of emergence, formation, and growth of environmental activism in Rojhelat.[2] It is worthy of note that many of these environmental activists were women, some of them occupying leading roles in the environmental NGOs.[3] Kurdish environmental NGOs enjoy widescale support from the local population. For instance, as one interviewee stated, the NGO Chya has roughly 14,000–15,000 members and sympathizers (author's interview with anonymous environmental activist, May 18, 2020). The main reason for such support for Chya is that "Chya provides a space for belonging to a serious organization that contributes to CoE, and warns the public of threats facing Kurdistan's nature, at a time where the natural environment of Kurdistan experienced massive and systematic devastation" (author's interview with Farzad Haghshenas, May 22, 2020).

Kurdish Environmental Activism (EA), which upholds an alternative to the entrenched approach of the state's environmental policy, offers a model of environmental governance referred to as "civic environmentalism." According to DeWitt John, this "features an emphasis on dealing with problems at state and local levels and involves a political process in which divergent values are recognized and many individuals and organizations work collaboratively to forge balanced, comprehensive solutions" (John 1994, 30). However, one element of John's civil environmentalism does not apply to Kurdish environmental NGOs' and activists' relationship with governmental institutions, which is conflictual rather than collaborative. In many regards, the values manifested in Iran's state-centric development are increasingly inappropriate for the social and economic realities of Rojhelat. Regarding the question of how to deal with environmental issues in Kurdistan, environmental NGOs and activists have argued that the government should adopt a system characterized as "Negotiate—Don't dictate" (Crowe and Shryer 1995, 28).

CHYA, ROJHELAT'S GREEN PLATFORM

The Kurdish word *chya/çiya* (mountain) is the name of Rojhelat's most prominent and proactive environmental NGO. Chya was founded in summer 1999 to conduct environmental and cultural activities in the area of Mariwan, a city and county in Rojhelat which borders Iraqi Kurdistan. Chya views the environment (*jinge*) as a wide range of biological and cultural elements, each of which in their own way shapes our thinking and lives (BN *Chya* 1, no. 1 [October 2008]: 1).

In the founding statement of Chya, it is emphasized that Chya is not a political organization, something repeatedly stressed by its members ("Shenakhti az Anjomen-e Sabz-e Chya" [An Introduction to Chya Green Association], BN *Chya* 1, no. 1 (October 2008): 1). Instead, members and sympathizers of Chya describe this NGO and its mouthpiece, *Chya*, as fulfilling a series of diverse, harmonious functions, such as being "a green tribune for 'green thinkers' (*sawzbir*)" (BN *Chya* 2, nos. 24–25 [July 2009]: 1–12), or "the voice of cultural, civic, and environmental activists" ("Chya "seday fe'alin-e ferhengi, medeni, ve zist mohiti" [*Chya*, the voice of cultural, civic, and environmental activists], Monthly *Chya* 1, no. 1 [November 2015]: 1). Regarding the role of *Chya* magazine as a platform that allows the voice of the marginalized to be heard, Amin Azizi refers to the magazine as "the voice of the oppressed" (Azizi 2011a, 3). Aside from a few short interruptions, the regular appearance of *Chya* magazine over a decade (2008–2018) has been a unique achievement. From 2008 to 2013, Chya published seventy biweekly "Special Environmental Newsletters," available to the public for free (Monthly *Chya* 2, no. 14 [December 2016]).

As shown in my previous study (Hassaniyan 2020a), EA in Rojhelat has become immensely securitized. While jail sentences, persecution, and assassinations of environmental activists have resulted from this securitization, self-censorship and enduring fear and anxiety are other elements of the everyday lives of environmental activists and their families. Chya has not been an exception from this state of securitization. Chya has spoken out vocally against it, as expressed in a statement published in *Chya* protesting the "unexplained arrests" of its members Sharif Bajwar, Behroz Darwand, and Mohammad Iraj Qaderi by the Islamic Revolutionary Guard Corps (IRGC) intelligence service, *Itellat Spah*, on July 23, 2011 (BN *Chya* 3, no. 65 [August 2011]: 2). Nevertheless, the list of arrests, torture, and killings of environmental activities in Rojhelat has become longer in succeeding years (author's interview with Haghshenas).

The Iranian state has also tried to limit the impact of NGOs, even on a local scale, through more institutionalized ways. From July to December 2010, Chya monitored 886 cases of wildfires in the forests of Mariwan and Sanandaj (BN *Chya* 3, no. 64 [June 2011]: 2). Chya's public denunciations of the local authorities' mismanagement of wildfires in the 2010s can be identified as the catalyst of tense relations between Chya and governmental institutions (BN *Chya* 2, no. 45 [August 2010]: 1–8). As a result of this antipathy, the local authorities have excluded Chya from coordinated activities aimed at solving Mariwan's environmental challenges.

CHYA'S DIVERSIFIED FOCUS

Culturalization, or the establishment and spreading of a culture that considers the sustainability of the environment by individuals, households, businesses, industries, and institutions, has been among the key tasks of Chya in Kurdistan. Despite strong public support for the project of CoE, this has not been an easy task. On multiple occasions, powerful forces, such as state institutions and opportunist local figures, have stood in the way of Chya's realization of CoE. However, some examples of Chya's activity toward this objective reveal the emergence of some new, and noteworthy, trends in the community that has been Chya's operational base.

The motto "We are an element of nature, and not its owner" has been a central value in the practice of CoE in Kurdistan (Monthly *Chya* 1, no. 9 [June 2016]: 6). Based on this and similar mottos, such as "I think green, that's why I exist," awareness campaigns have been mobilized around the threats facing the environment (Azizi 2009, 1). Promoting a culture of coexistence between nature and human beings has been a principal element of this project. Highlighting potentially destructive approaches to economic growth and development, and the unsustainable extraction and overconsumption of natural resources as major root causes of environmental destruction, environmental activists in Rojhelat have encouraged policy-makers and the public to rethink their approaches to, and interaction with, nature. On the communal level, a wide range of activities has been organized by Chya as elements of CoE, including clean-up activities and training the community in waste management and recycling. For instance, the frequent clean-ups of cities and neighborhoods by members of the public have become routine in Mariwan. According to Farzad Haghshenas, "This decade-old practice of cleaning cities and neighborhoods has now become a culture. This hopefully will spread to other parts of Kurdistan and will remain as part of Chya's legacy for future generations" (author's interview with Haghshenas). The endeavors of CoE are not limited to urban areas. As reported in *Chya* (under the section "Environmentalism in villages of Mariwan") (BN *Chya* 3, no. 49 [October 2010]: 5), Chya's educational teams travel to villages and remote areas where they offer activities. These environmental trips have offered environmental workshops, health and hygiene education, helped establish women's groups, and facilitated discussions of the everyday challenges faced by women in these communities.

Rojhelat is known for its four distinct seasons and moderate climate, and in recent years, its colorful natural sites have been increasingly popular tourist destinations for internal tourists from other parts of Iran. Mariwan has not been an exception. In particular, Zrebar (known, because of its ecosystem, as *negin-e sabz* or the "green jewel"), has become the region's most attractive tourist destination, due to its beauty and the joy it offers visitors.

However, Zrebar has also suffered from destructive activities (BN *Chya* 2, no. 36 (March 2010): 2). In addition to threatening activities such as the construction of an oil refinery, the activities of private investors, land speculators, governmental institutions (for instance, military institutions such as the IRGC and the army), and touristic activities have each, in different ways, caused damage to Zrebar's ecosystem. The IRGC and the army's sometimes irregular and illegal construction of residential, commercial, and military compounds, for example, threaten the life and sustainability of Zrebar (BN *Chya* 1, no. 23 (July 2009): 3).

Recognizing and promoting the role of women in different aspects of life, not least in solving Kurdistan's environmental challenges, has been another element of CoE. *Chya* has covered topics including "women's role in the interaction between environment and economy" (Koneposhi 2009, 4), "women and the environment" (Kurdistani 2016a, 3), and "women's leading role in the utilization and protection of environment" (Rafi'i 2017, 6). Environmental activists claim that through human history, women have played an important role in agriculture and care for the environment, and have thus been in close interaction with nature. The focus on gender diversity has meant the encouragement of women and securing them a place in any activities organized by Chya. In this regard, Chya has established the Women's Committee of Chya, tasked with "assessing and promoting the role of women in CoE" (BN *Chya* 3, no. 49 (October 2010): 2). However, while women are well-represented in Chya's activities, there have been no women on its leadership board. On multiple occasions, Chya has co-organized activities with environmental NGOs run primarily by female environmental activists. For instance, Chya hosted a major tree-planting event in Mariwan, in 2010, when 500 members of the public planted over 1,200 trees (BN *Chya* 2, no. 36 [March 2010]: 2); the *Anjomen-e zanan-e rah-e sabze Kordestan* (Women's

Association of the Green Road of Kurdistan) participated in the event (BN *Chya* 2, no. 36 [March 2010]: 2).

Highlighting the links between Kurdish national identity and Kurdistan's nature (such as forests, mountains, and rivers) is a clear feature of EA in Rojhelat; *Chya*, therefore, reminds its audience that "environmentalism is a task of patriotism" (BN *Chya* 3, nos. 58–59 [February 2011]: 7). By drawing a link between Kurdish national identity and Kurdistan's natural environment, and spreading awareness among the Kurdish people of the threats facing this environment, one writer for *Chya* points to Kurdistan's forest as a national treasure, stating that "as a repressed and deprived nation, nature is our only capital" (Anwar Rewshen, "Darestan samaneki neteweyi" [Forests, a national treasure], BN *Chya* 1, no. 18 [May 2009]: 3).

As mentioned above, the Monthly *Chya* has also been a platform for denouncing the Iranian regime's discriminatory policies toward Kurds in Rojhelat. *Chya*'s criticisms of environmentally damaging state activities—such as deforestation due to industrial timber-cutting (Qaderi 2015, 3), the IRGC's counterinsurgency measures,[4] trans-Iranian transportation of Kurdistan's water resources to central Iran, and the state's exploitation of Kurdistan's mineral and other natural resources—have adopted a terminology in which Rojhelat is described as an internal colony of the Iranian state (Hassaniyan 2020, 362). *Chya* has described the massive destruction of Kurdistan's environment as "eco-terrorism," with the state as a "co-mafia" (Monthly *Chya* 3, no. 27 (January 2017). For instance, Mas'od Binande has claimed in *Chya* that the Kurds in Rojhelat are the subject of a "state of exception and unequal center-peripheral relations, in which Kurdistan is a peripheral and isolated region captured in an unpredictable and exploitative economic system. The production system in Kurdistan is the product of a centralized economic system, imposed on Kurds, resulting in massive deprivation" (Binande 2017, 1–2). The mismanagement of wildfires in the forests of Rojhelat, and the northern regions of Iran, have been linked to the state's discriminatory policies toward Kurds (Nezeri 2010, 8).

The sociopolitical and economic disenfranchisement of Kurds in Rojhelat, reflected in the destruction of its environment and exploitation of its natural resources, has meant that Kurdish environmental activists have raised questions about the state-centric and elitist definition of "security and development" in Iran. For instance, in articles published in *Chya*, Kurdish environmental activists and intellectuals argue for "the need for deconstructing the concept of development" (Hamid Koneposhi, "Bazsazi-e mafhom-e amniyet" [The need for deconstructing the concept of development], Monthly *Chya* 1, no. 1 (November 2015): 6), "development based on indigenous values and premises" (Daswar 2016), and the standpoint that environmental security is "more important than national security" (Kurdistani 2016b, 3).

They ask for a sustainable economic and political system with respect to Rojhelat's environment and natural resources, taking into account the rights and wishes of its population. By challenging the state's developmental activities, such as the construction of dams (Azizi 2011b, 3) and oil refineries, they demand that any developmental activities in Kurdistan should consider the environment and advocate the adoption of "a model of [environmental] security defined by this region's population" (Azizi 2016, 1).

The campaign for education in Kurdish and mother languages other than Persian in Iranian schools is a topic that has been systematically covered in *Chya*. Each edition of *Chya*, features Amin Azizi's column entitled "Language Politics," in which the writer argues for the importance of educating in mother languages, including examples of countries with successful bilingual or multilingual education systems.

Azizi condemns political regimes that deny people the right of speaking and receiving education, in their mother tongue, equating its denial with "racism and genocide" (Azizi 2017, 5).

CONCLUSION

This study has argued that EA in Rojhelat is a nascent trend, dating back to the late 1980s. The underlying motivations for its occurrence are the protection of Rojhelat's natural environment from the Iranian state's multifaceted destructive military, economic, and developmental activities, spreading popular consciousness of environmental concerns, through CoE, and promoting "principles of direct democracy, gender equality and ecological well-being in a needs-based economy" (Hunt 2019). Kurdish environmental activist groups from different parts of Kurdistan, despite differences in the time of their formation and specific practices, share fundamental values, such as viewing environmental and ecological struggles as the touchstone for the liberation of all humanity (Hunt 2019, 7), campaigning for direct democracy, and viewing humans as a part of nature rather than its owner (Monthly *Chya* 1, no 9. [June 2016]: 6).

Through the deforestation and wildfires of recent decades, resulting in part from the IRGC's shelling of highlands and national parks (Hassaniyan 2020b), the Islamic regime as part of its counter-Kurdish movement efforts, has escalated its war on Kurds in Rojhelat. Kurdish civil society, with its environmental activists on the frontline, has been the main, if not the only, defender of Kurdish nature, and the main opposition to the regime's brutal policies in Rojhelat. Taking into account the emergence, evolution, practice, and discourses of Kurdish EA, as a form of civilian resistance, this trend should arguably be viewed within the context of the Kurdish liberation movement in Rojhelat and other parts of Kurdistan, in addition to its place within the global environmental movement.

NOTES

1. To avoid any confusion, references to the magazine will appear as *Chya* (italicized); non-italicized "Chya" should be understood as referring to the NGO itself. The biweekly newsletters (BN) and monthly editions of *Chya* are available respectively at http://fa.chya.ir/?page_id=122522 and http://fa.chya.ir/?page_id=66. Due to the large numbers of the editions of *Chya* included in this study, specific references to authors of pieces in *Chya* are made only in the notes, omitted from the bibliographic entries.

2. Other than Chya, environmental NGOs based in Rojhelat include (listed with date established): Pajin, the People's Association for Protecting the Environment in Bane (1999); Wllat, the Bokan Institute of Environmental Defenders (1999); the Green Road of Sanandaj (2001); Hewazo, Diwandere (2001); Ilam, Kabir Kuh Green Association (2002).

3. The following are some of the female CEOs of environmental NGOs in Kurdistan: Fahime Qadem Khayrian (*Anjomen-e sabbz-e hefz-e mohit-e zist*, Sanandaj, 2001); Ferokhloqa Mo'temedwaziri (*Jam'yet-e Kordestan-e sabz*, Sanandaj, 2001), Snor Khaledi (*Hewazo*, Diwandereh, 2001), Fateme Ardelan (*Anjomen-e zenan-e zamin ve tose'ey-e paydar*, Sanandaj, 2001).

4. Especially during the past two decades, part of the IRGC's counterinsurgency policy has been to shell and burn down forests in Rojhelat, using the justification that they are hiding places for Kurdish fighters.

REFERENCES

Afrasiabi, Kaveh L. 2003. "The Environmental Movement in Iran: Perspectives from Below and Above." *Middle East Journal* 57, no. 3 (Summer): 432–48.

Azizi, Amin. 2009. "Men sawz bir dekemewe boye ham" ["I Think Green, That's Why I Exist]." *BN Chya* 1, no. 22 (July): 1–8.

Azizi, Amin. 2011a. "Belavoki Chya; Hawari denge Khenkawekan" [Chya is the Voice of the Oppressed]. *BN Chya* 3, no. 62 (May): 1–8.

Azizi, Amin. 2011b. "Bendawekan: gesh-e u peshkawtoyi, yan wehm u goman" ["Dams: deveLopment and Progress or Questionable Steps"]. *BN Chya* 3, no. 67 (September): 1–8.

Azizi, Amin. 2016. "Amiyet ve hoqoq-e meliyetha ["Security and the Rights of Ethnonational Groups"]. *Monthly Chya* 1, no. 6 (April): 1–8.

Azizi, Amin. 2017. "Siyaset-e zebani ve amozesh-e chand zebanegi" [Language Policy and Multilingual Education]. *Monthly Chya* 2, no. 17 (March): 1–8.

Binande, Mas'od. 2017. "Ma hemegi kolbar-e waz'iyet-e wejeyi hastim" [We Are All Kolbars in a State of Exception]. *Monthly Chya* 2, no. 21 (July): 1–8.

Crowe, M. Douglas, and Jeff Shryer. 1995. "Eco-colonialism." *Wildlife Society Bulletin* 23, no. 1: 26–30.

Daswar, Adnan. 2016. "Tawse'-e ber paye-y engarehay bomi" [Development based on indigenous values and premises]. *Monthly Chya* 1, no. 9 (June): 1–8.

Doyle, Timothy, and Adam Simpson. 2006. "Traversing More Than Speed Bumps: Green Politics Under Authoritarian Regimes in Burma and Iran." *Environmental Politics* 15, no. 5: 750–767.

Fadaee, Simin. 2011. "Environmental Movements in Iran: Application of the New Social Movement Theory in the Non-European Context." *Social Change* 41, no. 1: 79–96.

Gurses, Mehmet. 2012. "Environmental Consequences of Civil War: Evidence from the Kurdish Conflict in Turkey." *Civil Wars* 14, no. 2: 254–271.

Hassaniyan, Allan. 2020a. "Environmentalism in Iranian Kurdistan: Causes and Conditions for its Securitisation." *Conflict, Security & Development* 20, no. 3: 355–378.

Hassaniyan, Allan. 2020b. "The Gains and Risks of Kurdish Civic Activism in Iran," *MERIP* 295 (Summer): https://merip.org/2020/08/the-gains-and-risks-of-kurdish-civic-activism-in-iran/.

Hunt, E. Stephen. 2019. "Prospects for Kurdish Ecology Initiatives in Syria and Turkey: Democratic Confederalism and Social Ecology." *Capitalism Nature Socialism* 30, no. 3: 7–26.

Koneposhi, Maso'd. 2009. "Neqsh-e zenan dar rabete-y bayn-e tabi'at ve eqtesad" [Women's role in the interaction between environment and economy]. *BN Chya* 1, no. 22 (July): 1–8.

Koneposhi, Hamid. 2015. "Bazsazi-e mafhom-e amniyet" [The need for deconstructing the concept of development]. *Monthly Chya* 1, no. 1 (November): 1–8.

Kurdistani, Kawe. 2016a. "Zan ve mohit-e zist" [Women and the environment]. *Monthly Chya* 1, no. 6 (May): 1–8.

Kurdistani, Kawe. 2016b. "Balater az amniyet-e melli" [More important than national security]. *Monthly Chya* 2, no 13 (November): 1–8.

John, DeWitt. 1994. "Civic Environmentalism." *Issues in Science and Technology* 10, no. 4 (Summer): 30–34.

Marquart-Pyatt, Sandra T. 2012. "Explaining Environmental Activism Across Countries." *Society & Natural Resources* 25, no. 7: 683–99.

Nezeri, Maryam. 2010. "Inja Iran ast, ve ba'zi az ostanha inja ostanterend" [This is Iran, and some provinces are even more peripheral]. *BN Chya* 3, no. 52 (November): 1–8.

O'Riordan, Timothy. 1981. "Environmentalism and Education." *Journal of Geography in Higher Education* 5, no. 1: 3–17.

Qaderi, Iraj. 2015. "Atesh-e Jangelhay Zagros zir khakester-e so'alat" [The mysterious wildfires in the forests of Zagros]. *Monthly Chya* 1, no. 1: 1–8.

Rafi'i, Narges. 2017. "Naqsh-e mohem-e zenan dar behrebardari ve negehdari-e mohit-e zist" [The importance of women's role in the utilization and protection of the environment]. *Monthly Chya* 2, no. 20 (June): 1–8.

Rose, Bird Deborah, Thom van Dooren, Matthew Chrulew, Stuart Cooke, Matthew Kearnes, and Emily O'Gorman. 2012. "Thinking Through the Environment, Unsettling the Humanities." *Environmental Humanities* 1, no. 1: 1–5.

Chapter 12

The Kurdish Freedom Movement and Gezi

Strategic Reluctance and Tactical Ambiguities

Kumru Toktamış and Isabel David[1]

Social ecology has constituted a significant component of the Kurdish freedom movement's political agenda, especially in the twenty-first century (as the present collection demonstrates). Yet the Kurdish participation and presence during the Gezi protests of 2013, which started as an environmentalist resistance against urban redevelopment of a park at the center of Istanbul, has been a controversial topic, marked by strategic reluctance and tactical ambiguities.

The Gezi resistance is considered a pivotal political moment in Turkey and particularly during AKP (Justice and Development Party) rule. What started as an encampment in late May 2013 against the destruction of one of Istanbul's few remaining green spaces—Gezi Park—soon became a mass protest across several cities around the country against AKP rule after the police forcefully evicted the initial occupiers (see David and Toktamış 2015; Hemer and Persson 2017). The original environmental protests amalgamated with the rejection of AKP's rising authoritarianism and increasing encroachment on personal lifestyles (like bans on alcohol consumption and the introduction of religious courses at school). Indeed, since the party reached power in 2002, AKP has been implementing a hegemonic project of cultural, political, and economic transformation making use of wealth transfer, clientelism, restrictions on fundamental freedoms, and the control of the state apparatus and media (Sayarı 2011; Buğra and Savaskan 2014; David 2016; Esen and Gümüşçü 2016). In the weeks that followed the initial occupation, the Gezi protests mobilized 2.5 to 3 million people in seventy-nine out of eighty-one cities across Turkey (İnsan Hakları Derneği 2013).

During the Gezi protests, Istanbul's Taksim Square was turned into a space of communes and collectives for almost two weeks, occupied by a diverse array of societal groups: religious minorities such as Alevis, Armenians, Kemalists, nationalists, Sunni Muslims, Kurds, feminists, Christians, LGBTI (Lesbian, Gay, Bisexual, Transgender and Intersex), and football fans (David and Toktamış 2015). Among these groups, the participation of Kurds is of particular importance because of their critical role for the democratization process in Turkey. This process is directly and critically linked to Kurdish politics.

However, the presence and participation of Kurds during the two months of the Gezi protests all around Turkey has been a controversial subject since the early days of this pivotal collective action (Krajeski 2013). While the political and massive absence of the Kurdish freedom movement (KFM) has been noted, there was no denying that the Kurds, local pro-Kurdish political groups, and their leaders were visibly and actively present, despite political hesitation during the early days. Even more important is the frequent suggestion that the pro-Kurdish party HDP (People's Democratic Party) seems to have emerged as the true heir of the Gezi spirit with its subsequent reorganization (Bettoni 2015; Goyette 2015; Göksel and Tekdemir 2018). This reorganization extended the party's strategies of diversity and multi-vocality in forging alliances of dissent in Turkey.

The conversion of the previous Kurdish party BDP (Peace and Democracy Party) into HDP was a political decision that correlates with the aftermath of Gezi protests (Toktamış 2019). Yet little has been written on Kurds and Gezi and, in particular, on the relations between the HDP and Gezi (see Göksel and Tekdemir 2018). While the role of Kurds during the Gezi protests is commonly downplayed by the mainstream participants, such as Kemalists, as they failed to participate in expectedly large crowds, several important personalities from the KFM were present at Gezi, like BDP, starting with the critical role of Sırrı Süreya Önder at the beginning, and later presence of other party officials, such as Istanbul representative Sebahat Tuncel, within the first week, as well as a prominent ecology and Gezi activist, Professor Beyza Üstün, who was elected as an HDP representative in 2015. More importantly, the statement by the BDP co-chair Selahattin Demirtaş on June 6, [2013] illustrated the strategic reluctance and willing support of the KFM for Gezi. Clearly, Kurdish participation and presence at Gezi were heterogeneous, with multiple affiliations.

In the 2014 presidential elections and in the June 2015 legislative elections, HDP officials sought consistently to attract Gezi activists, evoking the resistance as a seminal moment in Turkish history (Göksel and Tekdemir 2018, 389–390). On the other hand, there was the drastic inability of the HDP to connect with the Istanbul electorate, illustrated by the meager 4.84 percent vote (420,000 votes) of their mayoral candidate, Önder, in the March 2014 local elections.

The KFM was a critical political actor that understood the significance of the Gezi protests, yet the historical circumstances of the so-called peace/resolution process hindered effective political mobilization. It displayed strategic reluctance in its inability to mobilize against a negotiation partner, the AKP, while agreeing with the larger democratic claims of the protesters (even those whose nationalistic sentiments were of deep concern). Tactical ambiguities ensued, as seen in its lack of centralized action and fragmented responses from the party officers. This implicitly cautious, yet visible, massive presence, created critical opportunities and obstacles for Kurdish politics.

This chapter, which intends to be an exploratory study, delves into the strategically reluctant, tactically ambiguous pathway that connects the Gezi protests and the emergence of the HDP as viewed by protagonists from the KFM and members of other political parties. We conducted interviews and reviewed testaments of several HDP officials, and some non-HDP activists, and reviewed the İmralı minutes (Öcalan 2015) to collate a narrative of HDP's emergence as an originally wavering, yet transformed political actor within the political opportunities created by the Gezi protests. Interviewees include two HDP members, one member of the CHP (Republican People's Party), who is a founding member of the Taksim Solidarity initiative, and one member of the Future Party. We do not disclose their names, given the political climate in Turkey. Interviews were conducted via Zoom or by telephone between the summer and fall 2020. We had originally intended to conduct further interviews. However, these plans were thwarted by the imprisonment of potential interviewees during the writing of this chapter.

The chapter is organized as follows. First, we address the Gezi protests. Next, we present the findings of our interviews and archival materials on the participation of Kurds. We then conclude the chapter.

THE GEZI PARK PROTESTS AND HDP: RENEGOTIATING NATION AND DEMOCRACY

The redevelopment plans for the area of Taksim Square approved in 2009 were one among multiple gentrification projects implemented in Istanbul during the past decades (see Rivas Alonso 2015). From November 2012, Gezi Park became one of the main sites of resistance against the redevelopment plans, as activists (mobilized by environmental and urban social movements such as Taksim Solidarity) started to guard the park and held common breakfasts every Sunday from January 2013, together with vigils every Saturday, from 3:00 p.m. to 6:00 p.m., in order to prevent construction (Sendika 2012).

Unlike previous environmental and urban struggles in Istanbul against gentrification projects, the Gezi protests became a landmark in Turkish politics and society.

A first explanation is related to the symbolic meaning of the space. As a site of struggles and demonstrations, Taksim Square, one of the most vibrant centers of Istanbul, has been a major locus of contention in Turkish history, where the meanings and the identity of the Republic, the socialists, student movements, and labor have been repeatedly (re)negotiated (see Walton 2015; Toktamış 2015). On May 28, 2013, police violence used to evacuate the park, occupied by around fifty youths, added new symbolism to Taksim and Gezi, in particular. Indeed, disproportionate violence used against protesters (dubbed as looters—*çapulcu*—by then prime minister Recep Tayyip Erdoğan) and the evocation of majoritarianism (Hürriyet Daily News 2013) by Erdoğan as a legitimation argument in favor of the redevelopment plans prompted massive protests that extended to seventy-nine out of the eighty-one cities in Turkey, mobilizing between 2.5 and 3 million people with the help of social media, as the traditional media censored news of the protests. This reaction launched the contentious processes in the subsequent weeks that have continued to shape Turkish politics and society since, with the unraveling of AKP's authoritarianism and the ongoing negotiations of the interconnected meanings of democracy and of Turkish identity.

For more than two weeks, Gezi Park turned into a commune space with collective libraries, kitchens, seminars, concerts, math, and yoga classes, interspersed with occasional clashes with the police, who were using gas canisters to attack unarmed civilians. Inspired by Gezi, similar collective initiatives of public forums emerged in several cities, becoming a locus of protest and negotiation on neighborhood redevelopment and democratic participation, marking a cultural resistance against the authoritarian forms of political activism in Turkey (David and Toktamış 2015). While the coalition of protesters at Gezi has been analyzed as being mainly composed of the urban, educated, working and middle classes (Özden and Bekmen 2015), participants hailed from a wide variety of societal groups and identities: Alevis, Islamists (including anti-capitalist Muslims and revolutionary Muslims), Kurds, feminists, Christians, LGBTI (Lesbian, Gay, Bisexual, Transgender and Intersex), Kemalists, nationalists, environmentalists, and football fans. Several studies have analyzed the sociological change at and through Gezi, as these previously antagonistic groups developed relationships of cooperation and mutual understanding (and, among some, prejudice reduction) while cohabiting the same space and discussing politics, society, and religion (Acar and Uluğ 2016; Göksel and Tekdemir 2018). Of particular importance was the bridging of the divide between Turks and Kurds, the symbolic event of which was the killing of Medeni Yıldırım in Lice by the Turkish military, during a demonstration against the construction of a new gendarmerie outpost. Some, if not all, Gezi activists immediately appropriated Yıldırım as a hero of the resistance against state violence, chanting slogans in Kurdish (Karakayalı and Yaka 2014, 126).

The shared experience of being subjects of state violence, irrespective of social, economic, ethnic, and religious affiliation, created the consciousness that all forms of oppression and injustice are connected, allowing them to overcome otherization (see Toprak 2009) and develop a shared sense of identity, in a process whereby minorities shed their "negative historicity" (Tambar 2016). This was enabled by the mechanisms and processes embodied in the practices[2] and infrastructures[3] of the protest (Karakayalı and Yaka 2014, 124–25). As violence unfolded at Gezi and beyond Istanbul, activists actively negotiated the meaning of democracy as dialogue (across cultural, social, political, and ethnic divides) and coalition-building (as opposed to majoritarian conceptions forwarded by Erdoğan), leading to a "recomposition of the political subjects" and creation of a new sense of people, beyond the binomial hegemonic nationalist and Islamic identities (Karakayalı and Yaka 2014). This process is linked to another transformation: the renegotiation of Turkishness as *Türkiyelilesme* (a conception of citizenship that moves away from the Republican conception based on the uniformity of ethnicity and religious origin).

It is precisely these renegotiated meanings of democracy and Turkishness that were consciously appropriated by HDP.[4] When the party was formed in 2012, it was little more than a collection of dispersed political parties and groups with a traditional left-wing ideology (equality, anti-capitalism, radical democracy) (Tekdemir 2016; Göksel and Tekdemir 2018; Celep 2018). On the other hand, since 1990, the pro-Kurdish parties,[5] including BDP, had never managed to surpass 6 percent of the votes, which was insufficient to pass the 10 percent threshold to enter the Turkish parliament. This changed after Gezi. The starting point was the integration of the BDP into HDP in 2014,[6] followed by the HDP candidacy of Selahattin Demirtaş to the presidential election of 2014, where he obtained 9.7 percent of the votes, launching the June 2015 legislative elections, where HDP gained 13.1 percent of the votes. The HDP party program and manifesto explicitly defend the concept of Türkiyelilesme, reaching out to the excluded identities of Turkey (ethnic, religious, economic, social—which are reflected in party candidates to elections) and articulates it with participatory democracy and environmentalism—the banners of Gezi (Tekdemir 2016; Grigoriadis 2015). The historic result of June 2015 prevented the AKP from obtaining an absolute majority, the first time since 2002. Since then, HDP has always passed the 10 percent threshold. Furthermore, HDP's implicit support for the mainstream opposition CHP (Republican People' Party) candidates in local elections in 2019 (HDP did not nominate mayoral candidates in the major Western cities) was fundamental for the victories of Ekrem Imamoglu in Istanbul and Mansur Yavas in Ankara. These results have transformed HDP into the pivotal opposition party in Turkey and, consequently, a fundamental actor in the democratization process. Had the HDP not embodied the diverse Gezi spirit, its struggles, reflections, renegotiations, and contributions, we argue, this recomposition of the opposition would not have happened.

KURDS AT GEZI?

The controversial connection between Kurdish politics and Gezi goes beyond the confinements of an ecology movement. There is no denying that the priority of Kurdish politics at the time of the Gezi protests was the ongoing so-called peace/resolution process; that is multi-layered negotiations centered around İmralı Prison where captive PKK leader Abdullah Öcalan was meeting with Kurdish politicians from BDP (Toktamış 2019). During our interviews and archival studies, we have established that Kurdish presence and participation at the Gezi protests were strategically reluctant because of the ongoing peace/resolution negotiations and, therefore tactically ambiguous, since Kurds were actually present at many protests as members of the grassroots groups and as selected political leadership. Yet, they avoided larger participation and visibility. It is evident, however, that HDP has been an inadvertent product of the political and cultural opportunities created by the Gezi protests due to the following: (a) the role of the BPD Istanbul branch; (b) Demirtaş's (albeit belated) endorsement of the Gezi protests, and the ensuing assessments of the KFM; (c) continuity between Gezi ecological groups and HDK[7]-HDP; and (d) the emergence of the idea of HDP as a party for all segments of people living in Turkey at the İmralı negotiations.

BDP'S HESITATION AND ITS ISTANBUL
OFFICE'S ACTIVE PARTICIPATION

BDP's response to the Gezi protests was hardly uniform and centralized, especially during the early days. At the time, there was the ongoing Peace-Reconciliation process between the governing AKP and the Kurdish political leadership that included clandestine negotiations with the PKK forces located in Qandil, the PKK leader imprisoned on the İmralı Island, and the BDP leadership. Unlike the BDP Central Office, the BDP Istanbul Office was an active partner in the Taksim Solidarity group that was created to oppose the urban redevelopment project, according to one of our interviewees from

HDP (interview 1). Sırrı Süreyya Önder, a member of the parliament from BDP (Istanbul Office), famously arrived at Taksim declaring himself as "the representative of the trees" of the park (Tariq 2018). His attempt to stop the bulldozers from demolishing the trees on the first day of the protests was widely publicized.

Interviewee 1 and other activists present at the protests regarded their activism as an extension of the already ongoing HDK (Halklarin Demokratik Kongresi, n.d. a) activism. Founded in 2011, the HDK is an alliance of multiple groups, organizations, and networks of civil rights and resistance in Turkey, some, but not all, of which are affiliated with the HDP. Since its foundation, ecological concerns, in addition to women's and workers' movements, constituted a formative component of their collective actions. In that sense, it was not surprising, albeit politically controversial, when a prominent member of the parliament, Önder, became a very early supporter of the protests and the target of brutal attacks by the police.

As explained by interviewee 1:

In spring 2013, for weeks we were busy having possible coalition and alliance meetings with different [smaller] political parties and civil society groups such as LGBT or People's Houses in our offices right by Taksim. When Gezi started, I called two other party members and told them that something is going on in Taksim. They were preoccupied with our HDK meetings and party affairs. I said "fuck the meetings, I shall go to Taksim." . . . Three days later, I remember saying this: "Whatever we are trying to establish here with HDK is already happening there at Gezi."

Even after Önder was hospitalized on May 31, after being hit by a gas canister, BDP party officials were expressing their distrust of the emerging protests. Such distance and the political critique of the events were clearly expressed by, among others, İdris Baluken, one of the party leaders in parliament who resolutely expressed his distrust. On June 3, he stated that "slogans and images that are reinforcing the status quo have become the dominant subjects," and "BDP would not stand side by side with these racist, nationalist, sexist, uniform and militaristic sections of the society" (Haberler.com 2013).

The hesitation and diverse response within the BDP were also observed by other organizers of Gezi protests. Our interviewee from CHP observed:

There were two groups within BDP at Gezi. On one corner, there were people whose only subject was rising Öcalan posters. The other group was trying to communicate or get together with other groups, carrying BDP signs. There were different people within the party. (Interview 3)

Our interviewee from the Future Party also spoke about the diverse responses coming from the BDP and Kurds in favorable terms:

> There are all kinds of Kurds; conservative, leftist, PKK-oriented, AKP-oriented. The Kurds at Gezi were young Kurds. Many were at Gezi because it was a freer public sphere—to enjoy, express themselves, to connect with others who were not like them. There were some politically oriented individuals, the majority of whom were direct or indirect followers of BDP. They were more prudent. Most were sympathizers of BDP. Kurds from Istanbul are less militant than [those] in Mardin; they have a more emancipated public sphere; they are freer in their life sphere, they are freer from BDP [central decision-making]. (Interview 4)

DEMIRTAŞ'S "BELATED" ENDORSEMENT AND PKK'S ACKNOWLEDGEMENT OF MISTAKES

During May 2013, Demirtaş was co-president of the BDP and one of the active negotiators of the peace/resolution process as he was regularly attending the meetings with the PKK leader Öcalan at the İmrali prison. PKK had declared a cease-fire on March 23, as requested by Öcalan and the immediate agenda of the KFM was the ongoing peace/resolution process. A month later, at a widely attended press conference in Qandil, Karayılan, a prominent PKK commander, announced that the PKK would begin to withdraw from Turkish territory by the first week of May. The general understanding among the researchers of Kurdish politics was that

> the Gezi events [. . .] spread rapidly throughout the country and put the Kurdish political movement in a difficult position. This was because the events targeted the AK Party government: the partner of the Kurdish resolution process. Therefore, the Kurds felt that participating in the Gezi events could have a negative impact on the resolution process. This was the reason behind the often-pronounced call: "where are the Kurds?" (Görmüş 2019)

However, by June 6, Demirtaş, following another visit to İmralı island, announced that "Öcalan was enthusiastically saluting the Gezi resistance," stating:

> I consider the resistance meaningful. [. . .] Most certainly, this position has created a new political rupture. Yet, no one should allow themselves to be manipulated by nationalist, putschist forces. The democratic, revolutionary, patriotic and progressive forces of Turkey should not allow this mobilization to be controlled by such politics. (Radikal 2013)

Demirtaş aptly explained BDP's position later in July, stressing the significance of the Gezi protests: "We always supported the democratic demands towards freedoms, but we were cautious about forces that were aiming to manipulate the mobilization towards a *coup d'état*" [against the elected government] (CNN Turk 2013).

While such endorsement was significant, the hesitation of the KFM was later in August explained by Cemil Bayık, a PKK commander in Qandil, as a mistake. According to Bayık, there were two considerations for this hesitation: one was that Kurdish participation would make the Turkish state more aggressive toward the protests, and second, the process led by the PKK leader Öcalan would be harmed. He stated that Gezi was a mobilization that opened up democratic political action. For that reason, it was an action that served the resolution process. Not participating, hesitating was a mistake. Bayık further specified that having posters of Atatürk and Öcalan side by side at Gezi was a positive development because "this is an indication that the Turkish society is expecting a resolution" (Hamsici 2013).

Our second interviewee from HDP explained this hesitation as follows: "The protests did not seem like an opportunity for the peace negotiations. Kurds could see that there were legitimate claims, but they could not fathom if these claims would actually support the ongoing process." The interviewee further highlighted the same dilemma addressed by Demirtaş on the sixth day of the protests: "If Kurdish participation were massive, the protests would be marred as PKK actions, which would end up being delegitimized by the government and also participants themselves. . . . There was no material Kurdish presence, but Kurds participated."

Interviewee 1 from HDP also expressed critique of this strategic reluctance:

> Understandably, Kurds were cautious about the overwhelming presence of the nationalist segments of the society. At times, older nationalists were asking us about the Kurdish youth with statements like "our youth do not know how to clash with the police, yours are more experienced, where are they?" There was a political will of not revealing their actual participation. Hence Kurds were not taken seriously, not seen as credible by the majority of the Gezi activists. This obviously culminated into a lost political opportunity.

The dynamic nature of the Kurdish participation in Gezi was also noted by the other activists. As the CHP interviewee explained:

> A few days later, they [BDP] realized that Gezi could not be stopped. The effects of Gezi were bigger than they imagined. Most BDP members then started to support us and they withdrew the initial declarations. This was very important because there are different aspects in Gezi: it was an environmental

movement, a humanitarian movement. It is a universal cause that needs the support of all political parties, from the Left, to Right, from all parts of Turkey. These are universal causes.

Our interviewee from the Future Party expressed a positive approach regarding the presence of the Kurdish activists at Gezi as follows:

> From a sociological point of view, [Gezi] was positive. From a political point of view, it was positive in the first ten days, but afterwards it became an agenda of the leftist radicals. BDP never behaved like radicals. They were more centre. BDP got sympathy from everyone. They did not misuse Gezi, like the radical Left.

CONTINUITY BETWEEN GEZI ECOLOGICAL GROUPS AND HDK-HDP

The relationship between the Gezi protests and the emergence of HDP as a political actor is best expressed by the close and active connections with the ecologists and the green movement. Two aspects of this relationship are the presence of the HDK ecologists in Taksim and later their political participation through HDP in national politics.

Both interviewees from HDP indicated that there was a tent with a BDP flag at the entrance of Gezi, and HDK and BDP elements were present at the ecology tent from early on. Interviewee 2 also indicated that one of the first tents that was burned down by the security forces belonged to the HDK ecologists.

Veteran ecologist and a professor of environmental engineering, Beyza Üstün, who has always been a staunch critic of the commercialization of natural resources, was at Gezi Park since the early days of the protests and later was elected as an HDP representative to parliament. According to Professor Üstün, the Gezi protests were "defending our common and collective habitat, as well as our bodies." She identified the collective action in and around Gezi as follows:

> When we observe the Gezi process, this is what we see: various struggles that have already been going on in the valleys are carried into the city, the struggles against urban redevelopment that were separately going on in different neighbourhoods have united. Furthermore, the struggle against the commercialization of health care, the student struggles against the commercialization of education and struggles for all other freedoms are expressed at Gezi as body politics. [. . .] The distinctive quality of the Gezi process was all struggles for freedom assembled together in the urban setting. This is how we need to describe Gezi. (Yesil Öfke 2013)

Üstün, who was interviewed several times during Gezi, at times immediately after the attacks of the security forces (Yesil Bülten 2013), argued that these protests were not only about protecting a certain number of trees in Taksim, but a culminating moment of all resistances that were going on throughout the country, such as against the commodification of the hydro-electrical infrastructure around the Black Sea region since the early twenty-first century (Sayan 2019; Yaka 2020), or local women's resistance against the privatization of agricultural commons or deforestation in Eastern Turkey (Göner and Rebello 2017; Shahpurwala 2019), or the community resistance against gold-mining activity in Western Turkey, Bergama, over several decades (Çoban 2007).

With her comprehensive contextualizing and critical approach to the ecological struggles in Turkey, Üstün soon drafted HDP's ecological program and was elected as one of the Istanbul representatives of the party (Yeşil Bülten 2015). She was imprisoned in September 2020 (Reuters 2020) and released in June 2021, as the case against HDP officials, including Prof. Üstun, is still pending at the time of writing.

In addition to Sırrı Süreyya Önder, journalist Ahmet Şık, and ecologist Beyza Üstün, seven other Gezi protesters became candidates for HDP during 2015 June and 2018 general elections, declaring a new Gezi-like collective action (ANF 2015). Among those, Gülsüm Ağaoğlu and Ahmet Şık were elected in 2018; Arife Çınar became a member of HDP's Central Executive Committee in 2016.

HDP AS AN EMERGING IDEA AT THE İMRALI NEGOTIATIONS DURING THE GEZI PROTESTS

It is hardly challenged that HDP was activated during Gezi at the İmralı negotiations where the larger strategic issues, such as disarmament, conflict in Syria, Kurdish rights as citizens, were discussed (Toktamış 2019). Kurdish politicians who were meeting almost once a month with Öcalan and functioning as interlocutors with the Kurdish population were officers of BDP, which was a party concentrating its efforts to mobilize electoral participation in regions where the majority of the population were Kurds.

However, during their June 2013 meeting, Öcalan indicated that transforming the ongoing HDK coalitions with diverse social and political groups into HDP was "the correct project," and they can "make good use of the post-Taksim winds by introducing a new party more in line with the spirit of the times." He assured the BDP officials that this would not make them lose their BDP base, which was a political one, yet urged them to unify diverse segments of the society to "stage the expected birth of the new party in the aftermath of Taksim, which will help them to overcome the 10% threshold as well" (Öcalan 2015, 95).

A month later, during their July meeting, he reiterated his desire to unify the forces of the Left in Turkey, in response to Demirtaş's question regarding entering the elections as BDP:

> I have been saying for over 20 years not to ignore the Turkish left. We also have responsibilities towards the Left. The legal elements should act together. Look at the way CHP is deceitfully working with Alevis. That is unacceptable. We need to be responsive. Your inability to develop new tactics was obvious during the Gezi events. I, myself, can achieve things with only five Turkish friends. . . . Now, we have thousands of Turkish friends, right? Maybe it is a bit delayed, but appropriate; [. . .] this can be done in the West with HDP. There are certain tactical efforts that are needed to be done immediately. When the parliament re-opens, there is a need to convert into a party of all Turkey. . . . Thus, all the representatives become representatives of all Turkey instead of only representing the Kurds. . . . You need to move rapidly, take striking actions, now that there are tons of intellectuals [around us]. (Öcalan 2015, 103)

Among other topics that were crucial during these negotiations, Öcalan continued making strategic and tactical suggestions for the reorganization of the Kurdish representation in parliament with an umbrella party unifying diverse segments in the entire country. In October 2013, three BDP representatives, Ertugrul Kürkçü, Sebahat Tuncel, and Önder (who were only able to be elected as independents), formed the first HDP group in parliament with the participation of independent Levent Tuzel. While Öcalan was adamant about uniting the leftist groups in Turkey, he also understood that the active Kurdish participation in such a union was only possible if Demirtaş, then the leader of BDP, was in the leadership (Öcalan 2015, 268). During the presidential elections of 2014, Demirtaş was the candidate of the HDP and all left opposition groups and parties. By 2015 June, HDP entered the legislative elections as the party uniting many social, political, and cultural groups in Turkey and managed to overcome the 10 percent threshold, becoming the third largest party in the Grand National Assembly of Turkey, preventing the AKP from maintaining its absolute majority (Yildiz 2015).

CONCLUSIONS: STRATEGIC RELUCTANCE AND TACTICAL AMBIGUITY AT GEZI AS AN OPPORTUNITY

The Gezi protests, triggered as a defense of an urban park, were indeed a pivotal culmination of the opposition against commercialization and commodification of habitat and against the increasingly authoritarian AKP regime.

Paradoxical as it may be, seeking for a resolution to further seal its political power by disarming the PKK, the AKP was the only available negotiation partner for the KFM (Toktamış 2018). Under the circumstances, the expectation of a peaceful resolution prevented different segments of the KFM from providing resolute and centralized support to the Gezi protests. At the same time, a massive mobilization of the Kurds at Gezi would alienate scores of Kemalist nationalists. Yet, as Demirtaş and often others, including some of our interviewees and analysis of archival transcripts, clearly expressed, the demands of the Gezi protesters concurred with the demands of the Kurds as citizens, as workers, as women, and in defense of their environment. Kurds were historically cautious and politically concerned yet not unwilling to participate in the protests. As a matter of fact, not only the political and cultural claims of Gezi, such as embracing diversity and inclusiveness, resisting authoritarianism, and commodification of the commons and collective habitat made their way to the HDP program, but also the very activists who themselves were the embodiment of these values and demands found a home in the party.

NOTES

1. We are dedicating this chapter to the Gezi activist and one-time HDP member of parliament Professor Beyza Üstün, and the other activists who were arrested in September 2020 for protesting against the Turkish government's implicit support for jihadist groups in Syria against the Kurdish resistance in Kobanî in 2014.

2. Examples of these practices are: resisting police violence (erecting barricades and devising strategies against tear gas learned from Kurds), cultural performances, small assemblies, abolition of money, free goods, free medical aid, exchange of knowledge, the protection of religious protesters by the nonreligious during prayers, iftar meals attended by the religious and nonreligious groups (see Acar and Uluğ 2015; Göksel and Tekdemir 2018).

3. Namely the small size of the park, enabling the development of a commune.

4. On HDP becoming a party of Turkey, instead of solely a pro-Kurdish party, see Celep 2018.

5. People's Labor Party (HEP), Freedom and Equality Party (ÖZEP), Freedom and Democracy Party (ÖZDEP), Democracy Party (DEP), People's Democracy Party (HADEP), Democratic People's Party (DEHAP), Democratic Society Party (DTP).

6. Another part of BDP formed the Democratic Regions Party (DBP).

7. People's Democratic Congress.

REFERENCES

Academics for Peace. 2020. "Free Peace Defenders in Turkey." *Barış İçin Akademisyenler.* https://barisicinakademisyenler.net/node/2109.

Acar, Yasemin Gülsüm, and Özden Melis Uluğ. 2016. "Examining Prejudice Reduction Through Solidarity and Togetherness Experiences Among Gezi Park Activists in Turkey." *Journal of Social and Political Psychology* 4, no. 1: 166–79.

ANF. 2015. "HDP'nin Gezici Adayları." *Nor Zartonk*, May 20. https://www.norzartonk. org/hdpnin-gezici-adaylari-7-haziran-yeni-bir-gezi-olacak/.

Bettoni, Dimitri. 2015. "Gezi and HDP: A Wedding Never Celebrated." *Osservatorio: Balcani e Caucaso, Transeuropa,* June 15. https://www.balcanicaucaso.org/eng/ zone/Turkey/Gezi-and-the-HDP-a-wedding-never-celebrated-162227.

Buğra, Ayşe, and Osman Savaşkan. 2014. *New Capitalism in Turkey. The Relationship between Politics, Religion and Business.* Cheltenham: Edward Elgar.

Celep, Ödül. 2018. "The Moderation of Turkey's Kurdish Left: The Peoples' Democratic Party (HDP)." *Turkish Studies* 19, no. 5: 723–47.

CNN Turk. 2013. "Demirtaş Interview: We Support the Demands of the Gezi Protests." *Ankara Güne Doğru.* July 30, 2013. Video. https://www.youtube.com/ watch?v=zJU0fFg3sUM.

Çoban, Aykut. 2004. "Community Based Ecological Resistance: The Bergama Movement in Turkey." *Environmental Politics* 13, no. 2: 438–60.

David, Isabel. 2016. "Strategic Democratisation? A Guide to Understanding AKP in Power." *Journal of Contemporary European Studies* 24, no. 4: 478–93.

David, Isabel, and Kumru F. Toktamış. 2015. *'Everywhere Taksim': Sowing the Seeds for a New Turkey at Gezi.* Amsterdam: Amsterdam University Press.

Esen, Berk, and Sebnem Gümüşçü. 2016. "Rising Competitive Authoritarianism in Turkey." *Third World Quarterly* 37, no. 9: 1581–1606.

Göksel, Oğuzhan, and Ömer Tekdemir. 2018. "Questioning the 'Immortal State': The Gezi Protests and the Short-Lived Human Security Moment in Turkey." *British Journal of Middle Eastern Studies* 45, no. 3: 376–93.

Göner, Özlem, and Joseph T. Rebello. 2017. "State Violence, Nature and Primitive Accumulation: Dispossession in Dersim." *Dialectical Anthropology* 41: 33–54.

Görmüş, Alper. 2019. *Efforts to Solve the Kurdish Question: The Standpoints of the Parties and the Opposition (2002–2019).* London: Democratic Progress Institute. https://www.democraticprogress.org/publications/research/assessment-report- by-alper-gormus-efforts-to-solve-the-kurdish-question-the-standpoints-of-the- parties-and-the-opposition-2002-2019/.

Goyette, Jarred. 2015. "Meet the Young Activists Who Upended Turkish Politics and Want a New Model for the Middle East." *Public Radio International,* June 11. https://www.pri.org/stories/2015-06-11/meet-young-activists-who-upended- turkish-politics-and-want-new-model-middle-east.

Grigoriadis, Ioannis N. 2016. "The Peoples' Democratic Party (HDP) and the 2015 Elections." *Turkish Studies* 17, no. 1: 39–46.

Haberler.com. 2013. "Taksim Gezi Parki'ndaki olaylar." 3 June. https://www. haberler.com/taksim-gezi-parki-ndaki-olaylar-4694678-haberi/.

Halkların Demokratik Kongresi. n.d. a. "HDK Diyor Ki." https://www.halklarin demokratikkongresi.net/hdk/hdk-diyor-ki/52.

Halkların Demokratik Kongresi. n.d. b. "Halkların Demokratik Kongresi 1. Kentsel Dönüsüm Çalistay Raporu." https://www.halklarindemokratikkongresi.net/halklarin-demokratik-kongresi-1-kentsel-donusum-calistay-raporu/644.

Hamsici, Mahmut. 2013. "Cemil Bayık: Gezide Yanlışlar Yaptık." *BBC Turkish Service.* August 29, 2013. https://www.bbc.com/turkce/haberler/2013/08/130828_cemil_bayik_3_gezi_cemaat.

Hemer, Oskar, and Hans-Ake Persson, eds. 2017. *In the Aftermath of Gezi: From Social Movement to Social Change?* London: Palgrave Macmillan.

Hürriyet Daily News. 2013. "Turkish PM Erdoğan Calls for "Immediate End" to Gezi Park Protests." June 7, 2013. http://www.hurriyetdailynews.com/turkish-pm-erdogan-calls-for-immediateend-to-gezi-park-protests-.aspx?PageID=238&NID=48381&NewsCatID=338.

İnsan Hakları Derneği. 2013. "Gezi Parkı Direnişi ve Sonrasında Yaşananlara İlişkin Değerlendirme Raporu." Ankara: İnsan Hakları Derneği.

Karakayalı, Serhat, and Özge Yaka. 2014. "The Spirit of Gezi: The Recomposition of Political Subjectivities in Turkey." *New Formations* 83: 117–38.

Krejeski, Jenna. 2013. "In Taksim Square Where are the Kurds?" *New Yorker,* June 11, 2013. https://www.newyorker.com/news/news-desk/in-taksim-square-where-are-the-kurds.

Öcalan, Abdullah. 2015. *Demokratik Kurtuluş ve Özgür Yaşamı İnşa (İmralı notları).* Neuss: Weşanen Mezopotamya.

Özden, Barı Alp, and Ahmet Bekmen. 2015. "Rebelling Against Neoliberal Populist Regimes." In *'Everywhere Taksim'. Sowing the Seeds for a New Turkey at Gezi,* edited by Isabel David and Kumru F. Toktamış, 89–103. Amsterdam: Amsterdam University Press.

Radikal. 2013. "Demirtaş: Öcalan Gezi Parkı Direnişini selamlıyor," June 7, 2013. http://www.radikal.com.tr/turkiye/Demirtaş-Öcalan-gezi-parki-direnisini-selamliyor-1136729/.

Reuters. 2020. "Turkey Orders 82 Arrests," September 25, 2020. https://www.reuters.com/article/turkey-security-kurds-int/turkey-orders-82-arrests-including-kurdish-opposition-members-over-2014-protests-idUSKCN26G0VW.

Rivas Alonso, Clara. 2015. "Gezi Park. A Revindication of Public Space." In *Everywhere Taksim': Sowing the Seeds for a New Turkey at Gezi,* edited by Isabel David and Kumru F. Toktamis, 231–48. Amsterdam: Amsterdam University Press.

Sayan, Ramazan Caner. 2019. "Exploring Place-Based Approaches and Energy Justice: Ecology, Social Movements and Hydro-Power in Turkey." *Energy Research & Social Science* 57.

Sayarı, Sabri. 2011. "Clientelism and Patronage in Turkish Politics and Society." In *The Post-Modern Abyss and the New Politics of Islam: Assabiyah Revisited,* edited by Faruk Birtek and Binnaz Toprak, 81–94. Istanbul: Istanbul Bilgi University Press.

Sendika. 2012. "Taksim nobeti gunluğu—Taksim Dayanışması." November 8, 2012. http://www.sendika.org/2012/11/taksim-nobeti-gunlugu-taksim-dayanismasi/.

Shahpurwala, Aiman. 2019. "Conflict, Narratives and Forest Fires in Eastern Turkey." LUP Student Papers.

Tambar, Kabir. 2016. "Brotherhood in dispossession: State Violence and the Ethics of Expectation in Turkey." *Cultural Anthropology* 31, no. 1: 30–55.

Tariq, Haniya. 2018. "It Started with the Trees." *Friday Times,* June 1, 2018. https://www.thefridaytimes.com/it-started-with-the-trees/.

Tekdemir, Ömer. 2016. "Conflict and Reconciliation Between Turks and Kurds: The HDP as an Agonistic Actor." *Southeast European and Black Sea Studies* 16, no. 4: 651–69.

Toktamış, Kumru F. 2015. "Evoking and Invoking Nationhood as Contentious Democratisation." In *'Everywhere Taksim'. Sowing the Seeds for a New Turkey at Gezi,* edited by Isabel David and Kumru F. Toktamış, 29–44. Amsterdam: Amsterdam University Press.

Toktamış, Kumru. 2018. "A Peace That Wasn't: Friends, Foes and Contentious (re-)Entrenchment of Kurdish Politics in Turkey." *Turkish Studies* 19, no. 5: 697–722.

Toktamış, Kumru. 2019. "(Im)possibility of Negotiating Peace: 2005–2015 Peace/Reconciliation Talks Between the Turkish Government and Kurdish Politicians." *Journal of Balkan and Near Eastern Studies* 29, no. 2: 286–303.

Toprak, Binnaz. 2009. *Being Different in Turkey: Religion, Conservatism and Otherization.* Istanbul: Open Society Foundation/Boğaziçi University.

Walton, Jeremy. 2015. "'Everyday I'm Chapulling!' Global flows and Local Frictions of Gezi." In *'Everywhere Taksim': Sowing the Seeds for a New Turkey at Gezi,* edited by Isabel David and Kumru F. Toktamis, 45–57. Amsterdam: Amsterdam University Press.

Yaka, Ozge. 2020. "Justice as Relationality: Socio-Ecological Justice in the Context of the Anti-Hydropower Movements in Turkey." *Journal of the Geographical Society of Berlin* 151, no. 2–3: 167–180.

Yeşil Bülten. 2013. "Gezi Parkı Direnişi 4. Gün: Interview with Beyza Üstün." June 10, 2013. https://www.youtube.com/watch?v=5kEIVAYDLAU.

Yeşil Bülten. 2015. "HDP'nin Ekolojisi: Interview with Beyza Üstün." April 21, 2015. Video. https://www.youtube.com/watch?v=qmWww7aKTXU.

Yesil Ofke. 2013. "Beyza Üstün: Bir Manifesto Gibi Söylesi." *Yesil Direnis,* September 26, 2013. https://www.yesildirenis.com/en/2020/09/26/prof-dr-beyza-ustun/.

Yildiz, Güney. 2015. "Turkey's HDP Challenges Erdoğan and Goes Mainstream." *BBC World News,* June 8, 2015. https://www.bbc.com/news/world-europe-33045124.

Chapter 13

Hasankeyf, the Ilısu Dam, and the Kurdish Movement in Turkey

Laurent Dissard

Once located on the shores of the Tigris River in the province of Batman in Southeastern Turkey, Hasankeyf managed to capture the world's attention before its flooding in early 2020 by the rising waters of the Ilısu Dam built some fifty kilometers downstream.[1] The small medieval town would feature regularly in the *New York Times* and *National Geographic*, oftentimes in these somewhat exaggerated terms: "a 12,000-year-old village . . . something out of a surreal fairy tale . . . a place with a palpable feeling of historical continuity and survival . . . one of [the country's] most mythic places" (Hansen 2018). There has been a general tendency to romanticize Hasankeyf in these journalistic accounts as this dreamlike, even spiritual place, stuck in a faraway Orientalist past. (Visiting the town the last time, I was told, was like saying goodbye to a loved one you will never see again.) If mentioned at all in these media reports, its residents were often portrayed as the innocent (Kurdish) victims of another state-funded (Turkish) infrastructural monstrosity, while their many profound disagreements about the dam itself usually remained unreported.

Despite all of the efforts by archaeologists to salvage its archaeology, the pressure of environmental activists and human-rights advocates to stop the Ilısu Dam, the support of celebrities like Turkish pop-star Tarkan or Nobel Prize laureate Orhan Pamuk, the countless number of protests and Global Action Days both locally and internationally, and the general agreement worldwide, except perhaps in Hasankeyf itself, that the town should be saved, Hasankeyf, in the end, was submerged by Turkey's latest, state-of-the-art engineering feat in the early months of 2020. The imminent destruction of its historical monuments and archaeological ruins, as well as the flooding of its surrounding natural landscapes, had sparked the initial media buzz. Hasankeyf, however, would soon become more than the sum of its cultural and natural heritage and quickly turn into a site of contestation for the environmental movement in Turkey, as well as a potent symbol of Kurdish resistance against the Turkish state.

> Regarding the dam construction, all of the historical and cultural heritage, especially those of Hasankeyf, has been protected with the utmost care. (Recep Tayyip Erdoğan during the inauguration of the Ilısu Dam and Hydroelectric Power Plant on May 19, 2020)

This chapter begins with Turkish president Recep Tayyip Erdoğan's inaugural speech at the Ilısu Dam on May 19, 2020, during which he lists all of the supposed benefits of the controversial mega-infrastructure and proclaims that everything has been done to preserve the heritage site of Hasankeyf. It then relates some of the successes, but also the ultimate failure, of the anti-dam initiatives to protect Hasankeyf in order to explain its relationship to the broader Kurdish movement in Turkey. Finally, the chapter examines the perspective from the people in Hasankeyf themselves, many of whom perceived the construction of the Ilısu Dam, as well as the attempts to organize against it, both of which were intended to help them supposedly, as unwanted interventions into local affairs from the outside.

"PROTECTED WITH THE MOST UTMOST CARE"

The Ilısu Dam and Hydroelectric Power Plant were officially inaugurated by Turkish president Recep Tayyip Erdoğan on May 19, 2020.[2] Turkey's fourth-largest dam is part of the Southeast Anatolia Project (*Güneydoğu Anadolu Projesi* or GAP), a development project implemented in 1984 to improve, according to its planners, the economic and social conditions of the country's impoverished, mostly Kurdish, southeastern provinces. During the opening ceremony of the Ilısu Dam, President Erdoğan declared that all historical and cultural heritage threatened by the rising waters of the recently built mega-infrastructure, including the medieval town of Hasankeyf, had been "protected with the utmost care."[3]

The opening ceremony took place inside the hydroelectric power plant itself. Two red mushroom-like buttons, ready to be pushed in order to ignite one of the eight Austrian ANDRITZ Hydro turbines, were placed for the occasion on top of a futuristic-looking table bracketed by two white podiums on a large aqua blue stage. A short promotional film projected in the background during the inauguration showed high-flying cinematic drone shots of the mega-infrastructure and its recently filled reservoir, as well as scenes of water gushing down abundantly from the dam's flood gates, high voltage switchyard and transmission lines sending electricity across the country, and boys and girls running through irrigated fields of pesticide-covered vegetables. The clip then ventured inside the belly of the beast with stylized shots of pristine cables and wires, close-ups of rapidly spinning turbines, and a quick peek inside the control room's state-of-the-art monitoring equipment. High-energy technology aestheticized to its fullest shown in perfect harmony with its surrounding environment. Nature, in other words, tamed for the good of the nation, or at least for its political and economic elite.

The event was aired live on most government-controlled news channels, including TRT HABER, as well as on President Erdoğan's Facebook page. Not present inside the power plant amid the COVID-19 outbreak, Erdoğan delivered the inaugural speech from his office in Ankara by videoconference with a small Turkish flag pinned on his pickle green, plaid suit and the official seal of the Presidency (a sixteen-pointed sun surrounded by sixteen five-pointed stars standing for the sixteen Great Turkic Empires as mythologized in pan-Turkish ideology) behind him. The ceremony provided Erdoğan with an occasion to rewrite the past by placing his Justice and Development Party's reign (the *Adalet ve Kalkınma Partisi* or AKP) in power since 2002 at the pinnacle of Turkish history. During the AKP's eighteen-year rule thus far, Erdoğan reminds his audience, Turkey has gone from 276 to 861 dams, from 97 to 778 power plants, from 202 to 587 irrigation ponds, and from 84 to 331 drinking water facilities and other facts. He announces during the same speech the future construction of one dam per month for a total of seventeen in the next year or so. "In energy and water, do you know what this means?" Erdoğan concludes, "Turkey can be counted as one of the top revolutionary countries in the world in renewable energy."

His AKP, Erdoğan claims, has done more since its ascension to power in 2002 than any other party since the founding of the Republic in 1923. This might sound like pumped-up electoral rhetoric, but it also indicates how Erdoğan places himself historically above Mustafa Kemal Atatürk, the founding father of the Turkish Republic. Elsewhere, Erdoğan reminds us all how his rule has not only boosted the country's democracy and economy, improved education, health, justice, safety, and transportation but also satisfied the needs of local farmers whose soils were hungry for the dam's water. Ilısu, he adds, will not only irrigate fields but these farmers' hearts as well. Elsewhere, he calls the newly built infrastructure a necklace adorning the Tigris River. It not only embellishes the river and its surroundings, according to Erdoğan, but will also blow a wind of peace, brotherhood, and wealth that will be felt in the region for centuries. Finally, he claims, the magnificence of the dam will crush terrorism. It is the best answer, he adds, to anyone in Turkey who complains about their country to foreigners.

In this inaugural speech, President Erdoğan not only delves into what Brian Larkin (2013) has called the politics and poetics of infrastructures but also lists some of the dam's more technical attributes (for example, it is 135 meters high and possesses a reservoir capacity of more than 10 billion m3 and a total established power of 1,200 MW). The Ilısu Dam, combined with the soon-to-be-built Cizre Dam further south near the Syrian-Iraqi-Turkish border, will have a total energy capacity of 1.1 billion KW/h and irrigate at least 765,000 hectares of land. According to the speech, it will make 2.8 billion Turkish liras every year. Much of the speech, in fact, consists in enumerating all of the dam's benefits in comparison to its cost. He states, for instance, how

> the cost of the Ilısu Dam, including the resettlement, the protection of historical and cultural heritage, the construction work and other spendings, reached a total of 18 billion liras. Regarding the dam construction, all of the historical and cultural heritage, especially those of Hasankeyf, has been protected with the utmost care. Just for the works at Hasankeyf, a total of 200 million liras was used.

This is a rare occurrence in the speech where Erdoğan refers to Hasankeyf flooded just a few months prior to the inauguration. Here, however, the medieval town represents just another cost for him. More generally, the protection of cultural heritage is equated to so many million Turkish liras, but how exactly this large sum was used is not detailed. For Hasankeyf, the amount itself (200 million liras) seemed to be proof enough that the town had indeed been saved. The sum is not negligible of course, but it is relatively small compared to the 2.8 billion liras the dam will bring each year. What is more important for Erdoğan, however, is how the money spent on the protection of cultural heritage allows him, in the end, to make the claim, true or false, that indeed Hasankeyf "has been protected with the utmost care."

HASANKEYF IS (NOT) KURDISH

Anti-dam activism to Save Hasankeyf first began in the late 1990s. If these different "Save Hasankeyf" initiatives were able to slow down the construction of the Ilısu Dam over the years, they ultimately failed to rescue the medieval town from its rising waters in early 2020. Led by different groups and individuals, who do not all openly declare themselves pro-Kurdish for fear of being labeled "terrorists" by Ankara, the "Save Hasankeyf" initiatives are nonetheless constitutive of a small part of what the broader Kurdish movement is in Turkey today.

One of the earliest initiatives to protect the town of Hasankeyf from the Ilısu Dam was the UK-based "Ilısu Dam Campaign" led by the Kurdish

Human Rights Project (KHRP), Corner House, Friends of the Earth UK, and the comedian and activist Mark Thomas. This first campaign was able to stop the construction of the dam in 2002 by pressuring banks and companies in the Swiss, German, Austrian, French, Italian, British, and Turkish consortium to withdraw from the project (Ronayne et al 2005, 67). This was only a temporary setback for the dam's proponents, however. In August 2006, a groundbreaking ceremony is held by then prime minister Recep Tayyip Erdoğan at Ilısu, as another Swiss, Austrian, German, and Turkish consortium reinitiated construction work. Meanwhile, opposition grew and the longest-running campaign against the Ilısu Dam, the "Initiative to Keep Hasankeyf Alive" (*Hasankeyf'i Yaşatma Girişimi* or HYG), was launched the same year. Over the next decade and more, its main activities would consist in carrying out fact-finding surveys and preparing monitoring reports, organizing local and international protests, as well as informing the most vulnerable people in the threatened areas about their rights.

Not alone, the HYG coordinated its efforts with other organizations such as "Hasankeyf Matters," a group of volunteers working for the protection of the medieval town, and the Istanbul-based environmental organization *Doğa Derneği*, which carried out its own "Save Hasankeyf" campaign between 2007 and 2013. Among other achievements, the latter was able to establish an information center in Hasankeyf as well as attract the support of Turkish singers Tarkan and Sezen Aksu, and renowned author Yaşar Kemal. Meanwhile, the Manfred-Hermsen Stiftung, a German foundation dedicated to nature conservation and environmental protection, also played a key role in the "Stop Ilısu—Save Hasankeyf" campaign begun in 2006. Their efforts brought the suspension of the Ilısu Dam again in June 2009 when most European banks withdrew their support after learning that no adequate Environmental Assessment Plan (EAP) existed for the dam. Another setback for Ankara, but again a brief one. Construction work was reinitiated in 2010 after the government called upon a new consortium to step in, an alliance now reduced to Turkish and Austrian construction firms and several Turkish banks (Halkbank, Akbank, Garantibank) still willing to finance the project.

Work is again interrupted, first in 2013 following a successful lawsuit filed by Turkey's Chamber of Architects and Engineers (Doğa Derneği 2013), and a last time in 2015 after the PKK kidnaps two head workmen from the construction site. These interruptions are again only momentary. In spite of protests organized after 2015 near Ilısu, as well as several Global Action Days for Hasankeyf coordinated across the world by a new umbrella group called the Mesopotamian Ecology Movement (Hunt 2019), the cause seemed lost at this point. The rekindling of the conflict between Turkey's military and pro-Kurdish armed groups after 2015 served as the opportunity Erdoğan needed to finish the dam. A state of emergency is declared in 2016, which allowed the army to increase its presence around Ilısu. Construction work could now continue uninterrupted as security forces encircled the dam rendering it no longer accessible.

Anti-dam activism did not stop then, nor did it stop after the complete inundation of Hasankeyf in early 2020. The HYG, for instance, is still active despite the fact that the town it set out to save no longer exists. It is still made up of a coalition of more than eighty-eight environmental organizations, human-rights groups, heritage specialists, professional associations, and trade unions, Kurdish and non-Kurdish alike, for the most part based in nearby Batman or Diyarbakır. At the international level, HYG is not afraid to label itself as a pro-Kurdish coalition. On its English website, for instance, it describes itself as an "environmental movement in Turkey's Kurdish provinces." (The use of these last three words—Turkey's Kurdish provinces—is enough to label someone pro-Kurdish in Turkey.) At the national level, however, HYG is more careful so as to not jeopardize its existence. Its Turkish website omits terms such as Kurds, Kurdish, or Kurdistan. If HYG's association to the broader Kurdish movement brings it more credibility, or perhaps more symbolic capital, in a foreign context, its members almost never display their sympathy to the Kurdish cause in Turkey for fear of being attacked or closed down.[4]

This is one of the many traits of the Kurdish movement in Turkey, which is composed of activists who, for many of them, need to hide their allegiance to the movement itself for fear of reprisals from the state. The environmental activists, human-rights organizers, and heritage specialists who participated in the different "Save Hasankeyf" initiatives, for example, distanced themselves from the Kurdish movement, while simultaneously giving it substance and meaning through their activism and political engagement. Hasankeyf is not "Kurdish" per se, unlike the city of Diyarbakır nearby, let us say, which is considered to be Kurdistan's cultural capital by many. Conquered, ruled, and inhabited by different Arabic, Turkish, and Kurdish tribes, Hasankeyf's population has spoken a combination of Arabic, Turkish, and Kurdish over time. More recently, and more significantly perhaps, residents in Hasankeyf have not voted in mass for pro-Kurdish parties. In 2019, for example, the local candidate from Erdoğan's AKP won the municipal elections with 60 percent of the votes over the 34 percent of the pro-Kurdish Peoples' Democratic Party (*Halkların Demokratik Partisi* or HDP).

So, why did Hasankeyf become a site of Kurdish contestation against the Turkish state, as well as a powerful symbol of resistance for the Kurdish movement itself? Always in search of consensus among its followers and popular support beyond Kurdistan, the Kurdish movement found in Hasankeyf an opportunity to raise several "Kurdish" issues (such as environmental degradation, human-rights abuse, military violence, and heritage destruction) without ever having to refer to them as solely "Kurdish."

One issue in particular strikes a chord in Kurdish activism. Hasankeyf has been the victim of forced evacuation long before the Ilısu Dam flooded its residents out of their homes.[5] Even though again Hasankeyf is not intrinsically Kurdish, the town nonetheless resonates in people's minds as another example of (Kurdish) villagers evicted from their homes by force. In the end, it operates as a powerful symbol for the Kurdish movement because it echoes the fate of so many other places forcefully evacuated by Turkey's military in its fight against Kurdish "terrorism."[6]

"YOU CAN GO TO JAIL FOR PICKING UP A STONE"

The "Save Hasankeyf" initiatives attracted international attention without ever gaining full support at the local level. Hasankeyf's residents, who were not for the construction of the dam, found it difficult to protest against it during marches for example. In the end, many people in Hasankeyf perceived both the dam construction, as well as its protests, as unwanted foreign intrusions into their lives.

No one in Hasankeyf ever seemed to know exactly when the dam would be finished. I often heard people tell me "in five or six years," which indicated that the state body, in this case the State Water Works (*Devlet Su İşleri* or DSI), responsible for the building and upkeeping of the country's dams, either did not know or kept the information to itself. One of my informants, a school teacher born in Hasankeyf who lived in the nearby city of Batman with his wife and two children, told me how his parents had been approached by representatives of the DSI (at least, this is what the parents understood them to be), who promised them that their new house would be like paradise (*cennet*) once the dam was built. The son was attached to his hometown, of course, but did not necessarily feel the need to "save" it from the Ilısu Dam he had heard about all of his life. He even admitted to me that people would be "relieved" when Hasankeyf finally disappeared because it would allow them to mourn a place that many felt was lost already. Just a few days after the visit of then prime minister Recep Tayyip Erdoğan to Ilısu for the groundbreaking ceremony of the dam in August 2006, I asked him what he thought of the Ilısu Dam and the protests organized against it:

> I don't want this dam, of course, nobody wants this dam. . . . The bones of my grandparents buried in their graves do not want this dam. . . . People come to protest. They want to save (*kurtarmak*) Hasankeyf, he told me.
>
> Even famous people come to Hasankeyf, I added.

He took a whiff from his cigarette and continued pensively.

We are grateful of course. I thank them of course . . . everyone is welcome here, but we have lost our town already.

He was not alone in expressing this. Hasankeyf was "lost" long before the dam because it had been left to deteriorate. Turkey's so-called modernization had taken its toll on historic Hasankeyf. There had been a highway built across the old town in 1964, the forced evacuation of people from their cave dwellings a few years later, the systematic use of concrete for any new constructions, and the armed conflict between the PKK and the military since 1984. The town's urban fabric and architectural heritage had been significantly damaged as a result. The irony was that Hasankeyf was a protected site supposedly. A first decision to preserve such historic towns in Turkey was taken by the High Commission for the Preservation of Cultural Entities and Monuments in 1978 (Ronayne 2005, Shoup 2006). Hasankeyf then became a first-grade archaeological site in 1982 and excavations are launched soon after to document and preserve its archaeology. On paper at least, Hasankeyf was protected. On the ground, however, it was simultaneously being destroyed. The irony behind the fact that it was the state responsible for both (protecting and destroying) did not escape my informant, who later told me half-jokingly: "You can go to jail for picking up a stone here, but there's nothing wrong with destroying the entire city with a dam."

- What about the protests? I continued. They were able to stop the dam before.
- Recep [referring to Prime Minister Erdoğan] came here, didn't he, so it will be built. . . . Look, it's a small place here. Everyone knows each other. Everyone meddles with everyone . . . "yap, yap, yap" [He makes a hand gesture imitating a mouth talking]. If you participate in the protests, people are going to say you are into this or that [referring to the PKK], but there is no *terör* here. . . . You came here, right? It is safe. Nothing bad happens. . . . There is nature, history. It is calm, clean.

He then said something I didn't need to write down in my notebook to remember.

- The AKP comes with buses, plays some music, and then leaves. . . . The protesters too, they play some music and leave. . . . Erdoğan doesn't like the protesters, they are terrorists for him. The protesters don't like Erdoğan, he is a dictator for them. . . . In the end, we will lose Hasankeyf and both of them will be wrong.

An over-simplified opposition between a good "Kurdish" Hasankeyf and a bad "Turkish" Ilısu Dam hides too much of the complex reality on the ground where things are never so clearly divided. My informant was not for the dam, of course, but he could not bring himself to protest against it either. According to him, this is what outsiders from Diyarbakır or Istanbul, or even Europe, did instead, but certainly not him. Speaking against it was possible, but taking part in a collective march, oftentimes organized by foreigners, seemed more risky. This did not mean that he swore allegiance to the AKP either. Instead, he was scared like many others of being associated with the PKK. Protesting meant putting your reputation, your family, and your life under unwanted scrutiny. As a result, the anguish of residents about to lose their home in Hasankeyf did not necessarily translate into active participation in the anti-dam campaign, which some even perceived as being masterminded by the PKK.[7]

CONCLUSION

A new town has been built a few kilometers away from the Tigris River with houses for some of the displaced families, a recently opened state-of-the-art Hasankeyf Museum, as well as the few medieval monuments relocated at the last minute from the rising waters. There, it would seem that old Hasankeyf had indeed been "protected with the utmost care" as President Erdoğan declared in his inaugural speech. The different "Save Hasankeyf" initiatives only managed to postpone the town's destruction for a few years. If not all of them were openly pro-Kurdish, they nonetheless constituted a small part of what the Kurdish movement is in Turkey today. Hasankeyf, which is not a "Kurdish" town per se, has nonetheless become a powerful symbol of resistance for the Kurdish movement in the last few decades. The place served especially as a strong reminder of all the (Kurdish) villages forcefully evacuated and destroyed by Turkey's armed forces during its war with the PKK.

Similar mega-dams and hydroelectric power plants on the tributaries of the Tigris and Euphrates, as well as on Turkey's many other rivers, have already inundated or will soon destroy other places of historical and cultural significance. Despite the lack of active participation from the majority of local people, anti-dam activism at Hasankeyf did manage to shine a spotlight on the city for a while at least. Unlike so many other desecrated town or village that most of us will never know about, Hasankeyf attracted international attention mainly through the efforts of anti-dam activists. The Kurdish movement as a whole also managed to seize the occasion that the flooding of Hasankeyf presented in terms of visibility in order to raise other pressing issues in the region. The now flooded town thus became a resource for activists (Kurdish or not) used to shine a light on human-rights abuse, military violence, economic poverty, and cultural and environmental destruction.

Finally, Hasankeyf has not just played a symbolic role in both the Kurdish and environmental movements in Turkey but also became a key testing ground for the Kurdish movement's recent turn to political ecology in the last two decades.

NOTES

1. Before its destruction, Hasankeyf comprised "a lower, modern town on the riverbank and an ancient, ruined citadel (kale) atop a high bluff . . . it feature[d] a 12th-century bridge that was once part of a major silk route to east Asia, the rare and beautiful 14th-century minaret of the Sultan Suleyman mosque, and, at its outskirts, the tomb of Zeynel Bey, a Selçuk-era monument with a striking blue tile exterior. The citadel of Hasankeyf [sat] atop a cliff ascended by a narrow stone stairway that passe[d] through an ornately carved 13th-century gate. The cliff top, about 600 m long and 250 m wide, [was] covered almost entirely with ruined buildings, including the 12th-century Great Mosque (Ulu Camii), a large medieval cemetery, a church, the palaces of the medieval Artukid and Eyyubid rulers, and a wide variety of domestic buildings dating from the Middle Ages to the early 20th century. . . . The site ha[d] a dramatic view of the valley of the Tigris and the hills around, dotted with tombs, caves, and shrines" (Shoup 2006, 243–4).

2. May 19th is celebrated in Turkey as the Commemoration of Atatürk, Youth and Sports Day (*Atatürk'ü Anma, Gençlik ve Spor Bayramı*), a national holiday commemorating Mustafa Kemal's landing at Samsun and the beginning of the Turkish War of Independence.

3. All quotes (translated by the author) in this first part are from Recep Tayyip Erdoğan's May 19, 2020, *Ilısu Barajı'nda Enerji Üretimi Başlama Töreninde Yaptıkları Konuşma* [Inaugural speech at the start of energy production at the Ilısu Dam], available online at https://www.tccb.gov.tr/konusmalar/353/120267/ilisu-baraji-nda-enerji-uretimi-baslama-toreninde-yaptiklari-konusma.

4. For example, the *Kurdî-Der Batman*, a research association on the Kurdish language based in Batman, which was accused of having ties with the PKK in 2013 (https://anfenglish.com/news/case-filed-for-closure-of-kurdi-der-8977), is the only member of the HYG that has the word "Kurdish" in its name.

5. "The traditional inhabitation within caves in the old fort was moved to the side of the highway down below by the decision of the governor's office in 1967"(Başgelen 2006, 115). Three years after the construction of the highway across the old town in 1964, this relocation would cause further damage to the town's urban layout.

6. See van Bruinessen (1995) and Jongerden (2010) for more on the military evacuations of Kurdish villages in Eastern Turkey during the 1990s.

7. The ongoing conflict between Turkey's armed forces and the PKK has brought Turkey's successive governments to associate too quickly and unfairly anti-dam activists with PKK supporters (Hommes et al. 2016, 14).

REFERENCES

Başgelen, Nezih. 2006. "Gün saymaya başlayan eşsiz bir dünya mirası: Hasankeyf ve Dicle Vadisi." *Arkeoloji ve Sanat* 17: 114–119.

van Bruinessen, Martin. 1995. "Forced Evacuations and Destruction of Villages in Dersim (Tunceli) and Western Bingol, Turkish Kurdistan." Netherlands Kurdistan Society.

Doğa Derneği. 2013. "Controversial Ilisu Dam on Hasankeyf Halted by Turkish Court." International Rivers website, January 10, 2013. https://archive.internati onalrivers.org/resources/controversial-ilisu-dam-on-hasankeyf-halted-by-turkish-court-7787.

Hansen, Suzy. 2018. "In Turkey, a Power Play will Leave Ancient Towns Underwater." *National Geographic* (November 2018). https://www.national geographic.com/magazine/2018/11/turkey-flooding-dams-displaced-antiquities-mesopotamia/.

Hommes, Lena, Rutgerd Boelens, and Harro Maat. 2016. "Contested Hydrosocial Territories and Disputed Water Governance: Struggles and Competing Claims over the Ilisu Dam Development in Southeastern Turkey." *Geoforum* 71: 9–20. https://doi.org/10.1016/j.geoforum.2016.02.015.

Hunt, Stephen E. 2019. "Prospects for Kurdish ecology initiatives in Syria and Turkey: Democratic Confederalism and Social Ecology." *Capitalism Nature Socialism* 30, no. 3: 7–26. https://doi.org/10.1080/10455752.2017.1413120.

Jongerden, Joost. 2010. "Village Evacuation and Reconstruction in Kurdistan (1993–2002)." *Études rurales* 186: 77–100.

Larkin, Brian. 2013. "The Politics and Poetics of Infrastructures." *Annual Review of Anthropology* 42, no. 1: 327–343. https://doi.org/10.1146/annurev-anthro-092412-155522.

Ronayne, Maggie, Rochelle Harris, and Kerim Yildiz. 2005. *The Cultural and Environmental Impact of Large Dams in Southeast Turkey: Fact-Finding Mission Report.* London: Kurdish Human Rights Project [and] Galway: National University of Ireland.

Shoup, Daniel. 2006. "Can Archaeology Build a Dam? Sites and Politics in Turkey's Southeast Anatolia Project." *Journal of Mediterranean Archaeology* 19, no. 2: 231–258.

Chapter 14

The Kurdish Ecology Movement and Human Rights

Marlene A. Payva Almonte and
Thomas James Phillips

Human rights have become an international lingua franca for diverse social movements around the world, and their broad appeal can strengthen social movements. In the last decade, efforts have been made to "green" them in recognition of the existential threat of climate change. This chapter takes a critical look at human rights with a view to understanding what role they might play in the long struggle for democratic confederalism, with a particular focus on its ecological dimension. We argue that juridical and rhetorical human rights might be useful in actual conjunctural struggles over, for example, the construction of dams and the right to water, whereby, human rights can help to frame Kurdish realities in judicial claims, which may also increase the prospects of success within courtrooms and beyond, as a rhetorical tool for the dissemination of the Kurdish ecology movement. However, human rights could come with significant risks to the strategy of spreading democratic confederalist ideas beyond their moorings in parts of Kurdistan. Having identified some of those risks, we reflect on how the Kurdish ecology movement can and does seize the language of human rights and rearticulate it from a revolutionary perspective, concluding that human rights might play an important role in the struggle. We begin by outlining some aspects of the relationship between ecological harm and the Kurdish question before briefly describing the Kurdish ecology movement's main theoretical positions.

THE RELATIONSHIP BETWEEN ECOLOGICAL
HARM AND THE KURDISH QUESTION

There is a two-way relationship between the oppression of the Kurds and ecological harm. The following is an attempt to partially sketch its outlines in relation to two particular human rights, namely, the right to water and the right of self-determination.

Starting with the contested right to water, it is well known that the region is nourished by the Tigris and Euphrates Rivers. The latter, to which Turkey supplies 89 percent of the total flow, sustains the food, water, and energy needs of 60 million people and is the main source of water for 27 million people across Turkey, Syria, and Iraq (Shamout and Lahn 2015). Tensions over the health of this shared source of water are not new; they date back to the 1960s when states began to initiate hydroprojects including the construction of dams (Bilgen 2020). Between 1972 and 2020, the river lost as much as 45 percent of its flow and water salinity reached double its natural level (Shamout and Lahn 2015, 19–20). The consequences for the peoples of the region are severe (Solomon and Pitel 2018). But hydropolitics does not tell the whole story. The consequences of climate change make the problem more acute with increasing temperatures, which lead to increasing evaporation and decreasing precipitation, which could result in a 30–40 percent drop in rainfall over the river basin (Shamout and Lahn 2015, 21).

The water crisis connects ecological damage with the Kurdish question on several levels beyond its direct impact. Turkey legitimizes construction projects, such as the hydroelectric Ilisu Dam, as a way of meeting the country's climate obligations under the Paris Agreement (Hockenhos 2019), but such projects serve the additional purposes of displacing Kurds, constraining the PKK, and rearticulating the Kurdish question in racialized terms of capitalist development (Turks) versus backwardness (Kurds) (Jongerden 2010). The most prominent illustration of this is the conversion of Hasankeyf into a drowned world and the dislocation of its inhabitants to other parts of Turkey, even as it is claimed that the project will be a developmental boon to the region.

Viewed from the other end of the relationship, Turkey's conversion of the Kurdish question into a security question, and its deployment of large-scale violence in order to deny the Kurds a share in the right of self-determination, has significant environmental consequences. Just as the effects of climate change, including drought, contribute to the outbreak of wars affecting the Kurds (Malm 2017), so the outbreak of war exacerbates the impacts of climate change. Gurses points out that during the war of the 1990s, Turkey embarked upon "a systematic practice of deforestation to undermine rebels' capacity to operate" (Gurses 2012). This widespread destruction of invaluable carbon sinks is one example among many of the linkage between the securitization of the Kurdish question and ecological damage.

The Kurdish movement's answer to this multifaceted doom-loop lies in social ecology. In the next section, we will outline some marked features of democratic confederalism and how it seeks to overcome ethnic tensions, the logic of capitalist development, and the extractivist ideology in order to secure an ecologically just future.

THE KURDISH ECOLOGY MOVEMENT

The Kurdish ecology movement consists of a bouquet of organizations and grassroots projects with a penumbra of ecological concerns (Egret and Anderson 2016, 157–182). Its theoretical cues come from the works of the imprisoned head of the PKK, Abdullah Öcalan. Like most Marxist accounts of ecological damage, Öcalan roots the destruction of the biosphere in social relations; more specifically in the prevailing capitalist social relations, which give rise to what Foster, Clark, and York call a "metabolic rift" (2010) in the dialectical relationship between human beings and nature. As Öcalan puts it: "The ecological crises that erupted during the short history of civilization are the result of its destructive profit-oriented essence" (Öcalan 2020, 301). The first marked feature of the Kurdish ecology movement is therefore its anti-capitalism.

But the movement takes its analysis of environmental justice further than the familiar critique of capitalism. The American social ecologist Murray Bookchin has been a significant source of intellectual nourishment for Öcalan. In his works on social ecology, Bookchin emphasizes not only the centrality of capitalism to ecological crises but also the importance of hierarchy per se: "the very *idea* of dominating nature stems from the domination of human by human" (Bookchin 2015, 31). For Bookchin, this calls for a fundamental reconstruction of society that tackles hierarchies head-on and that transcends the nation-state. The nation-state is a core focus because it is, in Öcalan's view, "the fundamental form of power" that facilitates "the most far-reaching conquest and colonization that the society has ever experienced" (Öcalan 2020, 208). This critique translates into the Mesopotamian Ecology Movement's declaration of a "struggle against state sovereignty" (Mesopotamian Ecology Movement 2016). The second marked feature of the Kurdish ecology movement is therefore its anti-nation-statism.

These marked features are articulated via a revolutionary project called democratic confederalism, which is based on three pillars: direct democracy, ecological sustainability, and gender equality (Hunt 2019). In Öcalan's terms, it consists of multilayered, pluralistic political structures; direct democracy via local assemblies based on a federative principle; and "eco-industrialism," which reunites producers with the means of production via communal owner-ship (rather than state control) in a way that is strictly informed by ecological principles (Öcalan 2020). In theory, it is "democracy without a state" based on a model of direct, bottom-up participation (Öcalan 2017, 39). In order to deal with the problem of majority tyranny, which is often said to arise from direct (or populist) democracy, it institutionalizes a "revolutionary-consocia-tional regime" (Miley 2018).

It is worth noting that we have used the label anti-nation-statism instead of simply anti-statism because, as Burç puts it:

> The critique is . . . not made against the state as such but moreover against conceptualisations, in which nation and state territory are conflated and conse-quently lead to a political expression of a hegemonic relationship between the dominant nation and its minority. (Burç 2020, 328)

But one should bear in mind that such a system, by its very existence, would "challenge the legitimacy of the state and statist forms of power" and thereby "evoke increasing resistance from national institutions," and, as Bookchin saw it, such movements would face two choices, namely to "be radicalised by this tension" or "sink into a morass of compromises that absorb it back into the social order that it once sought to change" (Bookchin 2015, 18). One might therefore think of democratic confederalism as a form of dual-power (Burç 2020, 330) that would, if successfully implemented, hollow out the state over time and thereby present a challenge to states, even if the subject of its critique is nation-states.

Scholars have argued that democratic confederalism differs from the Marxist-Leninist inspired national liberation movements of the twentieth cen-tury insofar as there has been a shift from the language of collective national liberation to "a universal human rights paradigm" (Gerber and Brincat, 2018, 202). Indeed, Öcalan invokes a theory of human rights in his expostulation of democratic confederalism, namely, the theory of the inseparability of individual and collective rights (Öcalan 2012, 31). But to what extent can human rights be invoked to further a revolutionary project that is ecologi-cal, anti-capitalist, and anti-nation-state? It is to that question that the next subsection turns.

HUMAN RIGHTS: TACTICS AND STRATEGY

Rosa Luxemburg recognized that one cannot "contrapose the social revolu-tion . . . to social reforms," rather the latter offers the means of working in the direction of the former (Luxemburg 2006, 3). Extending this insight into the realm of legal struggle, Knox draws a useful distinction between tactics and strategy: the former concerns a revolutionary movement's conjunctural battles and the latter concerns its revolutionary objective. Tactical legal interventions, Knox reminds us, have strategic consequences, and unless one takes the fundamentals of the status quo for granted, it is necessary to pay close attention to them (Knox 2010, 193). When thinking about the invocation of human rights in the long struggle for democratic confederalism, one should therefore bear in mind that such rights might be tactically useful but strategically risky. In that spirit, we will consider how human rights might help in a conjunctural battle around the right to water and dam construction, before thinking about the possible strategic pitfalls of invoking human rights. The chapter will end by reflecting on how the Kurdish ecology movement can and does rearticulate human rights from a revolutionary perspective to avoid some of those pitfalls.

HUMAN RIGHTS TACTICS: THE RIGHT TO WATER

Despite the importance of water for human and nonhuman life and its interconnectedness with the ability to enjoy other human rights, the right to water has only recently started to be recognized and is still contested (Salman and McInerney-Lankford 2004). As McCaffrey notes, "the notion of a right to water emerged largely from non-binding documents adopted within the context of the United Nations, and . . . there does not yet appear to be a consensus among states on the existence of the right" (McCaffrey 2016, 232). In fact, there are no references to the "right to water" in the International Bill of Human Rights and main human rights instruments. Only the Convention on the Elimination of All Forms of Discrimination Against Women and the Convention on the Rights of the Child include references to the importance of access to water, "[but] neither expressly calls access to water a human right" (McCaffrey, 226). It could be said, however, that early in the twenty-first century, the "right to water" started its way toward recognition. General Comment 15 of the UN Committee on Economic, Social, and Cultural Rights (CESCR) acknowledges the right to water as "a prerequisite for the realization of other human rights." It establishes that "[the] human right to water entitles everyone to sufficient, safe, acceptable, physically accessible and affordable water for personal and domestic uses" and links the right to water with other rights. As CESCR General Comment 6 on the economic, social, and cultural rights of older persons previously recognized, General Comment 15 deems the right to water as "contained" in the right to an "adequate standard of living" (Article 11 International Covenant on Economic, Social and

Cultural Rights), and "indispensable" for its realization, "particularly since it is one of the most fundamental conditions for survival." In the same vein, General Comment 15 establishes that

> the right to water is also inextricably related to the right to the highest attainable standard of health . . . and the rights to adequate housing and adequate food. . . . The right should also be seen in conjunction with other rights enshrined in the International Bill of Human Rights, foremost amongst them the right to life and human dignity. (United Nations Committee on Economic, Social, and Cultural Rights 2003, 2)

Indeed, General Comment 15 acknowledges the interpretation of the International Covenant on Economic, Social, and Cultural Rights to cover the right to water and connect it with other rights, however, its absolute recognition can still be considered in progress. In fact, the challenges of the recognition of a right to water were acknowledged in General Comment 15 back in 2003 and, even today, despite its "[acknowledgement] in United Nations documents, [it] cannot yet be taken for granted" (McCaffrey 2016, 227). Yet, as Singh points out, "[while] water is yet to be explicitly recognised as an independent self-standing human right in international treaties, . . . international human rights law already entails specific obligations related to access to safe drinking water" (Singh 2019, 5). The right to water can, therefore, still be invoked in the context of Kurdish struggles related to access to water sources.

These struggles are exacerbated by climate change, which poses serious risks to the ability to access sufficient, safe, acceptable, physically accessible, and affordable water. The construction of the Ilisu Dam by Turkish authorities has major human rights implications in the southeast of the country as well as for Kurds in Iraq and Syria. It puts at risk a range of human rights such as, inter alia, the right to life, water, health, food, and an adequate standard of living as its construction directly impacts not only their private and family life, households, and means of income but also, for the risk of flood it poses, or even its potential misuse as a "weapon" (Marvar 2019). In fact, rights connected to the cultural heritage of Kurds and other populations affected by the flood of the ancient city of Hasankeyf, today submerged under water as a result of the opening of the dam, are also violated (Aykan 2018). While tensions around large energy projects are common worldwide, the long-lasting and multilayered nature of the tensions between countries co-beneficiaries of the Tigris and Euphrates' waters make the Kurds' position resulting from the dam construction particularly complex. Hence, it is important to note that water hegemony conflicts in the area, although a "central element of diverse conflicts" (Conde 2014) are not the only aspect or cause of them.

The human rights framework can be used as political rhetoric or in the judiciary to articulate Kurdish peoples' realities implying rights violations, particularly (for the purposes of this chapter) in relation to tensions with Turkey resulting from the Ilisu Dam construction. Given that Turkey has ratified the European Convention on Human Rights (ECHR) in 1954, lawsuits can be brought to the European Court of Human Rights (ECtHR), whose judgments have binding effect. However, this option is limited to Kurds in Turkey, even though the effects of dam construction are felt downstream by Iraqis and Syrians (Kurdish or otherwise). Nevertheless, beyond potential juridical gains, the ECtHR jurisprudence can provide guidance on the possibilities of invoking the right to water in connection with other rights, either to bring cases in other jurisdictions or to dress Kurds' claims with human rights rhetoric. The former might be valuable in purely legal terms as a way of mobilizing judicial decision-making in tactically useful directions. The latter could forge a connection point between Kurdish claims and other social movements via the shared language of human rights.

It should however be borne in mind that, while the ECtHR jurisprudence related to the right to water can provide some positive insights,[1] at the same time, recent developments on the matter, even if "unsuccessful," do provide, at least to some extent, grounds for auspicious prospects if human rights arguments are raised to address Kurds' concerns through litigation. In *Hudorovic and Others v. Slovenia*, touching upon the right to clean water and sanitation in relation to "vulnerable and disadvantaged members of society," the court takes a debatable approach to the application of Article 8 of the ECHR (right to respect for private and family life). Among other things, the court found "that access to safe drinking water is not, as such, a right protected by Article 8, in spite of the fact that without water the human person cannot survive." Probably having this in mind, the court established that in cases of "persistent and long-standing lack of access to safe drinking water," Article 8 would indeed be affected and, therefore, "when these stringent conditions are fulfilled . . . it may trigger the State's positive obligations under that provision." The court's view in this case hardly helps to back-up water rights claims of minorities and subaltern groups (David 2020). Yet, the court's finding provides room for the future development of case law seeking to identify the states' obligations related to the rights to water under Article 8 of the ECHR. Certainly, the court also found that the "[existence] of any such positive obligation and its eventual content are necessarily determined by the specific circumstances of the persons affected, but also by the legal framework as well as by the economic and social situation of the State in question." Accordingly, despite judicial defeats—including recent attempts before the ECtHR[2] "to link cultural heritage and human rights" (Aykan 2018) aimed to avoid Hasankeyf heritage loss—juridical and rhetorical human rights can provide a framework to articulate Kurdish claims in different forums at national and international levels. Besides, the increasing recognition of the linkages between human rights and climate change can potentially trigger an expansion of human rights in the future (Payva Almonte 2020). Developments in climate litigation can provide innovative rights-based arguments in which to ground Kurds' claims embedding, for example, the ecological ethos of democratic confederalism. Therefore, despite its incipient stage, the ability of the human rights legal framework to articulate the impacts upon rights resulting from climate change, which can translate into some progress in Kurdish "short-term" (judicial) struggles, should not be dismissed. Certainly, "losing cases" can also help to "garner media and public attention that elevate the political discussions on climate change" (Peel and Osofsky 2018, 67). Overall, as Blake argues, "despite its shortcomings, human rights law still constitutes one of the most effective available tools to curtail state dominance in the administration and implementation of international law because it is the only area of international law in which states have accepted fully binding obligations and allowed international regulation to interfere between the state and its citizens" (Blake 2011). Therefore, Kurds' conjunctural battles—around the right to water or otherwise—should still benefit democratic confederalist ideals—particularly those grounded on ecological concerns—from including considerations from a human rights perspective, particularly considering possible tensions arising from the recent construction of the Ilisu Dam, and the overall climate crisis.

HUMAN RIGHTS STRATEGY: POTENTIAL PITFALLS

Articulating Kurdish claims via human rights risks conceptually distorting some core features of democratic confederalism, namely, its ecological principles, its anti-nation-statism, and its anti-capitalism. Human rights have the power to bring about this distortion, because as well as promising to reduce suffering, they are productive of certain kinds of political subjectivities and political possibilities (Brown 2004).

First, human rights lack an ecological dimension, understood in Bookchin's terms as a concern with "the dynamic balance of nature, with the interdependence of living and nonliving things" (Bookchin 2005, 20). Insofar as human rights can be said to produce a certain view of society and nature, they have been described as the "most putatively anthropocentric of all legal orders of rights" (Grear 2011, 34). As presently constituted, they reflect a neoliberal notion of the environment whereby its value lies in its capacity to serve human needs and over which humans have hierarchical dominance (Morrow 2017). As Gearty puts it, a lack of interest in the nonhuman world is "built into the philosophy of human rights," which therefore provides governments with easy reasons for engaging in ecologically harmful activities (Gearty 2010, 9).

Second, human rights have a schizophrenic relationship with the state. Human rights seek to restrain the beast of state sovereignty in the name of human rights and, at the same time, rely upon the state as the omnipresent guarantor of those rights. As Kennedy puts it, human rights "[structure] liberation as a relationship between an individual right holder and the state" and thereby "equate the structure of the state with the structure of freedom" (Kennedy 2005, chapter 1). For Rajagopal, this "simply reproduces the same structures that have prevented the realization of those rights in the first place" (Rajagopal 2003). Human rights ideologically legitimize states as neutral entities without any kind of class content, which serves to undercut the kind of critique made by the Kurdish ecology movement. It is therefore difficult to make sense of human rights (at least in their orthodox understanding) from the perspective of a "democracy without a state" (Öcalan 2017, 39) because the very "structure of freedom" is missing.

Third, human rights discourses condemn certain forms of violence, such as torture, and approve others, such as the violence of the market (Rajagopal 2003, 196), which means that they do not tackle the root causes of ecological crises in capitalist social relations. In failing to forge the crucial link between capitalism and human rights violations, human rights can install a set of blinders on social movements, concealing from sight the structural violence of global capitalism. As Marks puts it, human rights can thereby "domesticate potentially radical demands on the social structure and bring with it the demobilization of oppositional activity" (Marks 2011, 77). In a slightly different vein, Whyte argues that hegemonic conceptions of human rights have provided the individualist moral framework conducive to the growth of neoliberalism (Whyte 2019). In the context of climate change, the UN Special Rapporteur on extreme poverty and human rights has tapped into this critique by sharply upbraiding international human rights organizations for their failure to recognize the need for "deep social and economic transformation" and their subsequent "deep denial of the real gravity of the situation" (UN General Assembly 2019).

These three strategic pitfalls add up to what Knox describes as "a key promise of human rights law," namely "that it is possible—without fundamental change—to reach a situation in which human rights are respected" (Knox 2020, 17). Human rights, in their institutionalized vernacular, encourage their users to buy into this key promise. The Kurdish ecology movement encourages the precise opposite. Relying upon the human rights framework in tactical struggles and contributing to its hegemony, therefore, comes with significant dangers in terms of crowding out the viability of the overall strategy.

There is, however, no getting away from the fact that human rights have become an international lingua franca for social movements and a way of framing their aims and demands in a way that is more likely to garner international support. In the next subsection, we will argue that the Kurdish ecology movement can and does rearticulate human rights from within the concrete struggle for democratic confederalism, using the right of self-determination as our example.

REARTICULATING THE RIGHT OF
SELF-DETERMINATION

The right of self-determination has, to an extent, been greened. It is widely recognized, for example, that rising sea levels put at risk the self-determination of peoples from small island states in the Pacific Ocean (Neroni Slade 2007). By analogy, one might argue that drought and desertification have similar effects for certain peoples of the Middle East. Indeed, scientific studies indicate that climate change could render parts of the region uninhabitable by the end of the century (Lelieveld et. al. 2016). But while acknowledging the grave symptoms of climate change, the right of self-determination—for all its undoubted importance—does not get at the roots of the problem. First, as Koskenniemi argues, it "acts as a justification of a State-centred international order" and during periods of political transformation works to "reconstitute the political normality of statehood" (Koskenniemi 1994). Indeed, the right was the legal vehicle through which the European nation-state form was instituted in former colonies, whose original intent was to use it as a stepping-stone to remake a deeply unequal international order (Getachew 2019). Second, insofar as the right has an internal dimension beyond statehood, it appears to be concerned with a right to thin forms of democracy (Franck 1992), and possibly liberal multiculturalism. This thin version of democracy combines a wide franchise with quite passive forms of citizenship (Wood 2016, 204–237), and while valuable and worth defending, it can typically be managed by global capitalism (Marks 2003). Third, common Article 1 of the two international human rights covenants, which sets out the right, provides that peoples may "freely dispose of their natural wealth and resources" subject to a set of broad qualifications that protect the rights of transnational corporations. Although this helps to ground the right to water in international law (among other useful purposes) (Gilbert 2013), it also reflects the extractivist ideology.

In contradistinction, the Kurdish ecology movement (and the KFM more broadly) conceptualizes self-determination in a way that eschews the nation-state and thin democracy in favor of "self-government on the basis of active citizenship and connectivity" (Jongerden 2017) and in a way that severs the link between national freedom and statehood (Miley 2018). The movement also replaces the extractivist developmental aspect of self-determination with the construction of "a functioning economy of eco-communities" where property belongs not to the extractivist state (to be disposed of for the benefit of peoples) or to corporations but to the communities that use it according to ecological principles (Öcalan 2020, 254–255). Self-determination on this rendering is not about buying into the ideology of institutionalized human rights and trying to ride it as far as one can rather it is about challenging the "political normality of statehood" (at least as statehood is commonly understood), challenging thin forms of democracy, and challenging the extractivist ideology on the terrain of human rights.

The Kurdish ecology movement's reconceptualization of self-determination demonstrates an important point, namely that social movements, as O'Connell puts it, "routinely engage in a sort of emancipatory or critical multilingualism" (O'Connell 2018a, 27) by mobilizing human rights claims alongside other, more radical languages. Human rights are contingent rather than set-in-stone, and social movements can develop what others have called "constitutive strategies" to "normatively reconfigure human rights" and enhance their emancipatory potential (Dehm, Golder and Whyte, 2020). The gist of our argument concerning the potential value of human rights to the Kurdish ecology movement is captured by Ed Sparer, who wrote:

> Various kinds of legal rights and entitlements may be used in a manner that helps to develop social movement, which in turn leads to expanded opportunities for a more humane society, or they may be used to help frustrate social movement by legitimating existing relationships. The meaning of a right or entitlement depends upon the way in which it intertwines with social movement. (Sparer, 1984)

While being wary of the pitfalls identified above, it is by embedding the language of human rights in more emancipatory projects, like democratic confederalism, that we might begin to put them to more meaningful use.

CONCLUSION

We have argued that in deploying human rights in conjunctural battles, movements for revolutionary change ought to pay close attention to the effects that such arguments could have on their long-term strategy. Having identified the potential tactical value of human rights relating to the right to water, we selected three marked features of the Kurdish ecology movement's approach to ecology, namely its ecological stance, its anti-capitalism, and its anti-nation-statism. We argued that institutionalized human rights could shape political subjects and open up political possibilities that diverge from these marked features and might crowd out the viability of the movement's overall objective. On the other hand, we recognized that the institutionalized, dominant narrative of human rights is not the only possible one. The movement can, and does, situate human rights discourse within its concrete struggle for democratic confederalism and rearticulates it from that perspective. This is to stress "the centrality of social struggle in shaping the concrete meaning of rights in specific contexts" (O'Connell 2018b).

Just as human rights can shape political subjects, political subjects can shape human rights. Given the existential threat of climate change, this is a challenge that needs to be met—and the Kurdish ecology movement is lighting a path.

NOTES

1. For example, in *Zander v. Sweden*, on water and property rights, the court was of the view that the ability to use drinking water within ones' property "was one facet" of the applicants' right of property, considered a "civil right" under Article 6.1 ECHR. *Zander v. Sweden*, Application No. 14282/88, ECHR judgment November 25, 1993, para. 27.

2. See, for example, *Ahunbay and Others v. Turkey*, Application No. 6080/06, ECHR judgment January 29, 2019 (communicated February 21, 2019); and *Ahunbay and Others v. Turkey, Austria and Germany*, Application No. 6080/06, ECHR judgment June 21, 2016.

BIBLIOGRAPHY

Aykan, Bahar. 2018. "Saving Hasankeyf: Limits and Possibilities of International Human Rights Law." *International Journal of Cultural Property* 25, no. 1: 11–34.

Bilgen, Arda. 2020. "Turkey's Southeastern Anatolia Project (GAP): A Qualitative Review of the Literature." *British Journal of Middle Eastern Studies* 47, no. 4: 652–71.

Blake, Janet. 2011. "Taking a Human Rights Approach to Cultural Heritage Protection." *Heritage & Society* 4, no. 2: 199–238. https://doi.org/10.1179/hso .2011.4.2.199.

Bookchin, Murray. 2005. *The Ecology of Freedom—The Emergence and Dissolution of Hierarchy*. Oakland: AK Press.

Bookchin, Murray. 2015. *The Next Revolution: Popular Assemblies & the Promise of Direct Democracy*. London: Verso.

Brown, Wendy. 2004. "The Most We Can Hope For . . . Human Rights and the Politics of Fatalism." *The South Atlantic Quarterly* 103, nos. 2–3: 451–63.

Burç, Rosa. 2020. "Non-Territorial Autonomy and Gender Equality: The Case of the Autonomous Administration of North and East Syria—Rojava." *Philosophy and Society* 31, no. 3: 277–448.

Conde, Gilberto. 2014. "El Agua entre Turquía, Siria e Irak: Barómetro de Conflictos?" *Regions & Cohesion* 4, no. 2: 81–100.

David, Valeska. 2020. "The Court's First Ruling on Roma's Access to Safe Water and Sanitation in Hudorovic et al. v. Slovenia: Reasons for Hope and Worry." *Strasbourg Observers,* April 9, 2020. https://strasbourgobservers.com/2020/04/09/the-courts-first-ruling-on-romas-access-to-safe-water-and-sanitation-in-hudorovic-et-al-v-slovenia-reasons-for-hope-and-worry/.

Dehm, Julia, Ben Golder and Jessica Whyte. 2020. "Introduction: 'Redistributive Human Rights?' Symposium" *London Review of International Law* 8: 225–32. https://doi-org.liverpool.idm.oclc.org/10.1093/lril/lraa018.

Egret, Eliza, and Tom Anderson. 2016. *Struggles for Autonomy in Kurdistan & Corporate Complicity in the Repression of Social Movements in Rojava and Bakur.* London: Corporate Watch.

Foster, John Bellamy, Brett Clark, and Richard York. 2010. *The Ecological Rift: Capitalism's War on the Earth.* New York: Monthly Review.

Franck, Thomas. 1992. "The Emerging Right to Democratic Governance." *The American Journal of International Law* 86, no. 1: 46–91. https://doi.org/10.2307/2203138.

Gearty, Conor. 2010. "Do Human Rights Help or Hinder Environmental Protection?" *Journal of Human Rights and the Environment* 1, no. 1: 7–22.

Gerber, Damian, and Shannon Brincat. 2018. "When Öcalan met Bookchin: The Kurdish Freedom Movement and the Political Theory of Democratic Confederalism." *Geopolitics*: 1–25. https://doi.org/10.1080/14650045.2018.1508016.

Getachew, Adom. 2019 *Worldmaking After Empire: The Rise and Fall of Self-Determination.* Princeton, NJ: Princeton University Press.

Gilbert, Jérémie. 2013. "The Right to Freely Dispose of Natural Resources: Utopia or Forgotten Right." *Netherlands Quarterly of Human Rights* 31, no. 3: 314–41.

Grear, Anna. 2011. "The Vulnerable Living Order: Human Rights and the Environment in a Critical and Philosophical Perspective." *Journal of Human Rights and the Environment* 2, no. 1: 23–44.

Gurses, Mehmet. 2012. "Environmental Consequences of Civil War: Evidence from the Kurdish Conflict in Turkey." *Civil Wars* 14, no. 2: 254–71. https://doi.org/10.1080/13698249.2012.679495.

Hockenhos, Paul. 2019. "Turkey's Dam-Building Spree Continues, at Steep Ecological Cost." *Yale Environment 360* website. October 3, 2019. https://e360.yale.edu/features/turkeys-dam-building-spree-continues-at-steep-ecological-cost#:~:text=Alamy-,Turkey's%20Dam%2DBuilding%20Spree%20Continues%2C%20At%20Steep%20Ecological%20Cost,completes%20the%20massive%20Ilisu%20Dam.

Hunt, Stephen E. 2019. "Prospects for Kurdish Ecology Initiatives in Syria and Turkey: Democratic Confederalism and Social Ecology." *Capitalism Nature Socialism* 30, no. 3: 7–26. https://doi.org/10.1080/10455752.2017.1413120.

Jongerden, Joost. 2010. "Dams and Politics in Turkey: Utilizing Water, Developing Conflict." *Middle East Policy* 17, no. 1: 137–43. https://doi.org/10.1111/j.1475 -4967.2010.00432.x.

Jongerden, Joost. 2017. "The Kurdistan Workers' Party (PKK): Radical Democracy and the Right to Self-Determination Beyond the Nation-State." In *The Kurdish Question Revisited* edited by Gareth Stansfield and Mohammed Shareef, 245–58. London: Hurst.

Kennedy, David. 2005. *The Dark Sides of Virtue: Reassessing International Humanitarianism*. Princeton, NJ: Princeton University Press.

Knox, Robert. 2010. "Strategy and Tactics." *Finnish Yearbook of International Law* 21: 193–229.

Knox, Robert. 2020. "A Marxist approach to *R.M.T. v the United Kingdom*" in *Research Methods for International Human Rights Law: Beyond the Traditional Paradigm* edited by Damian Gonzalez-Salzberg and Loveday Hodson, 13–41. London: Routledge.

Koskenniemi, Martti. 1994. "National Self-Determination Today: Problems of Legal Theory and Practice." *The International and Comparative Law Quarterly* 43, no. 2: 241–69. https://doi.org/10.1093/iclqaj/43.2.241.

Lelieveld, J., Y. Proestos, P. Hadjinicolaou, M. Tanarhte, E. Tyrlis, and G. Zittis, 2016. "Strongly Increasing Heat Extremes in the Middle East and North Africa (MENA) in the 21st Century." *Climatic Change* 137: 245–60.

Luxemburg, Rosa. 2006. *Reform or Revolution?* New York: Dover.

Malm, Andreas. 2017. "Revolution in a Warming World: Lessons from the Russian to the Syrian Revolutions." *Socialist Register* 53: 120–42.

Marks, Susan. 2003. *The Riddle of All Constitutions: International Law, Democracy, and the Critique of Ideology*. Oxford: Oxford University Press.

Marks, Susan. 2011. "Human Rights and Root Causes." *Modern Law Review* 74, no. 1: 57–78. https://doi.org/10.1111/j.1468-2230.2010.00836.x.

Marvar, Alexandra. 2019. "Turkey's Other Weapon Against the Kurds: Water." *The Nation*, November 11, 2019. https://www.thenation.com/article/archive/turkey-syria-iraq-kurds/.

McCaffrey, Stephen. 2016. "The Human Right to Water: A False Promise." *University of the Pacific Law Review* 47: 221–32.

Mesopotamian Ecology Movement. 2016. "The Aims of the Mesopotamian Ecology Movement." *New Compass* website. May 9, 2016. http://new-compass.net/articles /mesopotamian-ecology-movement-presents-its-aims.

Miley, Thomas Jeffrey. 2018. "The Perils and Promises of Self-Determination: From Kurdistan to Catalonia." In *Your Freedom and Mine: Abdullah Öcalan and the Kurdish Question in Erdogan's Turkey*, edited by Thomas Jeffrey Miley and Federico Venturini, 360–67. London: Black Rose.

Morrow, Karen. 2017. "Of Human Responsibility: Considering the Human/ Environment Relationship and Ecosystems in the Anthropocene." In *Environmental*

Law and Governance for the Anthropocene, edited by Louis J. Kotze, 269–88. Oxford: Hart.

Neroni Slade, Tuiloma. 2007. "Climate Change: The Human Rights Implications for Small Island Developing States." *Environmental Policy and Law* 37, nos. 2–3: 216–23.

Öcalan, Abdullah. 2012. *Prison Writings III: The Road Map to Negotiations.* International Initiative.

Öcalan, Abdullah. 2017. *The Political Thought of Abdullah Öcalan.* London: Pluto Press.

Öcalan, Abdullah. 2020. *The Sociology of Freedom.* Oakland: PM Press.

O'Connell, Paul. 2018a. "Human rights: Contesting the Displacement Thesis." *Northern Ireland Legal Quarterly* 69, no. 1: 19–35.

O'Connell, Paul. 2018b. "On the Human Rights Question." *Human Rights Quarterly* 40, no. 4: 962–88. https://muse.jhu.edu/article/708643.

Payva Almonte, Marlene. 2020. "Human Rights in the Context of Climate Change—Time for Evolution?" PhD Diss., University of Liverpool.

Peel, Jacqueline, and Hari Osofsky. 2018. "A Rights Turn in Climate Change Litigation?" *Transnational Environmental Law* 7, no. 1: 37–67.

Pouikli, Kleoniki. 2020. "From a Beacon of Hope to a Question Mark: Right to Clean Water and Sanitation across Europe in the Wake of the ECtHR Judgment Hudorovic and Others v. Slovenia." *Journal for European Environmental and Planning Law* 17, no. 3: 351–65. https://doi.org/10.1163/18760104-01703007.

Rajagopal, Balakrishnan. 2003. *International Law from Below: Development, Social Movements, and Third World Resistance.* Cambridge: Cambridge University Press.

Salman, M. A. Salman, and Siobhán McInerney-Lankford., 2004. *The Human Right to Water: Legal and Policy Dimensions.* Law, Justice, and Development series. Washington D.C.: World Bank.

Shamout, M. Nouar with Glada Lahn. 2015. "The Euphrates in Crisis: Channels of Cooperation for a Threatened River." *Chatham House* website. https://www.chatham house.org/sites/default/files/field/field_document/20150413Euphrates_0.pdf.

Singh, Nandita. 2019. *The Human Right to Water: From Concept to Reality.* Stockholm: Springer.

Solomon, Erika, and Laura Pitel. 2018. "Why Water is a Growing Faultline between Turkey and Iraq." *Financial Times*, July 4, 2018.

Sparer, Ed. 1984. "Fundamental Human Rights, Legal Entitlements, and the Social Struggle: A Friendly Critique of the Critical Legal Studies Movement." *Stanford Law Review* 36: 509–74.

United Nations Committee on Economic, Social and Cultural Rights. 1995. *General Comment No. 6: The Economic, Social and Cultural Rights of Older Persons.* https ://tbinternet.ohchr.org/_layouts/15/treatybodyexternal/Download.aspx?symbolno =INT%2fCESCR%2fGEC%2f6429&Lang=en.

United Nations Committee on Economic, Social and Cultural Rights. 2003. *General Comment No. 15: The Right to Water (Arts. 11 and 12 of the Covenant).* https://tb internet.ohchr.org/_layouts/15/treatybodyexternal/Download.aspx?symbolno=E% 2fC.12%2f2002%2f11&Lang=en.

United Nations General Assembly. 2019. *Report of the Special Rapporteur on Extreme Poverty and Human Rights.* https://digitallibrary.un.org/record/3810720? ln=en.

Whyte, Jessica. 2019. *The Morals of the Market: Human Rights and the Rise of Neoliberalism.* London: Verso.

Wood, Ellen Meiksins. 2016. *Democracy against Capitalism: Renewing Historical Materialism.* London: Verso.

Chapter 15

The Internationalist Project to Make Rojava Green Again

Stephen E. Hunt

Make Rojava Green Again (MRGA) was founded in January 2018, as awareness of the revolution's ecological dimension increased. Its name, of course, combines a mocking détournement of Donald Trump's nationalistic brand of exclusionist and anti-environmentalist populism, with a reconstructive vision of ecological transformation. As such, it is a microcosm that illustrates several of this collection's themes. Located at the Internationalist Academy in Cizîrê canton, MRGA is a key campaign of the Internationalist Commune of Rojava, aiming to provide practical mutual aid and to promote the aspiration to embed ecological solidarity within the paradigm of democratic confederalism. MRGA is a nexus that endeavors to connect several physical and cultural spaces: geographically spanning movement activists on the ground in northeastern Syria and the international solidarity community; intellectually and strategically linking the ecological pillar and other aspects of the Kurdish freedom movement's program; administratively it networks between international volunteers and relevant organizational bodies of AANES, such as the Ecology Committee of the Cizîrê Canton and the Committee for Natural Conservation; culturally it reaches out to the local Kurdish community, particularly to ecologically committed members of the youth movement, such as Rojava Fridays for a Future.

MRGA's practical projects consist of three areas of contribution: first, ecological restoration through creating tree nurseries, planting for reforestation—the Hayaka nature reserve is a priority project (Bance 2019)—and cleaning up rubbish in natural habitats; second, contributing toward agricultural production in rural cooperatives and working on community gardens and parks, and, finally, knowledge-sharing and awareness-raising. In this way, the group can help efforts to diversify food production away from monocultures (see Make Rojava Green Again 2019a) to extend food sovereignty and to enhance the quality of the soil, vegetation, and watercourses. Locally, the project supports the educational facility of the Internationalist Academy, helping to integrate ecological awareness within, and deepen the understanding of, the Kurdish movement's paradigm. In addition to participation in outreach events and social media, the main promotional project for MRGA's work has been the creation of an illustrated book, outlining the group's philosophy and projects, which has been translated into ten languages, now accompanied by an online reader (both accessible online).[1] With its foreword and endorsement by Murray Bookchin's daughter, Debbie Bookchin, to date, this is the most prominent publication about the social ecological aspect of the Rojava Revolution for an international readership.

Unfortunately, significant threats hedge MRGA's promise. There have been several incidents whereby state authorities have targeted individual Kurdish solidarity activists as they have returned from northeastern Syria. In the KRI, the police held Swiss MRGA volunteer Hawar Goran in custody for several days ("Make Rojava Green Again announces release of its Swiss member," *ANF News.* October 7, 2020), while Danish authorities confiscated the passport of another activist, Anne Dalum, to prevent her from traveling to volunteer for the project (Make Rojava Green Again 2019b). State repression has also resulted in the German government's ban on the Mezopotamya publishing house, leading to the seizure and confiscation of many pro-Kurdish movement titles, including 200 copies of the MRGA book (Make Rojava Green Again 2019c).

The small MRGA initiative, likely to be in double figures in terms of activist numbers, is at once vulnerable to repression, yet at the same time powerful in its potential to represent the KFM in an advanced form to the international community. MRGA's positive quantitative contribution to tree planting seems to have been a modest 2,000 trees per year over the short period since the group's launch. The challenge is to continue and endure, undertaking planting schemes from the production of seedlings to planting out and maintaining the trees to maturity, when there are formidable external factors that make its situation precarious. The most immediate threat is military aggression from Turkish armed forces and their proxy militias and mercenaries. MRGA's durability will also be dependent on its capability and capacity to reach out and respond beyond a dedicated but narrow demographic of young people from the ecological and anti-fascist direct-action movement, to draw in a wider cultural commitment to this project and to attract practical mutual aid in terms of material support and agroecological expertise. In long-term, peace-time circumstances, practical volunteer groups can have a substantial impact. The Scottish volunteer tree planting group Trees for Life,[2] for example, has planted two million trees to regenerate the remnants of the Caledonian Forest in the Scottish Highlands. The "Green Braids" initiative, an ambitious project at the University of Rojava, aims to plant four million trees within five years, a staggeringly bold objective, only achievable it is conceded if embedded in "societal culture" (Abdo 2021). Nevertheless, the great value of initiatives like MRGA is to root a political ecology project in real-world community restoration tasks, helping to encourage ecological regeneration and horticultural production. Other Western ecological projects for northeastern Syria include Slow Food International's "Gardens in Rojava" (Lomax 2020) and Kobane Roots (Associazione Ya Basta Bologna 2018). Such initiatives demonstrate a qualitative shift in mentality toward a green paradigm.

It is hoped that MRGA can act as a practical and intellectual exemplar that, if successful, could be scaled-up to a more quantitative impact in Kurdish areas and beyond, one based on expanding diverse, resilient spaces for food production and natural habitats counter to the commodification of the living world. Internationalist volunteer Viyan Qereçox reflects that small, solutions-focused projects have a vital role as spaces to create and pilot prototypes, providing opportunities for experiential learning that can ideally be translated to similar situations elsewhere ("Democratic Confederalism: Learning from Grassroots Democracy," Scottish Solidarity with Kurdistan webinar, May 19, 2020). While funding is urgently needed, it is also necessary to avoid capital-intensive approaches, nurturing such developments with a spirit of ingenuity and improvisation appropriate to local circumstances. Although aspects of MRGA have been on hold due to a deteriorating security situation and the impact of the coronavirus pandemic of 2020 onward, the emergence of the group is a significant one for the KFM's ecological dimension, and it is hoped that it will enjoy future flourishing.

NOTES

1. https://makerojavagreenagain.org/.
2. https://treesforlife.org.uk/.

REFERENCES

Abdo, Hogir. 2021. "Green Braids" a Campaign to Plant Four Million Trees in Syria's Northeast." *North Press Agency* website. February 23, 2021. https://npasyria.com/en/55005/.

Associazione Ya Basta Bologna. 2018. "Parte il progetto Kobane Roots! Sostieni un uliveto a Kobane." Associazione Ya Basta Bologna website. https://www.

yabastabologna.com/single-post/2018/11/28/Parte-il-progetto-Kobane-Roots-Sostieni-un-uliveto-a-Kobane.

Bance, Pierre. 2019. "Making Rojava Green Again: Annotations to the Booklet 'Make Rojava Green Again' by the Internationalist Commune of Rojava." *OpenDemocracy* website. June 21, 2019. https://www.opendemocracy.net/en/north-africa-west-asia/making-rojava-green-again/.

Lomax, Marcus P. 2020. "Permaculture in the Autonomous Administration of North and East Syria: Ecology as a Pillar in the Construction of an Alternative Political Society in Rojava." Masters dissertation, Leiden University. https://hdl.handle.net/1887/133483.

Make Rojava Green Again. 2019a. "Reber Apo is a Permaculturalist—Permaculture and Political Transformation in North East Syria." *Make Rojava Green Again* website. July 29, 2019. https://makerojavagreenagain.org/2019/07/29/reber-apo-is-a-permaculturalist-permaculture-and-political-transformation-in-north-east-syria/.

Make Rojava Green Again. 2019b. "Press release: Repression Against Ecological Campaign 'Make Rojava Green Again.'" *Make Rojava Green Again* website. April 18, 2019. https://makerojavagreenagain.org/2019/04/18/press-release-repression-against-ecological-campaign-make-rojava-green-again/.

Make Rojava Green Again. 2019c. "German State Confiscates "Make Rojava Green Again" Books." *Make Rojava Green Again* website. https://makerojavagreenagain.org/2019/03/07/german-state-confiscates-make-rojava-green-again-books.

Part IV

NATURE PROTECTION AND KURDISH ALEVISM

Chapter 16

Dersim as a Sacred Land

Contemporary Kurdish Alevi Ethno-Politics and Environmental Struggle

Ahmet Kerim Gültekin

A[1] mother uttered the following mournful and heartfelt words at the funeral of her young son, after he was killed as an TKP-ML TIKKO guerrilla, while on active service in the mountains of Dersim in 2020:

> He is my honor and pride. . . . I am a follower of his ideas, a supporter of his comrades. . . . They took my life. . . . They picked the flower of the Munzur. They killed the rose of the Munzur. . . . I am sure his comrades will continue fighting. . . . I entrusted them to the Munzur. (see Yenidemokrasi7 2020)[2]

This is a striking example of how Dersim (the sacred land) is being imagined and represented as a nonhuman entity, a powerful symbol that strengthens its roots in every one of such violent occasions.

In the last few decades, there has been a surge of academic interest in the contested issue of Kurdish Alevi ethnic identity among both Turkish and Western researchers (recently, for instance, see Gezik and Gültekin 2019; Çetin, Jenkins, and Aydın 2020). Intriguingly, this emerged at the crossroads of two separate, highly politicized, as well as continuously traumatized ethnic (which depends on spoken language) and religious ethno-political movements in Turkey, namely Kurdishness and Alevism (Gültekin 2019, 5). Although Kurdish Alevi identity has always been considered according to Turkish or Kurdish nationalist approaches, during the 1990s a vigorous challenge has emerged within a new socio-cultural sphere. That local awakening was mostly driven by a motivation to keep a clear distance from Kurdish nationalism, the Alevi movement, and the socialist movements alike. Predominantly intellectuals of Dersim (former members of various socialist fractions—see also Çetin, Jenkins, and Aydın 2020, 14), shaped a new way of understanding Kurdish Alevi identity (Gültekin 2020, 75–107). Seemingly, however, that awakening continues to relate to Kurdish politics, especially in the context of the HDP's rise since the 2010s. Nevertheless, tensions that have emerged between some leftist factions and the HDP during municipal elections suggest that essential differences remain (Orhan 2017, 201–14; Gültekin 2016, 175–91).

Within such rivalry, the emergence of new actors opens up a societal space for fresh political discourses. For example, when the customary political discourses of both the socialist organizations and the PKK lose their influence over Kurdish Alevis, a new political sphere appears in which different causes are raised, such as anti-dam protests, and other environmental issues which are also closely related to Kurdish Alevi belief and wider society. A new construction project of the state approved in 2019, for example, targets one of the most important pilgrimage sites, which is a large spring, a site known for miracles, where the sacred Munzur River is born. Currently, all Kurdish Alevi organizations as well as leftist movements are actively struggling against the project and sharing the same political discourse which effectively combines nature and the *Raa Haqi* religion (Duvarenglish 2020).

As awareness of Kurdish Alevism was raised—especially beliefs, rituals, and discourses relating to the natural world of Dersim[3]—independent observers more often noted this unique ethnic identity. During the early 2000s, this trend occurred alongside the growing environmental movements, which were becoming Turkey's dominant social movements (Aksu, Erensü, and Evren. 2016; Agar and Böhm 2018, 1228–49). However, as the tides of politics in Turkey change rapidly, such movements—especially those closely related to ethno-politics—have lost their early momentum. Nevertheless, defending nature, considered as a sacred being, against external threats, still forms part of the cultural identity in Dersim.

Kurdish Alevi culture has evolved a singular identity due to its unique and troubled history. The resilience of Kurdish Alevis, who have survived despite forced evacuations, massacres, and the devastation of their landscape, has partly been attributed to their deep-seated nature-based beliefs. A "sacred" and "impregnable" landscape full of mystic powers, as well as ancestors' souls, and a deep feeling of belongingness has been regarded as the secret power of Kurdish Alevism (Gültekin 2019, 10). The uniqueness seems to be found in mythological narratives regarding the landscape and its rebellious inhabitants throughout history. Such chronicles undoubtedly create a mental and psychological continuum in the present day when they continue to be fed by clashes between the guerrillas (not only the fighters of the PKK but also the members of several socialist groups) and the Turkish armed forces in the mountains of Dersim. Thousands of martyrs who fell on the soil of Dersim, and those who were killed on active service in the ranks of various socialist groups throughout Turkey and brought back to be buried in their land, raised the commitment of Kurdish Alevis to Dersim as a sacred land. That strengthens the perception of an endless cycle of a sacred war to defend the local population, which has always been defined by the notions of "to protect" and "being protected" by Dersim.

Additionally, as *Raa Haqi*[4] belief is spatially conceived in the form of a *jiare* (a sacred place or mystical-natural objects), the nature of Dersim becomes a sacred living-being. Such a perception also draws strength from beliefs regarding ancestors and martyrs. Every Kurdish Alevi individual— either a member of an *Ocak* (sacred lineage), or a *talip* (followers of Ocaks)— who lived and died in Dersim becomes a part of the same sacredness. They become a religious and personal binding element for living descendants. On the other hand, the endless—mythic or actual—cycles of struggles create martyrs. Every massacred civilian, marked and blessed as sacred by virtue of his or her innocence, contributes toward the understanding of the land according to religious and ethical levels of meaning. This heritage also shapes a unique connection to current political manifestos, predominantly related to Kurdish nationalism or some versions of socialism.

In keeping with this introduction, this chapter will be framed around the process of the reinvention of contemporary Kurdish Alevi ethnic identity, which seeks a consistent definition by drawing upon re-formulated specific geography, history, language, and religion. The chapter aims to explore the rising influence of nature-symbolism within the Raa Haqi religion, as a useful ethno-political example. First, for a better understanding of some emic terms, a brief introduction to the belief system will be given. Later, the chapter will evaluate the connections between environmental politics and natural aspects of the belief system. Last, the chapter will solidify the arguments through a recent example from Dersim.

JIAR U DIYAR (THE SACRED LAND)

Kurdish Alevis have long been considered as a part of Alevi[5] communities in Turkey, due to some similarity in the socioreligious organizations and theological discourses they share with other Alevi groups. They are distinguished, however, by their enclosed tribal organizations, inherited caste-like socioreligious positions, and a unique cosmology, transmitted orally over centuries. Hence some anthropologists have even argued that Kurdish Alevism constitutes an independent religion (Güler 2019, 45–113; Gezik 2014, 19–45).

The Raa Haqi belief has been practiced by Kurmanci as well as predominantly Kırmancki-speaking Alevi communities for centuries. It is rooted in enclosed tribal networks, which emerge through real and fictional kinship patterns, and can be understood through two underpinning aspects of the belief system. First is the *ikrar* (the sacred oath), a set of mandatory, sacralized social relations and institutions (Deniz 2019, 42–75). The ikrar legitimates and regulates socioreligious relationships between individuals and tribes.[6] Ikrar-based institutions, together with their religious and cultural heritage, preside over everyday social life. That caste-like implementation of the socioreligious order is also a distinctive feature of the Kurdish Alevi community (Gültekin 2019).

Second, the prominence of beliefs and rituals regarding jiares, which does not require any formal religious guidance or authority, allows more space for individual piety. Rituals at jiares rely on worshipping nature-based (living or nonliving) objects such as trees, forests, mountains, rocks, caves, rivers, lakes, fountains, fire, soil, wild animals, or the sun and moon. In everyday life, religious needs are supplied through jiare-based worship practices. While the ikrar-based organizations cover the social ritual sphere, the jiare cults give more independence to individuality in the sense of religious experience. That intriguingly creates a huge space for interpretation of new political discourses related to nature and religion. As Dersim is considered as a sacred land, the current ethno-politics—which has at times referenced ecology—easily conceptualizes a new nature-based religion (Gültekin 2019, 17–23).

There is a fascinating mythology concerning jiares. The collective memory is inherited and transmitted by a rich oral culture. Accordingly, two parallel universes exist: *Zahir*, the actual world where people live, and *Batın*, inhabited by deific creatures, sacred animals, mystic souls, or semi-deific individuals, such as prophets, *pirs*, or dervishes. Kurdish Alevis neither believe in Abrahamic heaven nor hell. Instead, they believe in metempsychosis. They consider the jiares as connection points between the two worlds. So, they spatially create and solidify sacredness as immanent. Each jiare has its deific characters. Lighting candles, sharing *niyaz* (a kind of bread), and sacrificing domestic animals (which is increasingly questioned by new generations) are the basic rituals (Gültekin, 2020; Çakmak 2013; Gezik and Çakmak 2010).

Jiares can be categorized by their natural characteristics (Gültekin 2020, 130–43; Çakmak 2013, 171–76; Gezik and Çakmak 2010, 97–98). This is essential to an understanding of how mythological narratives could easily be combined with modern environmental discourses and incorporated into the ethno-politics of Dersim during the anti-dam (hydropower plant) protests in the early 2000s. First, there are nonliving materials such as soil, rocks, water, and fire. Mountains, as pilgrimage sites; rivers, as living souls; particular places on the mountains, such as forests, a pile of rocks, or an uncommonly large cliff as living places of deific beings are all considered to be cornerstones of an extensive collective mind map. Much like the organizational model of a Kurdish Alevi tribe (Gültekin 2019, 13), there is a hierarchical and territorial demarcation among jiares. Thus, every square meter of Dersim directly relates to the perception of a sacred land, one deeply connected to individuals with (mythic or actual) ancestral ties. The sacred image of Dersim thus symbolically presents itself as the sum of socio-spatial borders marked by jiares. They are being transmitted across generations through oral traditions. Every phase of that history, which is strongly related to community survival, has been added to a layered, interconnected collective mind of mythological narratives regarding the sacred land. From mythology to actuality each war clearly reflects that understanding. Thus, there is an endless battle between those who want to destroy Dersim and those who protect it in the Batın world, reflecting a deep sympathy for the guerrillas in Dersim, who fight in hostile conditions against a powerful enemy.

Second, there are some personalized sacred objects retained by Ocaks. A glove, a cup, or some (un)natural, personified deific elements are regarded as relics of mythological ancestors and as beings able to create miracles. Third, there are living sacred creatures. Almost all of them are related to the wilderness. According to Raa Haqi belief, all living beings are reflections of souls, belonging to either humans or animals. In this context, wild goats, snakes, eagles, and fishes are prominent species. Contemporary perceptions of wild animals, therefore, are also crucial to understanding the direct relationship between Raa Haqi belief system, and modern environmental concepts. Protecting wild animals from hunting is regarded as a religious act in defense of sacredness, a stance in accord with ecologically aware attitudes beyond the religious community which value and respect the living world. Such interrelations play significant roles in uniting the defense of Dersim with ecology. Therefore, modern-intellectual actions or ideas, such as "saving nature" or "ecological awareness," connote a religious understanding for Kurdish Alevis. Besides, the wild animals are also regarded as living souls of former lives, who witnessed and suffered the atrocities in Dersim. Thus, they must be protected to defend cultural identity (Gültekin 2020, 130–37).

THE INSTRUMENTALIZATION OF NATURE WITHIN ETHNO-POLITICS OF DERSIM

Dersim (see Benanav 2015 for an outsider view of contemporary Dersim) is mostly portrayed as "a mountainous refuge" or "a natural fortress isolated by high mountains." Those narrations are deeply connected to a continuum of brutal violence during the twentieth century. The modern Republic brought a double burden for Kurdish Alevis, exposing the community to both "Turkification" and "Islamization" (Gezik and Gültekin 2019). The years between the "1921 Koçgiri Rebellion" in the western Dersim region and the "1937 Dersim Rebellion," in particular, witnessed a series of bloody conflicts. The last resistance culminated in the genocidal massacre of 1938, when tens of thousands were brutally killed, some notoriously with poison gas as they sought refuge,[7] there were forced evacuations, and the land was torched (Kieser 2011). Even though the consequences of such unsuccessful rebellions, defeats, and massacres were horrendous, the resistance shown in the defense of Dersim has subsequently been commemorated and honored as "defeated but never surrendered."[8] Defending home becomes defending sacredness, a concept deeply related to the natural world.

The second devastating event took place in the mid-1990s. All Alevis regarded the forced evacuations and environmental destruction of Tunceli province in 1994, due to intensified clashes between guerrillas and Turkish armed forces, as "the second 1938 Dersim massacre" (Jongerden 2010, 77–100). Many civilians (including children, women, and older people) disappeared (Göral, Işık, and Kaya 2013) or were publicly murdered. Most parts of Dersim were declared a "military forbidden zone"; a restriction that still applies to civilians.

During the early 1990s, when Dersim was facing such devastating atrocities, Kurdish Alevi associations were established in Turkey and throughout Europe (Keles 2014, 173–227; Le Ray 2015). These associations primarily aimed to raise awareness of both human rights violations and the strategic destruction of the environment for military objectives. The latter included setting fire to forests (or using bioweapons to destroy forests—Anfturkce 2016), destroying villages and agricultural lands, and killing animals. Kurdish Alevi associations primarily aimed to draw the attention of the international community to such crimes. There were two major strands to their campaigns, focusing on twin threats to the integrity of Dersim, which were to become prominent levers when raising awareness of further natural and humanitarian crises. The first was the re-definition of the "1938 massacre" as the "Dersim Genocide" and endeavor to establish international legal institutions' official recognition of such a designation. A related aim was to achieve acknowledgement of the "Dersim Genocide" by both the Turkish state and European states. Further demands have been to restore the name of "Dersim" to the province (instead of "Tunceli") and to make Kırmancki listed as an endangered language by UNESCO (with the name "Zazaki"). There is currently also an additional political campaign to add the Munzur River to UNESCO's World Heritage List (Dersimdeutsche 2019). In this way, the associations have aimed to legitimize the cultural and territorial identity of Dersim in the international arena, enabling it to check incursions by the Turkish authorities if necessary. A similar political approach continues in efforts to protect Dersim from the AKP's assimilation plans, by countering what they consider to be "culture tourism" (Pirha 2020).

The associations also aimed to mobilize an international ecologist movement against large-scale hydroelectric power projects, countering dams mostly intended to be constructed on the Munzur River for both economic and military reasons (Strasser and Akçınar 2014, 207–23). At this point, the success of the Kurdish Alevi diaspora's lobbying activities seemed to be vindicated, following the emergence of Dersim's contemporary popularity in international academic circles. The Kurdish Alevi associations were able to organize mass rallies in Dersim during the "Munzur Festivals," which is still being organized annually. They also recorded some legal successes in stopping some dam projects.

The festival's full name "Munzur Kültür ve Doğa Festivali" (Munzur Culture and Nature Festival) is a pioneering cultural-political event which started in the late 1990s. It has since encompassed substantial economic and social relations, connecting blockaded Tunceli city and a massive diasporic Kurdish Alevi community (Strasser and Akçınar 2016, 143–63). Its dynamism, which drew upon symbolic meaning regarding the survival of Dersim and brought together several social movements against state violence, led directly to the "Dersim awakening" (Çelik 2003, 141–57). The first festivals aimed to raise awareness of the destruction of nature as well as the substantial depopulation of the province due to intense clashes and heavy restrictions. Due to the ongoing limitations, from the mid-1990s it was, and remains, impossible to maintain a self-sustainable economy and accordingly a social life. Dersim had a population of 157,000 in the late 1980s, exceedingly small in comparison to other eastern provinces of Turkey, and accordingly had an underdeveloped economy; factories for instance were absent. Therefore, among Kurdish Alevis, such restrictions and enforced depopulation policies created a fear that the province could be extinguished. There were further reasons. At the time, the Turkish government was planning a series of hydroelectric dams on the main rivers of Dersim (Agar and Böhm 2018, 1228–49; Dissard 2017, 229–57). If these were constructed, the Munzur Valley and the Pülümür Valley, along with some other deep, narrow valleys, would have been submerged. The outcome would be the loss of the city of Tunceli's road connections; its only links to the province's other districts. Thus, the province would be dismembered, with the parts allocated to neighboring cities where the Sunni populations are in a majority and hostile to Kurdish Alevis. That is why the priority was given to a call to encourage Dersim people to settle back into their abandoned villages. By drawing on the psychological impact of gathering tens of thousands of people together at once in "the sacred land," the associations hoped to inspire the people to return and resist. To do so, the most frequently mentioned connection was surely the sacredness of the land (in terms of its natural environment) and the nobility of the duty to protect it.

The festivals also created an economic boom in the province, though it was temporary. They ignited the solidarity between Kurdish Alevis who had migrated and new generations, in the sense of reunification of a collective cultural identity. Furthermore, the festivals led to the spread of a new understanding of a modern Kurdish Alevi identity, which tends to maintain a distance from Kurdish nationalism, socialist movements, and, also, the mainstream Alevi movement in Turkey. They included a broad spectrum of intellectual activities, such as public discussions organized by academics or intellectuals from throughout Turkey and western Europe; documentary and movie screenings about historical and contemporary issues in Dersim; concerts and art exhibitions (Neyzi 2002, 89–109). There were also social and economic spaces, such as markets, where the new Dersim identity's cultural capital has been apparent in the form of various cultural consumer items, such as t-shirts, bandanas, re-stylized traditional folk clothes, and tattoos.

Importantly, however, alongside such festivals was the rise of criticism of clichés regarding the origins of the Kurdish Alevi identity. Books, journals, documentaries focusing on Kurdish Alevi ethnography and history were presented, and the events included performances of new ethnic-music streams, drawing upon recovered folkloric traditions. All of these contributed to the new understanding of contemporary Dersim symbolism associated with the need to protect "a sacred nature," and a unique, non-Islamic, egalitarian, peaceful religion, that is held to reflect the same nature (Gültekin 2019). Emphasizing a nature-based uniqueness can also at once be considered as a new way to reposition this movement among anti-state politics, while showing its difference.

The symbolic representation of nature as a living sacred being seems in perfect harmony with the new ethno-politics of Dersim in the 2010s. The festivals apparently created a new understanding of Dersim in the form of a re-imagined sacred homeland.[9] Relatedly, they were also in part conceived as an annual spiritual journey to the ancestral lands and the jiares. Here, it is important to mention again that the province is a homeland for the Ocaks of Kurdish Alevism. Dersim also hosts well-known jiares, as pilgrimage sites, such as *Kemeré Duzgı* (Duzgı mountain) and the Munzur River (Deniz 2017, 13–33). Thus, undertaking pilgrimages to jiares and performing rituals were among the festivals' most prominent demands. Consequently, within a few years, some well-known jiares in Dersim became popular new sites of pilgrimage (Gültekin 2020). *Cemevis* were quickly built on desolated jiares as indicators of Kurdish Alevi identity in the public sphere. Cemevis in general were also the new institutional instruments (Es 2013, 25–43) of negotiations with the states at both national and international levels for all Alevis in the 2000s.

In addition, as Dissard (2017, 20) argues, motivations underpinning the anti-dam struggles, to protect the Munzur and Pülümür Rivers, often depend on the symbolization of victimized ancestors, martyred revolutionists, and natural beauty rather than recorded historical-archaeological sites. When HDP or independent socialist mayors established jiares as public parks in the city center, they were drawing upon the collective memory as part of a renewed political project in Dersim. This was part of a new strategy, since it was necessary to communicate with the people, exhausted by conflict, in a way beyond old rhetorical approaches. For this reason, almost all the marches were taking place on the valley and were finishing at jiares in the 2010s, newly built by municipalities. In the present day, discourses regarding the defense of Dersim continue to originate from a religionized political under-standing of a unique nature as a living nonhuman entity.

If jiares are to be considered as nature-worshipping practices, since they are grounded in ritual practices in the material living world, then the Ocaks and the beliefs that surround mythical ancestors of *aşirets* (tribes) are related to "ancestor-cults." Each tribe, regardless of whether it is an Ocak, has mythi-cal ancestors. The majority, although not all, of such ancestors occupy physi-cal spaces, with graves regarded as jiares. For this reason, practicing rituals at jiares precisely entails worshipping actual and mythical ancestors. This function establishes strong relationships between individuals and the land from *hane* (household) to aşiret levels. Thus, the re-symbolization of nature as the reference point of the sacredness of Dersim is common (Gültekin 2020, 137–43).

The sacralization of nature as a jiare is also a factor in relation to many sites throughout Dersim, where brutal massacres took place in history as well as in the recent past. The Hamidian period, when non-Muslim communities, including Kurdish Alevis, were subjected to mass violence in the 1890s; the Armenian Genocide in 1915; the Dersim uprisings of 1916, 1918 to 1921, and 1937; the 1938 genocidal massacre; and the 1994 forced evacuations are the most prominent, although sadly not the only, late nineteenth and twentieth-century examples. To these must be added dozens of places where guerrillas as well as civilians died, mostly young people from Dersim, and other tragic incidents occurred. All of these contribute toward an understanding of Dersim as a site of sacred geography, where natural beauties are saturated with a col-lective memory of loss and martyrdom. It is important here to say that most of these places continue to be respected as jiares in the present. The victims of twentieth-century Kurdish Alevi massacres are direct and known ancestors of the present generations. The act of such defense powerfully emphasizes the symbolical ethno-politics centering around modern Kurdish Alevism.

The formation of such places as memorial sites in the mid-2010s, therefore, represents a continuity of religious understanding of nature-based sacredness and cultural identity; history and a collective memory in spatialized form, shaped by ethno-politics of victimization. The Munzur festivals were the first cultural events to make such connections visible. More widely, and surprisingly, some memorial places were quickly regarded as new jiares (İlengiz 2020; Orhan 2017, 193–214). That creation of a sacred place on a political level, and its acceptance in the everyday culture so quickly and unproblematically, is another striking example of the flexibility and adaptability of perceptions of modern Kurdish Alevi identity. This illuminates our understanding of the relationship between the instrumentalization of nature and environmental ideas and ethno-politics.

CONCLUDING REMARKS

While this chapter was being written, another struggle against a mining company in Dersim was underway. The *Milli Köyü* (Milli Village) of Tunceli has been trying to stop operations at the stone mine, located beside the village, for several years. Unsurprisingly, villagers' arguments emphasize both "the destruction of jiares" and a call "to defend the homeland." This is a typical example that illustrates the overlapping connections between spatialized sacredness and the modern Kurdish Alevi identity. The following are the villagers' declaration and a campaign poster:

> The stone mine has been working for 15 years. Our village, located among forests, is being destroyed. We want it to stop immediately! *Milli* is hosting several jiares, which are elements of the *Kızılbaş*[10] faith. There is also the grave of *Silê Qiz* on the one side, and another sacred graveyard of our precious, the *Mezela Dewres*, on the other side of the stone mine. Moreover, there is the *Vile Jêlê*—a jiare at another side of the mine. . . . The state-supported local companies' unlimited, greedy ambition to make money leads to a massacre of nature and puts a damper on our lives. With every destroyed piece of land, we lose our remembrance, culture, social memory. What would be left if one is forcefully taken from its nature and memory?! We have been raised by praying to those rocks and mountains. . . . We have always sought refuge and believed in nature, mountains, and saw them as members of our families. . . . Besides, the Coronavirus, which incarcerated humankind into small concrete cells, has shown that the mistreatment of nature and wildlife ends with catastrophe. If nature continues to be harmed, the next generation will only face death and destruction. (Change.org 2020)

The declaration and the poster are current examples from Dersim, which indicates the continued construction of identity that appears in the social reproduction of symbolization of the land, history, religion, ethno-politics, environmentalism, and the future. Milli village, like others in Dersim, shares

Figure 16.1 **"Stop destroying Milli Village" Campaign.** *Source*: © Caner Canerik.

several overlapping layers of symbolism. The first one seems to be sacredness, derived from being a part of a more extensive concept of Dersim. According to this understanding, temporary social concerns and a contemporary understanding of identity are rebuilt upon the notion of a historic, even perpetual spatial dimension. In the politicized creation of jiares, continuity is synthesized from layers of social history, in the form of collective memory and modern-day social life. This history is undoubtedly imbued by Raa Haqi belief and hence, unique. For this reason, the images of elders, and sacred natural objects, such as trees, mountains, rocks, or forests, are invoked in several posters. Thus, the image above (Figure 16.1) symbolizes history, the past, and the inherited land, the present which both need to be protected.

The call for protection is also strengthened by the appeal of well-known, much-respected characters, such as Silo Qiz, a well-known local poet and traditional folk singer, which emphasizes the cultural specificity of Milli Village. Here, the process of re-imagination, drawing upon connotations of "the sacred land" and Raa Haqi culture, is further distilled using actual characters who are regarded as symbols of ethno-politics. Silo Qız, sang in Kırmancki. His songs are regarded as the voice of the resistance of a dying language. Elders, at this point, are the symbols of the dead as well as living ancestors, which is an essential dimension of the beliefs and rituals of Raa Haqi. The reference to ancestor-cults openly places a religious duty upon new generations, creating a strong motive to act.

In another poster (Figure 16.2), which celebrates the 1st of May (Yıldırım 2020), the survival of cultural identity is politically voiced with reference to environmental struggles. This is precisely consistent with the politicization of the Kurdish Alevi identity with socialist ideology that has occurred for almost half a century. However, in this contemporary understanding of socialist strug-gle, there is an emphasis upon solidarity with Turkish and international environ-mental movements. Here ecological awareness, living in harmony with nature, anti-capitalism, and gender equality are explicit. As proactive actors, a new generation of activists is including the symbolic and ideological sacralization of nature along with historical victimhood, to sustain Kurdish Alevi identity.

The Raa Haqi belief system has been forged over centuries. Although it has no written sources, oral tradition brings together hundreds of mythologi-cal narratives relating to enclosed socioreligious interrelations, social insti-tutions, and rules for individual conduct, as well as accounts of the tribes' origins. Even though such historical facts are unclear, focusing on a century-long modern past—especially the last half-century when the modern Kurdish

Figure 16.2 Poster Which Links the Celebration of International Workers' Day on 1 May with Ecology. *Source*: © Ekoloji Birliği.

Alevi identity emerged—has led to many theories to explain the evolution of its social structures and collective mentality. As we have seen, the living world of nature is also intimately connected to ancestral lands and a mythical world inhabited by deific creatures, a spatial dimension which it is deemed to share and be continuous with. Each hane, graveyard, jiare, and historic site of charged collective memory is a place where the two worlds touch together. For this reason, every external threat to the land is immediately perceived as a threat to cultural survival.

The case of Milli Köyü reflects this deep connection between the natural environment and prevailing religious belief. While Raa Haqi arises from two main dimensions, *ocak-talip* relations along with some ikrar-based socioreligious institutions and jiare beliefs, the latter currently seems to dominate collective as well as individual devotion. Moreover, the whole of Tunceli, the heartland of Dersim's cultural-geography, long subjected to mass violence and suppression, is in a way represented as a jiare. It still represents resilience, providing shelter for a subordinated identity and hopes for its survival. However, the success of this deep connection and reciprocal relationship relies upon a particular dynamism. Cultural identity is deemed as dependent upon the protection of the natural environment, which in turn is reliant upon communities willing to mobilize for its preservation. The new understanding of nature is being forged in the Kurdish Alevis' struggles against incursions by dam projects, mining companies, tourism policies, and other threats. Within this developing discourse, the threat of genocide is expanded to encompass the accompanying and more recent concept of "ecocide" (Bilgin 2020). The older tradition that the sacred land, full of jiares, must be defended from external threats is now dramatically reinvigorated. In the present day, more local, enclosed, and isolated identity-agendas, rooted in a unique cultural and religious belief system, are updated by modernized, internationalist ecological assertions seeking to connect the struggle with a wider global network, reflecting a contemporary Kurdish Alevism, sustaining and reimagining itself for the twenty-first century.

NOTES

1. A note on "Dersim": Contested place-name for the city and province with a Kurdish Alevi majority population; known as "Dersim" historically and in Kurdish, renamed as "Tunceli" by the Turkish state in the mid-1930s, just before the last resistance of Kurdish Alevi tribes. Although Tunceli and Dersim are used synonymously, I use Dersim to refer to the cultural geographic area, that is, the area that Kurdish Alevis associate with their reinvented ethno-religious identity.

2. I read about this event that occurred on June 2, 2020, while writing this chapter, and include it as a dramatic example of my theme.

3. *Dersim* was a principality established by the Ottomans during the nineteenth century in eastern Anatolia. The core of the region, namely "inner-Dersim" (*iç-Dersim*), has always been dominated by Kurdish Alevi tribes, and over time the name Dersim has become a synonym for a region where Kırmancki- and Kurmanci-speaking Alevi tribes live. Long after the foundation of the modern Republic, in 1935, the name of the region was officially discontinued. Today, it is called "Tunceli," meaning "Bronze-Hand," symbolizing "the ultimate victory of the republic against barbaric tribes" (see Gültekin 2019, 10).

4. *Raa Haqi* (The Path of The Truth) used to be the definition of the religion, namely Kurdish Alevism (see Deniz 2019, 45–75).

5. For a general understanding of Alevis in Turkey, see Yaman and Erdemir 2006; Olsson, Özdalga, and Raudvere, 2005.

6. There are four socioreligious positions. First, *talips*, followers of *Ocaks*; Second, *raybers*, who can be considered as local guides on religious issues; Third, *pirs*, the spiritual guides of their followers and the embodiment of their Ocaks; Finally, *mürşids*, who hold a superior position with judicial authority among the Ocaks (see Gültekin 2019, 12–17; Deniz 2019, 42–75; Gezik 2013, 11–78).

7. Recently shared documents reveal the poisonous gas trade between Nazi Germany and Turkey (Westernarmeniatv 2019).

8. The motto can be seen on epitaphs of fallen revolutionists' graves. This is another example of the contemporary understanding of politics and religion in Dersim. There is a deep symbolic connection between those who have fallen while fighting to protect Dersim, and the massacred ancestors (İlengiz 2020).

9. The PKK declared a unilateral cease-fire in the early 2000s, followed shortly by unofficial negotiations with the AKP. Consequently, some restrictions were loosened in Dersim, triggering a gradual increase in visits to the home of Kurdish Alevis from the diaspora (Göner 2012, 317–28).

10. Kızılbaş (Qizilbash or Kızılbash) is a historical term that refers to certain Kurdish- or Turkish-speaking Alevi groups who fought against the Ottomans alongside Safavids during the fifteenth and sixteenth centuries. In the present day, the term is mostly used to define Kurdish Alevis.

REFERENCES

Agar, Celal Cahit, and Steffen Böhm. 2018. "Towards a Pluralist Labor Geography: Constrained Grassroots Agency and the Socio-Spatial Fix in Dêrsim, Turkey." *Environment and Planning* 50, no. 6: 1228–49.

Aksu, Cemil, Sinen Erensü, and Erdem Evren, eds. 2016. *Sudan Sebepler: Türkiye'de Neoliberal Su-Enerji Politikaları ve Direnişleri*. İstanbul: İletişim Yayınları.

Anfturkce. 2016. "Türk devleti, Ovacık'ta ormanlık alana tırtıl atıyor" (Turkish State Throws Caterpillar to Forests in Ovacık). Accessed August 24, 2020. https

://anfturkce.com/kurdIstan/turk-devleti-ovacik-ta-ormanlik-alana-tirtil-atiyor-7
3060.

Benanav, Michael. 2015. "Finding Paradise in Turkey's Munzur Valley." *New York Times.* Accessed August 24, 2020. https://www.nytimes.com/2015/06/28/travel/ finding-paradise-in-turkeys-munzur-valley.html.

Bilgin, Deniz. 2020. "Dersim'de Yeşil Soykırım." Yeni Özgür Politika. Accessed August 24, 2020. https://www.ozgurpolitika.com/haberi-dersimde-yesil-soykirim -276.

Çakmak, Hüseyin. 2013. *Dersim Aleviliği: Raa Haqi.* Ankara: Kalan Yayınları.

Çelik, Ayşe Betül. 2003. "Alevis, Kurds and Hemşehris: Alevi Kurdish Revival in the Nineties." In *Turkey's Alevi Enigma: A Comprehensive Overview,* edited by Paul J. White and Joost Jongerden, 141–57. Leiden: Brill.

Çetin, Ümit, Celia Jenkins, and Suavi Aydın, eds. 2020. *Kurdish Studies,* 8, no. 1 (May 2020), *Special Issue: Alevi Kurds: History, Politics and Identity.* London: Transnational Press.

Change.org. 2020. "Dersim'in Milli Köyü'ndeki Taş Ocağı Kapatılsın." Accessed August 24, 2020. https://www.change.org/p/m%C3%BCteahhit-erdal-g%C3 %BCnta%C5%9F-dersim-in-milli-k%C3%B6y%C3%BCndeki-ta%C5%9F-oca %C4%9Fi-kapatilsin.

Deniz, Dilşa. 2017. "Dersim Kutsal Tarihi: Çevre, Toplum ve İnancın Ortaklaştığı Kutsal Coğrafyada Yaşam ve Direnme Hukuku." *Toplum ve Kuram* 12: 13–33.

Deniz, Dilşa. 2019. "Kurdish Alevi Belief System, Reya Heqi/Raa Haqi: Structure, Networking, Ritual and Function." In *Kurdish Alevis and the Case of Dersim: Historical and Contemporary Insights,* edited by Erdal Gezik and Ahmet Kerim Gültekin, 45–75. Lanham, MD: Lexington Books.

Dersimdeutsche. 2019. "Der Munzur soll frei fließen." *Dersim Site Deutsche.* Accessed August 24, 2020. https://dersimdeutsche.wordpress.com/2019/02/27/ der-munzur-soll-frei-fliesen/.

Dissard, Laurent. 2017. "From Shining Icons of Progress to Contested Infrastructures: Damming the Munzur Valley in Eastern Turkey." In *Contested Spaces in Contemporary Turkey: Environmental, Urban and Secular Politics,* edited by Fatma Müge Göçek, 229–57. London: IB Tauris.

Duvarenglish. 2020. "Construction Kicks off on Alevi Sacred Grounds in Eastern Turkey." Accessed August 12, 2020. https://www.duvarenglish.com/culture/2020 /08/11/construction-kicks-off-on-alevi-sacred-grounds-in-eastern-turkey/.

Es, Murat. 2013. "Alevis in Cemevis: Religion and Secularism in Turkey." In *Topographies of Faith: Religion in Urban Spaces,* edited by Irene Becci, Marion Burchardt, and José Casanova, 25–43. Leiden: Brill.

Gezik, Erdal. 2013. "Rayberler, Pirler ve Mürşidler." In *Alevi Ocakları ve Örgütlenmeleri.* edited by Erdal Gezik and Mesut Özcan, 11–78. Ankara: Kalan Yayınları.

Gezik, Erdal. 2014. *Dinsel, Etnik ve Politik Sorunlar Bağlamında—Alevi Kürtler.* İstanbul: İletişim Yayınları.

Gezik, Erdal, and Ahmet Kerim Gültekin. 2019. *Kurdish Alevis and the Case of Dersim: Historical and Contemporary Insights.* Lanham, MD: Lexington Books.

Gezik, Erdal, and Hüseyin Çakmak. 2010. *Raa Haqi—Riya Haqi / Dersim Aleviliği İnanç Terimleri Sözlüğü*. Ankara: Kalan Yayınları.

Göner, Özlem. 2012. "*A Social History of Power and Struggle in Turkey: State, Memory, Movements, and Identity of Outsiderness in Dersim.*" PhD diss., University of Massachusetts.

Göral, Sevgi, Ayhan Işık, and Özlem Kaya. 2013. *The Unspoken Truth: Enforced Disappearances*. İstanbul: Hafıza Merkezi / Truth Justice Memory Center.

Güler, Sabır. 2019. *Ötekinin Ötekisi—Etno-Dinsel Bir Kimlik Olarak Alevi Kürtlüğün İnşası*. İstanbul: İletişim Yayınları.

Gültekin, Ahmet Kerim. 2016. "Counter-Publics and the Local Practices: The Struggle for Revolutionary-Popular Local Government." In *The Road to Gezi—Resistance and Counter-Publics in 21st Century Turkey*, edited by Gamze Yücesan-Özdemir, 175–91. Ottawa, ON: Red Quill Books.

Gültekin, Ahmet Kerim. 2019. *Kurdish Alevism: Creating New Ways of Practicing the Religion - Working Paper Series of the HCAS "Multiple Secularities—Beyond the West, Beyond Modernities.*" Leipzig: Leipzig University.

Gültekin, Ahmet Kerim. 2020. *Kutsal Mekanın Yeniden Üretimi—Kemeré Dızgı'dan Düzgün Babaya Dersim Aleviliğinde Müzakereler ve Kültür Örüntüleri*. İstanbul: Bilim ve Gelecek Kitaplığı.

İlengiz, Çiçek. 2020. "Magical #Afterlives in Post-Genocidal Turkey." *Allegra lab*. Accessed August 24, 2020. https://allegralaboratory.net/magical-afterlives-in-post -genocidal-turkey.

Jongerden, Joost. 2010. "Village Evacuation and Reconstruction in Kurdistan (1993– 2002)." *Études Rurales* 2, no. 186: 77–100.

Keles, Janroj. 2014. "The Politics of Religious and Ethnic Identity among Kurdish Alevis in the Homeland and Diaspora." In *Religious Minorities in Kurdistan: Beyond Mainstream*, edited by Khanna Omarkhali, 173–227. Wiesbaden: Harrassowitz.

Kieser, Hans-Lukas. 2011. "Dersim Massacre, 1937–1938." In *Online Encyclopedia of Mass Violence*. Article published July 27, 2020. https://www.sciencespo.fr/mass -violence-war-massacre-resistance/en/document/dersim-massacre-1937-1938.html.

Le Ray, Marie. 2005. "Associations de pays et production de locality: la « campagne Munzur » contre les barrages." *European Journal of Turkish Studies* 2. Accessed August 24, 2020. https://doi.org/10.4000/ejts.370.

Neyzi, Leyla. 2002. "Embodied Elders: Space and Subjectivity in the Music of Metin-Kemal Kahraman." *Middle Eastern Studies* 38, no. 1: 89–109.

Olsson, Tord, Elisabeth Özdalga, and Catharina Raudvere, eds. 2005. *Alevi Identity— Cultural, Religious, Social Perspectives*. İstanbul: Swedish Research Institute.

Orhan, Gözde. 2017. "The Production of Space in Dersim in the 2000s." PhD diss., Boğaziçi University.

Pirha. 2020. "Dersim'i Yok Etme Politikasına Karşı Çıkmak Her Dersimlinin Görevidir." Accessed August 24, 2020. https://www.pirha.net/dersimi-yok-etme-politikasina-karsi-cikmak-her-dersimlinin-gorevidir-220165.html.

Strasser, Sabine, and Mustafa Akçınar. 2014. "Dersim Dernekleri: Transnationale Betrachtungen Ppolitischer Flüsse' zwischen Europ und der Provinz Tunceli/ Dersim in der Türkei." In *Migration und Entwicklung. Neue Perspektiven.*

Historische Sozialkunde, Internationale Entwicklung 32, edited by Ilker Atac, Michael Fanizadeh, Albert Kraler, Wolfram Manzenreiter, 207–23. Wien: Promedia.

Strasser, Sabine, and Mustafa Akçınar. 2016. "Dersim Across Borders: Political Transmittances Between the Kurdish-Turkish Province Tunceli and Europe." In *Migration and Social Remittances in a Global Europe*, edited by Magdalena Nowicka and Vojin Šerbedžija, 143–63. London: Palgrave Macmillan.

Westernarmeniatv. 2019. "Turkey has Bought Chemical Weapons from Nazi Germany for Dersim Massacre." Accessed August 24, 2020. http://westernarmeni atv.com/en/44510/turkey-has-bought-chemical-weapons-from-nazi-germany-for-dersim-massacre.

Yaman, Ali, and Aykan Erdemir. 2006. *Alevism—Bektashism: A Brief Introduction.* İstanbul: Horasan Yayıncılık.

Yenidemokrasi7. 2020. "Yasaklar ve Engellemeler Sökmedi, Hasan Ataş Ailesi ve Yoldaşları Tarafından Toprağa Verildi" [Despite all Detentions Hasan Ataş was Buried by His Comrades and Family]. Accessed June 9, 2020. http://www. yenidemokrasi7.net/yasaklar-ve-engellemeler-sokmedi-halk-savascisi-hasan-atas-ailesi-ve-yoldaslari-tarafindan-topraga-verildi.html (site discontinued). Also accessed at Yeni Demokrasi Gazetesi (@yeni_demokrasi1). 2020. "Hasan Ataş'ın annesi: 'Onurumdur, gururumdur, ben oğlumun arkasındayım'" Twitter video, June 5, 2020. https://twitter.com/yeni_demokrasi1/status/1268879803891027968 ?lang=en.

Yıldırım, Barış. 2020. Facebook. Accessed August 20, 2020. https://www.facebook .com/photo/?fbid=1351712051688805&set=a.104690273057662.

Chapter 17

The Philosophy of Ecology and Rêya Heqî

Religion, Nature, and Femininity

Dilşa Deniz

RÊYA HEQÎ/KURDISH ALEVISM

This chapter investigates the deep roots of the relationship of Rêya Heqî with the natural world of its sacred land and particularly explores the prohibition upon the hunting of animals based on the myth of Gola Buyêr.

Unlike Islam, Rêya Heqî, or Kurdish Alevism,[1] does not have written texts, but a strong oral tradition based on myths and religious songs, which can be considered as the main sacred sources of the faith. The religious songs (*dêş*) lay out the philosophy and history of the religion. Myths, the archive of the collective memory, are understood and analyzed by every member of the society as prescribed. The philosophy and jurisdiction of the Kurmanc (pronounced like Kurmanj) community have been transmitted through these myths as part of a larger societal interaction.

As traditional collective stories that chronicle small changes within an unchanging framework, myths are often passed on orally from generation to generation (Deniz 2020b, 224) and provide a sacred prerogative that contains the institutions, traditions, and belief practices of a distinct region and function as a tool for their continuity and metamorphosis (Graves and Patai 2009, 23). As such, besides being the "bearers of important messages about life in general and life within society in particular" (Caldwell 1989, 4), myths connect to a timeless and infinite past in the present time and serve as vessels containing valuable information about social structures. Thus, this connection emanating from an undocumented time and place establishes an important interrelation with sacredness (Deniz 2020b, 225). This is the "principal structure on which the continuity of society is built as a fictive formation

243

encapsulating all elements necessary for the continuity of that society" (Deniz 2012, 86).

Marija Gimbutas thus argues that mythology, linguistics, ethnology, folklore, comparative religion, and historical documents can all be considered nonmaterial archaeological remnants (Murdock 2016, 11). Mircea Eliade, a philosopher and historian of religion, regards the inception of myth to be at a moment outside of time, in which stories of events occur in a more sacred period (1992, 43–44). As a special type of historical text, myths are mostly "sacred stories of religions" (Leeming 2003) and collectively collate a far past era still alive in the communal memory, thus providing a kind of archival technology and language. Claude Lévi-Strauss, the founder of structural anthropology, portrays myths as a language and thus a part of human speech to be known, and to be told (1963, 209) while Christopher Johnson thinks myths are "more than a language, . . . a kind of information technology, an instrument, a 'logical model' or 'logical tool'" (2003, 102). As a language and a kind of information technology, myths can make sense as "long as the speaker and listener, or the actor and the audience, share the same conventional thoughts and metaphysical time, space, and objects" (Leach 1976, 70). As such, the myths of Rêya Heqî in Dersim are shared by the speaker, listener, actor, and/or the audience of the same culture in the larger Dersim geography.

Rêya Heqî originated as an ancient Iranian religion and until the present has resisted unrelenting oppression from the hegemonic states of Abrahamic monotheism. As a survival strategy, Kurmancs (Kurdish Alevis) had to take refuge on the highlands of Anatolia and Kurdistan. As a result, teaching and generational transmission has been through the oral tradition, until recently. For instance, when forced by the Muslims to apply the Quranic teaching, Kurmancs would say *"Quran a me natiqe/* our Quran is oral," pointing to their religious songs and the distinctive and rich sets of myths that serve as their sacred texts of faith.

The myths of Rêya Heqî present a kind of functional and modular federative system (Deniz 2012, 53–54, 313). While these myths might seem to be independent stories, in reality each has a different task and together they form the total jurisdiction, philosophy, and culture of the religion. They are also the basis of the unique formation of the Kurmanc societal structure, a tribally based belief system (Deniz 2012, 71) by which duties and responsibilities are prescribed for groups and groups, groups and individuals, individuals and individuals. This is derived from the inherited authority of the Ocax tribal system. The Ocax system (Deniz 2019, 52–53), summed up by the phrase "hand to hand, hand to God/Heq," distributes rights and responsibilities among its groups and individual members in a partially vertical but overarching horizontal hierarchy in which all the religious and secular societal institutions are interconnected (Deniz 2012, 78).

Kurmanc identity is also based on this federative system. A Kurmanc is born into both an ethnic and religious identity that includes a certain geography regarded as sacred land. As the geography is an important part of one's identity "Kurdish Alevis still maintain close contact with their sacred and beloved land, Dersim" (Gültekin 2019, 4). Dersim (in its traditional extent) is the sacred land of Kurmancs, for whom the concept of land differs from traditional indigenous or modern capitalistic understandings. According to indigenous tradition, people consider themselves as a part of the land, thus the land owns the people along with all its content (Randall 2009, 2.38–2.42 sec), while the Western-centric, capitalistic understanding considers that land is owned by humans. In Dersim, the relationship between people and the land is understood as something in between; it is deemed to be closer to indigenous traditions but has shifted recently due to indoctrination through the Western-centric domination of understanding through education and urbanization.

Today, the land is both collective and individual property (Deniz 2012, 210–11), with an emphasis on collective. While individual property is enjoyed by individuals, all individual land is also part of the tribe's collective property. While individuals have land ownership, the land is also registered in the name of a tribe (Deniz 2012, 48). The entirety of Kurmanc land equates to a federation of tribes. Tribal ownership does grant the right of individual ownership, but with a limitation: individual land can be sold only to a member of the same tribe; its sale follows tribal jurisdiction based on primogeniture (closeness of kinship) and is assigned in order: children, siblings, paternal cousins, and so forth (Deniz 2012, 48–49).

Another wider collective land ownership occurs regarding sacred geography. The understanding of sacred geography is similar to the indigenous understanding of the land. In this understanding, the land owns the people; land that belongs to everyone and at the same time everyone belongs to that land as well. This perception of land as collective is explained by an Aborigine woman, as follows:

For Aboriginal peoples, country is much more than a place. Rock, tree, river, hill, animal, human—all were formed of the same substance by the Ancestors who continue to live in land, water, sky. Country is filled with relations speaking language and following Law, no matter whether the shape of that relation is human, rock, crow, wattle. Country is loved, needed, and cared for, and country loves, needs, and cares for her peoples in turn. Country is family, culture, identity. Country is self. (Korff 2020)

Despite the changes through the years, Kurmancs still have a similar perception of the land. The term *herd a Dewresh* means both that the land belongs to dervish, the holy divine, and/or the land itself is a dervish. The whole, the sacred entity is dervish, or the sacred land itself is connected to the creator, god *Xwade/Heq, Xwedê,* and *Heq* can be translated as the main holy power, the unity of the whole, the main form of a creator. *Xwedê/Xweda* in Kurdish means "the one who creates from itself," most likely a reference to Mother Nature, *Xwedawend (Itself/herself/himself* in Kurdish), means goddess, again means gives [birth] from itself but the suffixes of *"wend"* more likely means in secret, hidden. Thus, *xweda* also has a feminine character, although used for god, if considered in the context of giving birth. *Xwedawend,* the goddess, means born/gave from herself in secret, while *xweza* has a dictionary definition as "nature"[2] with a related meaning. The "dê/da" suffixes mean given [birth], in the context of divine power means reproduction, and "za" means born and refers to animals giving birth. Thus, reiterating that the *herd a Dewresh* is the body of the *Xweda[wend]/Heq* and most importantly *xweza,* nature as a whole is dervish, and thus xweza/nature itself is a dervish.

While the creator sounds masculine due to the recent dominance of patriarchal culture, the term itself is unisex. It is only by pronunciation according to the last letter of the spelling that one can make it masculine or feminine, its unisex root reflected in the unisex nature of both Rêya Heqî and goddess worship. We also know that before patriarchy took control, as Edwin Oliver James points out, the essential features of male and female elements in goddess worshipping and recurrent androgyne phenomenon in the Goddess cult were everywhere (1959, 245). Dersimians still have this connection as well. Thus, the *Xweda* is an androgynous term of the deity and still strongly retains this meaning in Kurdish. Eric Neumann states as a cult object the Great Mother's unshapely figures were not only women, but also men, and were representing archetypal symbols of fertility (1955, 129). Traditionally, the Rêya Heqî religion addresses everyone in a unisex form as "can." "Unlike Abrahamic monotheist religions, . . . gender duality not as an opposition (male vs. female) but androgynal co-constitutive symbolizing fertility" appears in more traditional ways of thinking in these old Iranian religions (Deniz 2020b). *Xwadê/Heq,* the main form of the creator as mentioned above, is androgyne in form, but *Heq/Haq* also means the truth, reality, and rights. These combined meanings in Kurmanc terminology are not a coincidence but signal a symbiotic relationship. *Heq/Haq* means rights/justice and *heqi/haqi* means truth/eternity/holy unity, which includes everything both organic and inorganic as part of the Universe as a whole, the holy body. A soul travels from body to body (transmigration) and form to form at most one thousand times. After 1,001 times, it can join eternally to the holy body, *Heq/heqi.*

Heqi, the eternal truth, is used to explain death. Because the faith believes in reincarnation (Deniz 2012, 346–347), the term *çuye heqiya xwa* (went to her/his truth) is used, according to the faith, no one permanently dies. Death is the beginning of rebirth. Changing the end of this term indicates joining with the eternal body of the god[dess]. Death is necessary for rebirth, eternal truth, a journey in the *Rêya Heqî* (the path of the truth), a journey toward eternity from *herd/hard a Dewresh* (Deniz 2012, 118–130) to the holy land (the body of the *Xweda[wend]/Heq*). Thus, local and sacred geographies are related through a special, sacred jurisdiction.

In Kurdish Alevism, *Xizir* is the most important sacred role after *Xwade/Heq* (Çem 2010, 42). As an intermediary between *Xweda[wend]/Heq* and the *taliws* (believers), *Xizir* can appear anywhere and anytime to help those in need but is mostly a protector of travelers. This is specific to Alevism/Rêya Heqî, as Alevis identify their belief as the Path (*Rê/Raa*, in Kurdish), meaning that the whole process is a kind of journey of both spiritual and physical traveling. Either you are traveling on your own path to the holy eternity through changing forms or you are trained to act as a *dewrêsh/seyid* to reach a perfect human level (*marifet*) via an internal philosophical journey that enables you to serve the *taliws* (believers) in religious activity. The emphasis is on the path itself. Traveling, and the journey or path, are important symbolic and sacred acts, and so, thus, are the travelers themselves. We should also remember that the religious service also supplied by Seyids via traveling among the taliw villages, homes.

In Dersim's Kurmanc pantheon, *Xwede* and *Xizir* are omnipresent, but the connection between the [sacred] land and human also occurs in specific locations. In the inner Dersim pantheon *Dizgun* and *Mizur*, the central symbols of that collectivity, are associated with the particular locations where they performed their miracles. One symbolizes a mountain, *Kevirê Dizgunê* (the Stone of Dizgun), where water is sparse. The other symbolizes the birth of a sacred river, *Mizur*.

Dersim society is organized according to a religious caste system in which the term Ocax, (literally, "fireplace"), is meaningful for a faith that considers it as sacred and central to its rituals. Relatedly, Ocax as an institution refers to a large, tribe-like "family" with an inherited position. The system is based on a set of ancestral myths rooted in miracles by location. Nearly all Ocaxs have founding myths and relate to miracles that occurred in certain places in Dersim. In their magical narratives, the Ocaxs' ancestors have an interactive relationship with nature or the land, such as walking on water, escaping fire, talking to animals, receiving power from trees and animals, or settling somewhere for magical reasons. This system is particularly crucial for the Rêya Heqî community as it is used to solve their daily societal problems, socialize each generation, define jurisdiction, and maintain social solidarity, as well as functioning as a mediation system.

Because of the importance of the Ocax system in the Rêya Heqî, it has been targeted by the Ottomans and, later, by the Turkish republic, as clearly stated in several reports made by state officials including prime ministers,

ministers, and military officers (Kaynak 2010 and 2011). These reports take the insulting position that the religion is primitive (Kaynak 2010, 243) or *heretic* (Ünlü 2014, 425) and use this as a rationale for assimilation policies (Deniz 2020a, 23–24). The designation as heretic has made Alevis open and defenseless to the organized crimes commissioned by members of the Sunni Muslim religious order. As such, through all the reports from the Ottomans onward, *seyids* (the religious cast of Kurdish Alevism) of the Dersim region have been regularly blamed and targeted (Bulut 2009).

In recent years, other methods have been used not only to assimilate Alevis but acculturate Alevism as well. As a result, a huge body of the nature-worshipping oral texts and practices of Rêya Heqî have been wiped out. The huge gap thus created in the religion has been filled with Islamic teachings that directly contradict the core beliefs of Rêya Heqî/Alevism. Rêya Heqî/Alevism's only connection is based on Alī ibn Abī Ṭālib (born c.600, Mecca, killed in 661, Kufa, Iraq, who was a cousin and son-in-law of Muhammad, the Prophet of Islam). He was assassinated and his family was subjected to genocide by the Muslim army. Most probably, due to a kind of empathy for being victims of such religious violence that Rêya Heqî faith strongly opposes, his personality was reconstructed according to the Alevi philosophy, as a miraculous, fair, and humble character and replaced in the cycle of Rêya Heqî reincarnation and used as a resistance symbol. Because of that, presently this is used for the Islamization of Alevism.

The myths of settlement are the texts for the sacred and secular history of Dersim. Most importantly, they reference the spiritual journey from the past that reaches into the present to provide holding rights to sacred lands. Religion, the history of the religion, the settlement of the region, past events—all have been transmitted through mythical stories from generation to generation. Myths of settlement also serve to authenticate collective land ownership in Dersim. They act as traditional deeds to houses, wells, nearby trees, and other features, to declare them as sacred and open to visitors; worshippers, as well as the *taliw*s (who receive religious services) of the relevant Ocax, have a spiritual share in such land. As such, targeting myths of settlement is a political strategy for annulling traditional claims of religious and social positions within Kurmanc society.

These sacred mythical stories connect the land and the human, the past and the present, the individual and the collective, through which society, faith, and nature are dynamically interconnected. As such, the Dersim Rêya Heqî system of sacred grand law carefully defined, designed, and interconnected the relationships between human and nature, human and animal, human and human, individual and individual, individual and group, group and group. From a tree to a deer, from an eagle to a snake, from water to a mountain, from tribes to their *Pir ocaxs*—everything has been positioned and their relationships formulated. In the collectivity, everyone has the right to enjoy everything as formulated "*heqê her tişti heye*/ everything/everybody has right in it." This is the main frame of the sacred law of Rêya Heqî faith. Structured in this way, the jurisdiction of sacred and secular matters has functioned effectively.

This has led Kurmancs to treat the land as an eternal ancestral relic and as a sacred and continual connection between their ancestors and themselves. As such, the land cannot be sold but is to be protected along with all its contents. Dersimian Kurmancs do not sell their land because it is not a commodity that belongs to persons, but to the ancestral lineage. Thus, while individuals have the right to benefit from residing on the land, they also have a duty to protect it. If a situation obligates a sale of land, it must be sold to one of the members of the tribe, one who is a descendant within the same genealogy and has kinship priority.

Geography and faith are constructed simultaneously and cannot be dissociated. Dersim is a land of sacred geography, an important base of the faith, and is understood as belonging to a lineage of ancestors—from the first till the last and including the future as well. This understanding of land ownership resists commodification, even into the twenty-first century. As the collectivity views the land as sacred, their relationship to it is to preserve, protect, and carefully consume it; in return, societal survival is assured. This relationship applies to every member of the society. The land is seen as the "vital one." Its stature as such is the reason society organizes for its collective and consistent protection; it is the relic of the ancestresses/ancestors and the most sacred entity to be protected.

Water is another sacred entity for the Kurmancs of Dersim. It is regarded as a monolithic entity that contains all the springs and other bodies of water in nature. There is an important emphasis on the sacredness of subterranean water, so that all springs are accepted as sacred, although some are more so than others. Traditionally, no one puts any pollutant in the springs. Wherever water exists, it is common practice to show one's respect although this occurs mainly at wells. For instance, any Kurdish person of Rêya Heqî faith of the old generation who passed by any spring would bend down and kiss a rock or a tree near that spring. At a village well, they would kiss the wall of the well to show their respect.

As a sacred entity, water is used for consecration or purification as well. For example, in *Hewtemal* (early spring, a process that starts March 7 in the Kurdish calendar, March 20 in the Gregorian calendar, since the Kurdish calendar is thirteen days behind) water is sprinkled in the house for purification. In the past, once a year, a Kurdish woman of Dersim about to use a brand-new churn in early spring would take it to seven wells and put water and pebbles from these wells, as well as seven types of wild flowers, in the churn for a night. The next day they would perform a ceremony in which the churn is emptied and rewashed. When a newborn baby is forty days old there is a ceremony, in which the baby is washed with water that contains forty seeds of wheat. At one time, a woman was not allowed to go out of the house until the fortieth day post-partum, at which time she would be taken to the village well for a ceremony. Similarly, a new bride made her first public outing to the village well for a ceremony. In Ovacix, all brides visit the Mizur wells to say goodbye (Deniz 2012, 275) as well as to be purified and welcomed. All these important "first" ceremonies are connected to the acceptance of water's sacred—thus protective and purifying—nature.

Polluting water resources, in particular the Mizur River, was regarded as a crime against the faith. Committing crimes against water resources was a capital crime, and thus incurred a heavy penalty. Committing a crime against one water source is equated to committing a crime against all water sources, springs, and wells collectively. Thus, such a crime is committed against the whole society and the society's sanctuary. This has been observed by outsiders who visited Dersim. For instance, an Ottoman pasha, Ziya Yergök, writes in his memoir that "no one commits any sin, or kills anyone near Munzur springs. If they had to take an oath about an important matter, they first get into the water and then take the oath. [It is believed that] the one who commits any sin and the one who lies will be punished in accordance" (Önal 2006, 165). An Armenian author and traveler, Yeritsyan Antranik, observes the same issue, stating the belief in water's sacred nature and its potency as a significant, infinite, "lively" power (2012, 78, 96, 97, 125). Nuri Dersimi also points out that sacred water sites are the main sites where the bloodiest conflicts were resolved. "Any decision made there, undoubtedly, every person had to accept, and did not find any courage to oppose" (Dersimi 1992, 9). For example, when the Dersimian tribes, including Seyid Riza, decided to resist Turkish army operations in 1937, the decision was made in Halbori springs, a part of the Mizur River, and was "signed" by handing over a stone that was then thrown into the Mizur River (Aygün 2009, 292).

Until twenty years ago, no pollutants were thrown into the river. Elders and others who did not have any connection to Turkish urbanization believed that all natural waters occurring in the wilderness belonged to the wild animals, believing every living being has a right to it. All these examples show the centrality and sacredness of water in Rêya Heqî belief in Dersim, as well as other Kurmanc sites.

The myth of the *Mizur* is one of the most important sacred texts of Kurmancs' belief, further emphasiêzing the constructed bond between nature and society in the belief system of Dersim. It is yet another component of the strong interactive bond between religion and nature that ultimately connects human and the history of the homeland through which people and geography meet in a single identity. Myths as sacred texts are the main vehicles that transmit the meeting of human and geography in a common sacred life philosophy. Myths as a whole form the Alevi ecological philosophy that shapes jurisdiction and grand law. They archive the philosophy, law, and history of Alevi society, deciphered and transmitted from one generation to the next, both synchronically and diachronically. The Gola Buyêr myth is one of the main texts in the archive and is deconstructed here for its content and discourse in relation to the Kurmancs/Rêya Heqî's ecological understanding of the societal function of the relationship between nature and humanity.

GOLA BUYÊR: THE SACRED
FEMININITY, LAW, AND LIFE

For Kurmancs, Rêya Heqî depends on its myths to transmit its philosophy, solidarity, jurisprudence, and societal norms, as well as their intra-functioning. The Gola Buyêr myth is one of the most important sacred texts in this regard to be transmitted from generation to generation. It contains a unique discourse of Rêya Heqî philosophy and its jurisdiction in connection to the feminine character of natural law. A summary of the story follows:

> There was a king whose name was Bask and he had two sons and five daughters. In that time there was a difficult initiation that male candidates for kingship had to pass. As Bask's sons approached adulthood they had to undertake this rite of initiation. The brothers were sent to the *Miran* upland meadows to hunt; however, they were unsuccessful and began the return home empty handed. On the way back they saw a mother deer with her fawn drinking water from a spring on the slope of the *Bedro* mountain. They hit the mother and the fawn with their arrows at the same time, killing the fawn straight away and badly wounding the mother.

> While they had now successfully fulfilled their initiation, they had, however, also committed a serious crime: killing a mother and her fawn while they were drinking water. Very happy with their hunting success, the two brothers approached their prey. The mother deer cursed them for killing her baby while it was drinking water. She reminded them they were the sons of the king and candidates for kingship, and that everybody knew that even a snake would not touch anyone drinking water. She damned them and died. At her death a storm descended from the *Bedro* mountain that sent gushing water everywhere. In the pandemonium the brothers drew their swords and cut off each other's arms and heads. As their heads tumbled down next to the body of the mother deer, the

brothers turned to stone. The violent flood waters continued to rise, creating a deep, bottomless lake where the brothers remain forever.

In the meantime, the five sisters learned about the events and rushed to help their brothers. When they arrived at *Bedro* mountain, a whirlwind gathered and lifted them up until they disappeared into the sky. Later the sisters were found turned to stone: one on *Kimsor* hill, one between *Balix* and *Maskan*, one in *Tirsuku* below the *Melkiş*, one in the *Mile* Well where she fell while trying to catch the head of her brother. The fifth sister was later found in the bottom of the lake next to her brothers with a brother's arm she had tried to catch, but had tumbled down with it to the bottom of the lake. (Deniz 2012, 166-67; Yücel 2003, 68–69)

The main theme of the myth is the violation of a law about the right to life. The myth focuses on the continuation of life through the mother, a fawn, and the necessity of drinking water. First, according to the myth-constructed law, a mother with a dependent baby cannot be killed. According to Alevi faith, if you kill a mother with a dependent baby, you have committed two crimes, in fact, even separating a baby from a mother is considered a crime (Birdoğan 1995, 181). Without a mother, a baby cannot survive. Killing mothers means killing the infants who are the next generation. This is a violation of the right to survive as a species as well. In the grand law, every living being has the right to survive and regenerate.

Second, no one can be touched or killed while drinking water, another important aspect of natural law prescribed in the myth. Water is essential for survival, not for just one species but for all living beings. Depriving anyone of it is a crime, which is a philosophical jurisdiction that has been followed by the Kurmancs/Kurdish Alevis for thousands of years. Alevism includes many basic human rights, both individual and collective, that the United Nations has recently recognized in contemporary times.

The myth also defines the law on the treatment of females. The female is declared to be untouchable, protected, not only because she has a baby but also because of her role in regeneration. The representation of the female as mother underlines the role of the female in maintaining life and regeneration. There is no such role attached to the male in this regard. In contrast to motherhood, fatherhood/masculinity is associated with unnecessary and unlawful killing for the practice of power as represented by the sons of the king, and of course the king himself. Their involvement in hunting contradicts Rêya Heqî philosophy and jurisprudence. This myth is the main reason why the Kurdish Alevis of Dersim (Deniz 2012, 341), and elsewhere, do not practice hunting (with the exception of the partridge), even when they suffer from hunger.

The key construction is the combination of the baby, mother, and water and their relationships. The baby is the future of the community, the mother is the source of regeneration, and water is the vital material for life. As understood from the myth, the fawn is young, unprotected, and vulnerable, and without a mother cannot survive and accordingly is the one who dies first. This emphasizes the importance of the role of the female/mother.

The fawn and the mother are placed at the site of a spring, constructing water as the main material of life and emphasizing its importance for all living beings and as a collective vital material that every living being has a right to receive. As such, it is imbued with holy stature. Water is also related to fertility. All water, particularly subterranean water, is associated with the womb, and thus fertility and birth. Anahita is the goddess of all subterranean waters. Ana Fatme seems strongly to be a variation on the name Anahiti (Deniz 2012, 232, 248), which is an important source of mineral water near the center of Dersim. Ana Fatme is also the name of the daughter of Muhammed, the wife of Ali, the mother of Huseyin and Hassan. Associating a water source with a mother demonstrates the strong traces of goddess worshipping in the region. The religious reference to a mother whose sons, husband, and family members were brutally killed by Muslims in their conflict during a power struggle also carries a double meaning in a site, the desert, where many of Huseyin's supporters died from thirst. Although this connection has not been supported by the written sources, but here it is further used to reinforce the main discourse, that is, nobody can touch any creature engaged in receiving water, a vital need. This has been transmitted from generation to generation and still has an important impact on daily practice among Alevi believers.

While water has been sanctified in the myth as a vital element for all, on the other hand, it has also been constructed as a punishing power for violators of the law. Water has been transferred from a source of life to a source of death, from a lifesaver to a killer, punisher, and threat. Note that its immense power is activated by the call of the mother/female. Following her call, this power appears with huge anger and rage. Its character turns, in contradiction to its life-giving role, into an inexorable, frantic, fatal force. It transforms from potential to kinetic, from passive to active, from a source of life to a means of death. However, in its form as a power of death, it simultaneously acts as a protection of life, in that its punishing power performs an act of justice that protects society from a violation. This grand rule of nature, not surprisingly, is reflected in goddess worshipping (Deniz, 2020b), indicating Rêya Heqî has many traces of the past before patriarchy became dominant.

The connection between the mother deer and the power of the water must also be considered. When her call activates the giant flood, it is the connection of two carriers of life to fight against the violators. This cooperation and solidarity between the life forces of animal and nature demonstrates they are on the same wavelength regarding sacred law and violations against it. This indicates a shared and eternal communications network capable of sensing and transmitting a violation in order to activate a deterrent force. At one end is the female source of life, at the other is a vital element characterized by necessity and power.

Another layer of meaning in the myth is the dichotomization of masculinity and femininity. Masculinity is associated with hunting, an unnecessary killing presented as an important crime against the right to life in nature. The connection of hunting to unnecessary killing constructs masculinity as a potential threat to the right to life. In contrast, femininity is defined as the source and protector of life. The sisters who tried to save the masculine perpetrators are also severely punished. They are considered as violating the law by trying to help their violator siblings. The helpers are dealt a severe punishment in line with that meted out to the perpetrators. This demonstrates that helping, depicted as an "undesirable femininity position," is an equally serious crime to be dealt with accordingly.

We should remember that sisterhood is an important position in the Dersim religious-cultural codex that traces back to matrilineality (Deniz 2012, 167) in its culture and history. In societies where a matrilineal tradition previously existed, sisterhood appears more strongly throughout the culture. For instance, Dizgun, the holy mountain mentioned previously, has three sisters—Munzur Çem (2010, 81) counts Buyêr (Gola Buyerê) as the mountain's sister as well. One of these sisters, *Xaskar*, is positioned as a cave on Dizgun Mountain from which a tiny amount of water runs, despite a high altitude of 2,090 meters. Gürdal Aksoy matches Xaskar with Anahiti, the goddess of subterranean water (2012, 98). The second and third sisters are *Zêl* and *Karsnî*; both are nearby mountain peaks. All of them are sacred and viewed as protectors.

In the myth, the sisters are blamed for violating matrilineal jurisdiction that entrusts females with the continuity of life. By helping their brothers in their power struggle for kingship, a system in which they as women have no agency, their violations are equally as serious. As previously noted, killing a mother means breaking the generational line, and therefore the continuity of a society. The sisters are illustrated as attached to a patrilineal and patriarchist system as daughters, and who have no rights to rule, but act only as the helpers of their brothers/males. Thus, the myth also should be considered as strongly opposing domination through patriarchal power.

Reading through, the myth suggests that a conflict exists between patriarchist powers and religious and social groups, by which we can argue that the myth records a colonial process that the local population, in accordance with its faith, resists a newly perpetrated, male-dominated power and its ruling system. It means this undesirable alien power has not been accepted or obeyed by the local population. Based on the analysis of the myth, we can argue that the matrilineal, goddess-worshipping society and culture had conflicted with a patriarchist, monotheist colonial power as well. Encoded in the myth is the message that the grand law portrayed here existed prior to the kinship and siblinghood of the brothers and sisters. It also negates the belief that killing another life proves adulthood and power and argues that such acts go against the jurisdiction of existing religion and society. Thus, it is perceived as a crime against natural jurisdiction and deserving of capital punishment.

It is important to note the sisterhood in the myth represents young women who have not yet been mothers. The perpetrators' family consists of a father, two sons, and five unmarried daughters. Interestingly, the family that does not have a mother figure has been destroyed by the call of a mother deer. While femininity is constructed as the main source of regeneration, this family, which is constructed without a mother, is depicted as aggressive, disrespectful to the sacred land, and a violator of grand law. On one side, there is a mother and a baby who are arbitrarily killed, on the other is a family with no mother figure that is fully involved in committing a capital crime. We can argue this represents a conflict between an old, matrilineal society and a patrilineal, patriarchist alien power that has pressured this society through physical power and killing. It also suggests that the matrilineal, goddess worshipping faith that put the female in a central role has been attacked by a patriarchist power structure where women are positioned only as helpers. The myth concludes that a mother is the guarantee of a community's safety and preserving such natural and societal orders prevents chaos and crime in society. This reference to a mothers' role in generational socialization is connected to mother/goddess worship. Fathers/masculinity are coded as focusing their power as aggression toward life and nature. It strongly condemns proving masculine power by engaging in hunting (arbitrary killing) and aggression toward others, activities that violate sacred law and geography.

The myth denies justification for an apparently newly defined role for the sisterhood of women as a typical sibling reflex. As helpers, as sibling conspirators, they are condemned, and thus severely punished. From this point on, we understand that no one can violate the grand rule and kill for pleasure. No matter who committed such an act, all members of society should stand against it. More importantly, the law supersedes any type of kinship relationship. This is a powerful way of protecting the functioning norms of society.

There is additional meaning in the punishment of the sisters by the storm. They do not die next to each other, but fall apart, far from each other— symbolic of the killing of sibling ties. The storm and ensuing dust cloud symbolically represent anger toward their actions. The sudden storm and flood that scatter the sisters to different places show a strong reaction to their acts. They are punished in accordance: capital punishment in the form of death by petrification as they turn into a stone under the water. All are symbolic according to the crime. No kinship relationship is more important than natural law. Separating the siblings from each other is also an important response to the nepotism at the root of their violation of the law.

It is also important to note the style of punishment of the perpetrators, particularly the brothers' chopping off each other's arms and heads. In this case, the controlling factor is not the brothers but their limbs that act independently, or, more accurately, uncontrollably, to punish each other. The limbs revert to nature's control and act out nature's retaliation, enacting the law of retaliation "an eye for an eye." In Rêya Heqî jurisdiction, no one can be killed irrespective of guilt. Thus, it is nature that metes out appropriate justice. The out-of-control arms of the two perpetrators cutting off each other's arms and heads can also be understood as punishment for the limbs that gave (heads) and fulfilled (arms) the order. In this way, no individual is involved in killing the guilty parties. The holy justice prescribed by powerful, natural law is carried out by some "guilty" limbs under nature's command.

The myth establishes a philosophy and jurisdiction concerning the value of human and animal life. Seven human beings are punished for killing a mother deer and her fawn, affirming that every life is sacred and valued by the faith and thus the culture and society, and should be respected. This is contrary to Abrahamic tradition, in which men (humans) are positioned on the top of a hierarchy and are the center of everything. Consequently, all animals are to serve or feed the "men." The myth negates the notion that killing animals is man's positional right. That the victim is female and the young male perpetrators—the potential leaders of the society—are punished eternally establishes the higher position and value of the female and animal.

No perpetrator in the myth has a grave. On the contrary, depending on the gravity of the crime, some are sent to the bottom of the newly formed lake and are thus removed eternally from the earth's surface. Some of the sisters remained on land in the form of stone, as markers to show the consequences of helping the perpetrators of a capital crime. As stone markers that remain in the same spot forever, they serve as signs of a strong warning, an educational tool through which crime and punishment are taught.

CONCLUSION

This myth depicts important ecological aspects of Rêya Heqî faith and philosophy. It presents a sophisticated jurisdiction that compares favorably over present-day individual and collective bills of rights. The main tenets of the myth can be summarized, as follows:

1) No living being should be arbitrarily killed.

2) No mother who has a dependent baby can be killed or touched.
3) No newborn baby can be killed.
4) Everyone has the right to receive the vital necessity of water.
5) Hunting is a man's power game, and thus it is a crime.
6) No crimes can be committed within sacred geography.
7) All living beings in this sacred geography are under the protection of divine/natural law and no crime can be committed against them.
8) Violation of these rules is a capital crime and will be severely punished by the divine/natural law of the sacred land.
9) Any crime against any beings in the sacred land is also an attack on the whole community and sacred land itself. Accordingly, in Rêya Heqî teaching, hunting is a prohibited act by grand and religious laws. Deer, mountain goats, and ibexes are regarded as the most sacred animals in this land, thus hunting them is an aggravated crime of the ninth rule. Dersimians hold feelings of love and respect for these species and never kill them.

The myth indicates that the region and its faith have been under the pressure of colonization, namely by the Roman Empire, Mongolians, Seljuks, Ottomans, and most recently the Turkish republic. Historically, invaders have been aggressive and disrespectful toward the belief system in the region, and this remains the case in the present day. For this reason, Dersimians have chosen to live away from any central government and its disrespectful attacks on their religion and the values of the community, which continue today. These attacks, as recorded in the myth, are a result of the contradictions between the colonial powers and the faith of the Rêya Heqî and Kurmancs, specifically the contradictions between the Alevi faith and Islam. While Islam glorifies man and his ability to kill under the name of Jihad, Alevism condemns any killing and prohibits the participation of anyone in any type of killing. The criminals who commit such crimes are never accepted into Rêya Heqî rituals.

The myth depicts hunting as an arbitrary killing and a symbol of the male competition for power, and thus a threat to surrounding life. As hunting is coded as a threat to life, it is a violation of societal norms, sacred jurisdiction, and the grand law, thus it is a capital crime. It is also marked as an act of disrespect against sacred land where no crimes can be committed. For instance, recently, an American trophy hunter was forced to cancel his hunting program in Dersim, due to social outrage (Squires 2020).[3] This was even though the state, likely motivated by an attack on *Rêya Heqî*, deliberately grants permits for such hunters, even though they are fully aware of people's beliefs and feelings.

This is the main basis of political ecology in Dersim and for the conflict between colonial Turkish-Islamic power and the Kurdish Rêya Heqî followers, indigenous people who want to save their sacred land. Rêya Heqî is a nature, woman, and animal-friendly faith with a unique philosophy and jurisdiction. As represented in the myth, Rêya Heqî faith can be said to have a strong connection with goddess worshipping that can be traced back thousands of years. As such, it has been targeted by the patriarchist, Abrahamic monotheist colonial powers from the past to the present. Despite such systematic pressure, enduring over centuries, the Rêya Heqî faith and its followers have long resisted such incursions; the preservation of the faith's nature worship has been a vital aspect of that resistance.

NOTES

1. "Alevism" (*Alevilik*) is a distinctively modern term and concept (Dressler 2008, 283). Before Turkish domination through urbanization and the forced teaching of the Turkish language, Kurdish Alevis did not define themselves with that term. During prayer, they referred to "Rêya Heqî," which means the path of the eternal truth. Therefore, more recently instead of "Kurdish Alevism," Kurdish scholars have used the term Rêya Heqî as the community understood it before the imposition of the Turkish language. The term "Kurmanc/Kirmanc," indicates both the ethnicity (Kurds) and the religion (Rêya Heqî). Throughout this article, therefore, instead of referring to Kurdish Alevis, the terms Kurmancs and the term Rêya Heqî will be used to refer to the religion and the Kurdish followers, since they use the term of Kurmanc to refer to themselves in both the dialect of Kurdish, Kurmanci (sounds like Kurmanji) and Kirmancki/Zazaki (Deniz 2012, 52, 335). Both terms are the same in meaning, only different pronunciations in these dialects.

2. https://ku.glosbe.com/ku/en/xwezan.

3. Unfortunately, the news story, by Nick Squires, in *The Telegraph* inaccurately defines the Kurdish Alevis of Dersim as "a Turkish ethnic group," and "Shia Muslim" (Squires 2020). Dersimians are not ethnically Turkish, but Kurdish, are not Shia Muslim but Alevis/Rêya Heqî's followers. The only common thing between Alevis and Shia Muslims is the name of Ali. Shia Muslim fully follow five Islamic pillars that Alevis do not.

REFERENCES

Aksoy, Gürdal. 2012. *Dersim, Alevilik, Ermenilik, Kürtlük*. Ankara: Dipnot.

Antranik, Yeritsyan. 2012. *Dersim Seyahatname*. Istanbul: Aras.

Aygün, Hüseyin. 2009. *Dersim 1938 Resmiyet ve Hakikat*. Ankara: Dipnot.

Birdoğan, Nejat. 1995. *Anadolu Aleviliği'nde Yol Ayrımı, İçerik-Köken*. İstanbul: Mozaik Yayınları.

Bulut, Faik. 2009. *Dersim Raporları: İnceleme Genişletilmiş Basım*. İstanbul: Evrensel Basım.

Caldwell, Richard. 1989. *The Origins of the Gods: A Psychoanalytic Study of Greek Theogonic Myth.* New York: Oxford University Press.

Çem, Munzur. 2010. *Dersim Merkezli Kürt Aleviliği.* Istanbul: Vate.

Deniz, Dilşa. 2012. *Yol/Re:Dersim İnanç Sembolizmi, Antropolojik Bir Yaklaşım.* İstanbul: İletişim.

Deniz, Dilşa. 2019. "Kurdish Alevi Belief System, Rêya Heqî/Raa Haqi: Structure, Networking, Ritual and Function." In *Kurdish Alevis and the Case of Dersim, Historical and Contemporary Insights,* edited by Erdal Gezik and Ahmet Kerim Gültekin, 45–73. New York: Lexington.

Deniz, Dilşa. 2020a. "Re-assessing the Genocide of Kurdish Alevis in Dersim, 1937–38." *Genocide Studies and Prevention: An International Journal* 14, no. 2: 20–43. https://doi.org/10.5038/1911-9933.14.2.1728.

Deniz, Dilşa. 2020b. "The Shaymaran: Philosophy, Resistance and Defeat of the Lost Goddess of Kurdistan." *The Pomegranate: The International Journal of Pagan Studies* 22, no. 2: 221–248.

Dersimi, Nuri. 1992. *Kürdistan Tarihinde Dersim.* Diyarbakır: Dilan.

Dressler, Markus. 2008. "The myths of Rêya Heqî as Sacred Texts of the Religion and Society Religio-Secular Metamorphoses: The Re-Making of Turkish Alevism." *Journal of the American Academy of Religion* 76, no. 2: 280–311. Doi:10.1093/jaarel/lfn033.

Eliade, Mircea. 1992. *İmgeler Simgeler.* Translated by Mehmet Ali Kılıçbay. Ankara: Gece Yay.

Encyclopaedia Britannica. "Al-Ḥusayn ibn ʿAlī, Muslim Leader and Martyr," by the Editors of Encyclopaedia Britannica, last updated January 1, 2021. https://www.britannica.com/biography/al-Husayn-ibn-Ali-Muslim-leader-and-martyr.

Gezik, Erdal, and Huseyin Çakmak. 2010. *Raa Haqi-Riya Haqi, Dersim Aleviliği İnanç Terimleri Sözlüğü,* Ankara: Kalan.

Graves, Robert, and Raphael Patai. 2009. *İbrani Mitleri, Tekvin-Yaratılış Kitabı,* translated by Uğur Akpur. İstanbul: Say.

Gültekin, Ahmet Kerim. 2019. "Kurdish Alevism: Creating New Ways of Practicing the Religion." HCAS *Multiple Secularities—Beyond the West, Beyond Modernities.* Leipzig University.

Halis Paşa. 2010. "Halis Paşa Raporu." In *Jandarma Genel Kumandanlığı Dersim Raporu,* 242–250. İstanbul: Kaynak.

James, Edwin Oliver. 1959. *The Cult of the Mother Goddess, An Archaeological and Documentary Study.* New York: Frederic A. Praeger.

Johnson, Christopher. 2003. *Claude Levi-Strauss, The Formative Years.* Cambridge: Cambridge University Press.

Kaynak. 2010. *Jandarma Umum Komutanlığı Dersim Raporu* [1932]. İstanbul: Kaynak.

Kaynak. 2011. *Genelkurmay Belgelerinde Kürt İsyanları I-II.* İstanbul: Kaynak.

Korff, Jens. 2020. "Meaning of Land to Aboriginal People." *Creative Spirits* website. Last updated May 21, 2020: https://www.creativespirits.info/aboriginalculture/land/meaning-of-land-to-aboriginal-people#What_does_land_mean_to_Aboriginal_people.

Leach, Edmund. 1976. *Culture and Communication, the Logic by which Symbols are Connected. An Introduction to the Use of Structuralist Analysis in Social Anthropology.* Cambridge: Cambridge University Press.

Leeming, David. 2003. *Myth: A Biography of Belief,* Oxford Scholarship Online. DOI: 10.1093/0195142888.001.0001.

Lévi-Strauss, Claude. 1963. *Structural Anthropology.* New York: Basic Books.

Molyneux-Seel, L. 1914. "A Journey in Dersim." *The Geographical Journal* 44, no 1: 49–68.

Murdock, Maureen. 2016. "The Goddess and Marija Gimbutas." *Jung Society of Atlanta* 11: https://www.academia.edu/31735390/The_Goddess_and_Marija_Gimbutas.

Neumann, Eric. 1955. *The Great Mother.* Princeton, NJ: University of Princeton.

Önal, Sami. 2006. *Tuğgeneral Ziya Yergök'ün Anıları; Harbiye'den Dersim'e (1890– 1914).* Istanbul: Remzi Kitabevi.

Randall, Bob. 2009. "The Land Owns Us." *Global Oneness Project,* February 26, 2009. Video. https://www.youtube.com/watch?v=w0sWIVR1hXw.

Squires, Nick. 2020. "American Trophy Hunter Scared Off by Turkish Ethnic Group that Considers Goats Sacred." *The Telegraph,* December 9, 2020. https://www.telegraph.co.uk/news/2020/12/09/american-trophy-hunter-scared-turkish-ethnic-group-considers-/.

Ünlü, Barış. 2014. "Kürdistan/Türkiye ve Cezayir/Fransa: Sömürge Yöntemleri, Şiddet ve Entellektüeller." In *Türkiye'de Siyasal Şiddetin Boyutları,* edited by Güney Çeğin and İbrahim Şirin, 403–34. İstanbul: İletişim.

Yücel, Müslüm. 2003. *Kürt Coğrafyasında Göl ve Irmak Efsaneleri.* Istanbul: Evrensel Basım Yayın.

Part V

CONFLICT AND ENVIRONMENTAL DESTRUCTION

Chapter 18

Forest Fires in Dersim and Şırnak

Conflict and Environmental Destruction

Pinar Dinc

In October 2020, forest fires broke out in the İskenderun and Arsuz districts of Hatay, a southern province in Turkey bordering Syria. A news article published in the pro-government newspaper *Daily Sabah*, about the devastating forest fires in the province, highlighted the government's effective action plans regarding its firefighting measures and strategies and reported that 166 fire engines, fifteen bulldozers, drones, and 700 workers worked to put the fires out (Daily Sabah 2020). The news piece also mentioned the Children of Fire Initiative, which was defined as a group with close ties to the Kurdistan Workers' Party (*Partiya Karkerên Kurdistan*, PKK). It was reported that the Children of Fire Initiative has glorified the fires in Hatay and that "four provinces around Turkey fell victim to the PKK's hatred of nature on Saturday, with almost simultaneous fires destroying forests in various parts of the country . . . the PKK either claimed or was accused of starting many forest fires in the past" (Daily Sabah 2020).

It was true that the Children of Fire Initiative issued a statement, published on the *Nûçe Ciwan* website, claiming that "the towns and cities of those who invade and violate all parts of Kurdistan will also burn in this fire" (Nûçe Ciwan 2020). The Press Center of the People's Defense Forces (*Hêzên Parastina Gel*, HPG, known as the military wing of the PKK), however, denied any links with the forest fires in a statement (ANF News 2020). The Peoples' Democratic Party's (Halkların Demokratik Partisi, HDP), which is often viewed as the pro-Kurdish political party in Turkey's parliament, had accused the Justice and Development Party (*Adalet ve Kalkınma Partisi*, AKP) government of deliberate ecological destruction. The HDP Ecology Commission's co-spokesperson Naci Sönmez stated that they believed these forest fires in Hatay, as well as in the provinces of Cudi/Şırnak (Şirnex) and Dersim, were "not ordinary forest fires" but intentional destruction of the ecosystem for the economic benefit of capitalist corporations (Halkların Demokratik Partisi 2020). Sönmez also declared that the government was intentionally manipulating the public by reporting that these forest fires were the result of terrorist activities (Halkların Demokratik Partisi 2020). Indeed, the "PKK burned Hatay" hashtag [#HatayıPKKYaktı] on Twitter was a trending topic in Turkey during the forest fires in October 2020.

The recent forest fires in Hatay and the discussions around it have once again shown how media becomes an effective source that disseminates contradictory narratives of competing actors and how the content of mass and social media representation can assume the form of propaganda, manipulation, misinformation, or at times censorship and disinformation. What is disseminated or concealed in the media is highly interconnected with the owners of the media channels and their political alliances, particularly in authoritarian settings where media freedoms are under attack. In light of this background, the aim of this chapter is twofold. First, it focuses on environmental destruction during times of escalating war and conflict through politically non-biased data[1] about fires and conflict events in Dersim and Şırnak, to show the changes in the number of fires and conflict in these provinces during the peace talks between Turkey and the PKK from 2009 until the summer of 2015. Second, it examines the forest fires and conflict nexus through a qualitative discourse analysis of selected print newspapers in Turkey. In doing so, the chapter focuses specifically on August 2015 and shows the role of the opposition media as well as social media in persistently highlighting the environmental destruction in Turkey's Kurdish regions as a state strategy, while the mainstream media turned a blind eye.

THE FIRE AND CONFLICT NEXUS

Environmental destruction has historically been used as a military strategy in times of wars and armed conflict (Austin and Bruch 2000). Armed conflicts are known to have significant impacts on the land systems. To give some varied instances, it was used by the British Army against Malayan guerrillas, by the US Army during the Vietnam War in Southeast Asia, during the armed conflict between Rwanda and the Democratic Republic of Congo, and during the Gulf War in Iraq. Central governments have also used environmental destruction as a military tactic to deal with insurgencies during ethnonationalist civil wars in geographically isolated places, such as against the Tamils in Sri Lanka and Chechens in Russia (Gurses 2012).

The Ottoman Empire, and the Republic of Turkey that followed, are no exceptions to this trend, as both used the destruction of forests and nature as a military tactic against their rivals. One example of this strategy is revealed in the diary of Yeritsyan Antranik, an Armenian traveler. In his description of the late 1800s, Antranik writes that the Ottoman Sultan was told that Dersim would remain undefeated as long as its forests remained, which led the emperor to order his army to set Dersim's forests on fire (Antranik 2012). Although the Sultan's attempt miraculously failed in the late 1800s (Antranik 2012), during the early Republican period (1924–1938), the military forces once again used forest fires as a tactic in Turkey's Kurdish regions (Arslan 2014). Similar strategies were later used during the armed conflict between Turkey and the PKK in the early 1990s, when the armed forces used forest burning as a counterinsurgency tactic (Jongerden et. al. 2007; Etten et al. 2008; Dinç 2020a; Dinc et al. 2021), leading to the destruction of forests, as well as agricultural lands, livestock, and villages.[2] Such attacks were accompanied by the forced migration of thousands of people (van Bruinessen 1994). Based on the Kurdish Human Rights Project report (2002), over 3,000 villages were evacuated and/or burned between 1991 and 2001, the majority of which took place between 1993 and 1994. Such destruction not only has severe humanitarian and economic consequences for the people in these conflict zones but it also damages "the prospects for building and sustaining the peace in the aftermath of war" (Gurses 2012, 268).

Although we cannot talk about an uninterrupted process, there had been intermittent dialogue between Turkey and the PKK in the 2000s. Backchannel communications between Turkey and the PKK in 2005 led to the Oslo Talks between 2008 and 2011 (Dicle 2017), followed by the peace process (also referred to as the Solution Process or İmralı Process) between 2012 and 2015. When the peace process came to an end in July 2015, urban warfare between the PKK and Turkey started in various Kurdish cities in Turkey. This urban warfare continued between August 2015 and March 2016, leading to the death of hundreds of civilians, human rights abuses, the internal displacement of thousands of citizens, and massive destruction of Kurdish cities (Crisis Group Europe 2016). The human rights abuses and shift from peace and democracy were not limited to the Kurdish cities and people. The aftermath of the peace process, combined with the alleged coup attempt against the AKP government in July 2016, resulted in intensified oppression, criminalization, and violation of freedoms against—but not limited to—the Kurdish political movement, politicians, journalists, academics, human rights defenders, and activists.

Even during the peace process, the so-called "development" and "security" policies of the state were being implemented in Turkey's Kurdish regions through dam projects, mining activities, fossil-fuel power station constructions, as well as the construction of fortress-military stations (*kule-kol*s) (Orhan 2013). Many (Dinc 2020b; Dinc 2020c; Hunt 2019; TATORT Kurdistan 2013, 145; Gurses 2012) describe the state actions as deliberate

Table 18.1 Number of Conflicts (in Turkey, Şırnak, and Dersim) and Fires during the Extended Dry Season (Şırnak and Dersim) between 2009 and 2015

Year	Number of Fires in Şırnak	Number of Conflict Events in Şırnak	Number of Fires in Dersim	Number of Conflict Events in Dersim	Number of Conflict Events in Turkey
2009	162	9	1	2	44
2010	215	7	7	3	64
2011	121	6	5	10	121
2012	115	14	8	9	129
2013	114	0	2	0	7
2014	101	0	5	0	7
2015	405	30	14	5	232

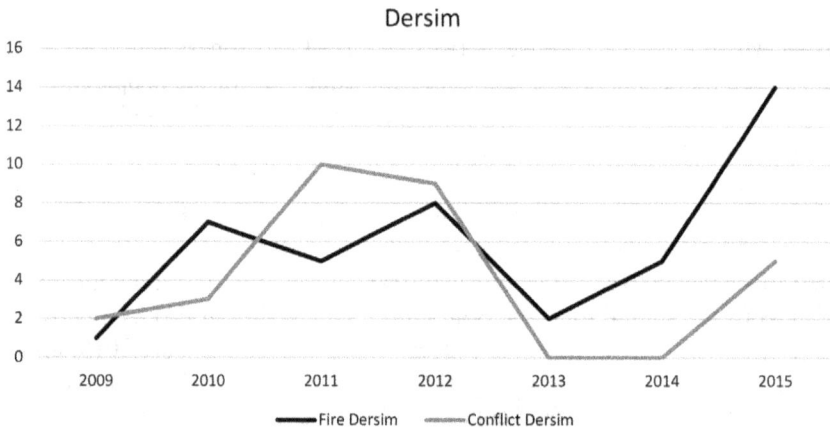

Figure 18.1 Rise and Fall in the Number of Conflicts and Fires in Dersim during the Dry Seasons between 2009 and 2015. *Source:* © Pinar Dinc.

and systematic ecological destruction. More recently, the environmental destruction in the area has been defined as a "neocolonial ecocide" (Cudi 2020). For the purposes of this chapter, I specifically focus on the fire[3] and conflict data for Dersim and Şırnak during the extended dry season—the six months between May 1 and October 31, 2015. Table 18.1 above shows a sharp increase in the number of conflict events[4] in Turkey between 2014 and 2015.

Conflict events in Şırnak and Dersim were reported as zero in 2014, whereas in 2015 the numbers were reported as 30 and 5, respectively. The fire data shows that the number of fires in Şırnak quadrupled and the number of fires in Dersim tripled from 2014 to 2015. Figures 18.1 and 18.2 display

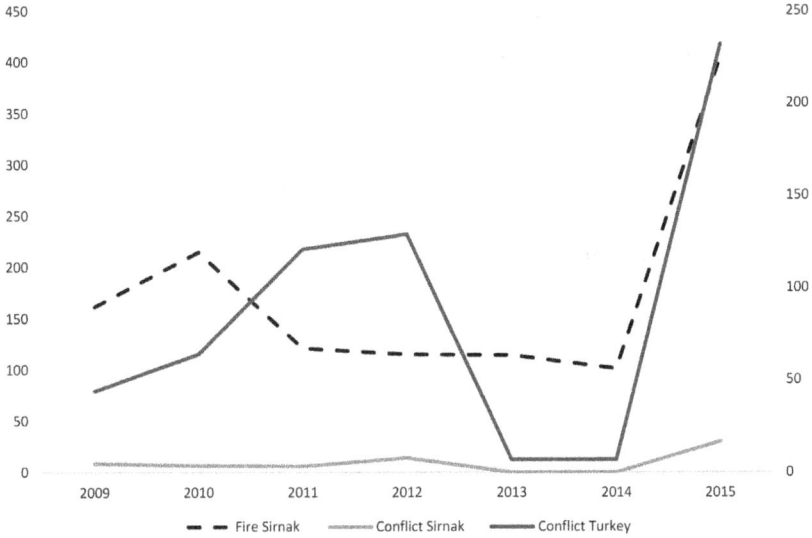

Figure 18.2 Rise and Fall in the Number of Conflicts (Turkey Plotted in Secondary Axis) and Fires in Şırnak during the Dry Seasons between 2009 and 2015. *Source*: © Pinar Dinc.

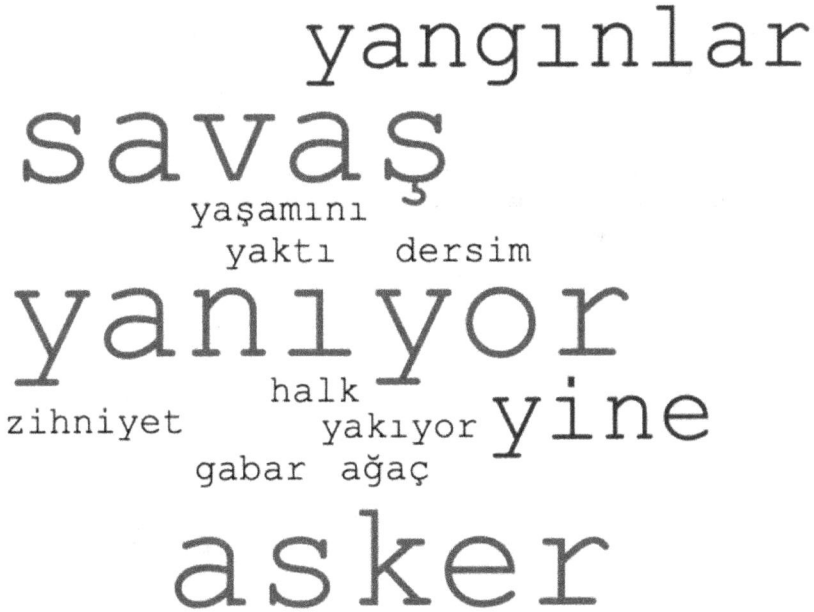

Figure 18.3 Word Cloud of News Article Titles in Özgür Gündem and Evrensel with Minimum Word Frequency of Two. *Source*: © Pinar Dinc.

the increasing trends in the number of conflicts and fires in both provinces in the aftermath of the peace process in 2015.

The spatial analysis of this data shows that in certain areas in Şırnak and Dersim, as the number of conflict events increases, the number of fires also increases. In other words, there is a positive correlation between the number of forest fires and conflict events in particular areas. This was also confirmed through the Bivariate Local Moran's Index (BiLISA), through which the correlation between fire and conflict variables were investigated (Dinc et. al. 2021). BiLISA analysis showed that the Cudi Mountains were located to the north of Silopi and the east of Cizre districts of Şırnak. The same applies to Dersim, where the Aliboğazı, Munzur Valley, Bali, Kutuderesi, and Geyiksuyu areas were the hotspots where both conflict and fire variables had high values (Dinc et. al. 2021).

MEDIA COVERAGE

Despite the positive correlation between the number of fires and conflict events in Dersim and Şırnak, media coverage of this phenomena is largely avoided, if not deliberately concealed. Table 18.2 shows the distribution of the seven newspapers analyzed in this chapter, together with their political leaning and the number of articles (including news pieces and opinion columns) published about the forest fires in Dersim and/or Şırnak during August 2015.

Table 18.2 shows that the mainstream newspapers did not report the forest fires, while left-wing newspapers were more attentive. *Evrensel* and *Özgür Gündem* were the two newspapers that recurrently published news articles and opinion columns about the forest fires not only in Dersim and Şırnak but also in other Kurdish regions in Turkey. A total of forty-eight news articles were published in *Evrensel* and *Özgür Gündem* in August 2015. Figure 18.3 displays a word cloud of the titles of all forty-eight articles, where stop

Table 18.2   Newspapers, Their Political Leaning, and the Number of Articles about Forest Fires in Dersim and/or Şırnak in August 2015

Newspaper	Political Leaning	No of Articles
Sabah	Mainstream, pro-government, right-wing	0
Sözcü	Mainstream, anti-government, Kemalist	0
Hürriyet	Mainstream, pro-government	1
Birgün	Left-wing	4
Cumhuriyet	Left-wing, Kemalist	4
Evrensel	Left-wing	19
Özgür Gündem	Left-wing, pro-Kurdish	29

var en değil
hemen yansın sizin
yanıyor daha pkk yine olsun
Kürt
arttırma yangın ateş
insan bizim bin
orman geliyor olmak sessiz
fenomen çok Allah
kapanmadan ses yok oldu
sonra size
ağaç veren şu devam yakar
su biz devlet Dersim
Lice TC Kürdistan
girmeyen yangını
şimdi yanan doğa
neden siz

Figure 18.4 Word Cloud of Tweets Shared on Twitter about Forest Fires in Dersim and/ or Şırnak between July 1, 2015, and December 31, 2015. *Source*: © Pinar Dinc.

words were removed[5] and the words that appeared at least twice were kept. The words burning [*yanıyor*], soldier [*asker*], and war [*savaş*] were the most frequently used (four times each), followed by again [*yine*], fires [*yangınlar*] (three times each); and its life [*yaşamını*], burned [*yaktı*], people [*halk*], mentality [*zihniyet*], putting on fire [*yakıyor*], tree [*ağaç*], and two place names, Dersim and Gabar in Şırnak (two times each).

Several issues were raised in the news articles on the environmental destruction in Turkey's Kurdish regions that were published in *Özgür Gündem* and *Evrensel* during August 2015. First, both newspapers placed a special emphasis on the importance of peace, their desire for the war to end, and the democratization of Turkey. Second, it is important to highlight that these two newspapers, and *Özgür Gündem*, in particular, used the original Kurdish names of the areas in which there were forest fires. This was not only a political choice but also a necessity considering their readership who will refer to these areas by their original Kurdish names. The news articles often used a language that highlights a sense of attachment and belonging to that geography, Kurdistan, but also "our trees" or "our nature." Third, both newspapers placed a special emphasis on the destruction of not only forests but the living space, including villages, pasture areas, and farming areas. Fourth, these forest fires were described as fires caused by the armed forces. *Özgür Gündem* in particular claimed that the soldiers started these fires and that forest fires were used as a war strategy against the Kurds. Fifth, the newspapers highlighted that the state was not making an effort to extinguish the fires, and it was also preventing civilians from taking part in firefighting. In relation to this, both newspapers reported on the protest activities against these forest fires and made a call to the wider society to not remain silent on this intentional environmental destruction. Finally, the systematic nature of these fires and the long history of environmental destruction in Turkey's Kurdish regions was emphasized, often with reference to the 1990s, which was also visible—albeit much fewer in number—in *Birgün* and *Cumhuriyet* as well.

Among the four news articles in *Birgün* that covered forest fires in Dersim and/or Şırnak, two drew links between forest fires, conflict, and the traumatic memory of the 1990s. On August 4 and August 14, 2015, *Birgün* published two news articles with the titles "What difference does this have from the 1990s?" and "Forest fires put out the peace." Three out of the four news articles published in *Cumhuriyet* on forest fires in Dersim and/or Şırnak were opinion columns written by Aydın Engin.[6] Engin also refers in these articles to the destruction of forests and forced evacuation of civilians in the 1990s and demands that the same mechanisms of the 1990s—meaning environmental destruction via the armed forces—should not be used once

Figure 18.5   Site of Forest Fire at Bingöl-Yayladere, in 2015. *Source*: © Mesopotamia Ecology Movement.

again by the AKP. However, it should be noted that there were no news articles that reported on the forest fires that took place in these provinces in August 2015.

The only article published in *Hürriyet* was on August 6, which reported a soldier who was "martyred in a forest fire." The article explains that the sergeant was taken to a hospital because he was affected by the smoke while tackling blazes in a forest fire in Şırnak.

Right-wing, nationalist newspapers, regardless of their position as pro- or anti-government, did not publish any news about the forest fires in Dersim or Şırnak. The news articles about the provinces mainly reported on "terror," "terrorism," "traitors," and the PKK members killed by the Turkish armed forces. While forest fires in other, mainly western, parts of Turkey were reported as fires that scared people living in these areas, provinces like Dersim and Şırnak were only visible in news articles that reported the killing of "traitors" in the military operations of the armed forces against the PKK. In August 2015, two news articles appear next to each other in *Sözcü* newspaper, where on the left column, the title reads "14 traitors have been killed in Tunceli and Şırnak" whereas, on the right column, a forest fire that scared the people in the Bursa province in western Turkey was reported. The pictures used in the newspaper are also noteworthy, as two military aircraft are seen in the news column about Dersim, whereas in the column about Bursa, two women are shown helping to extinguish the fires. What is noteworthy here is that this is not a one-time instance but a general tendency that can be seen in mainstream newspapers. News articles about Kurdish provinces are often related to "terror" events, while forest fires in these areas are not reported. This seems to be a deliberate choice, since forest fires taking place in other parts of Turkey are reported, while forest fires in conflict zones are not.

SOCIAL MEDIA

In the last decade, social media has started to replace traditional media in disseminating information and creating public participation in a way that includes different voices that struggle against hegemonic narratives (Khaldarova and Pantti 2016). Twitter and other participatory online platforms have become spaces in which activists, journalists, and the general public express themselves, engage in politics, and disseminate silenced news stories. In Turkey, too, social media plays an important role in protest movements: the Gezi Park protests of 2013 are one of the most recent and significant examples of such movements. However, as Ozduzen and McGarry (2020) underline, social media is used by a diverse group of actors with opposing views, which are contested during and after the course of events.

The analysis of posts about forest fires in Turkey's Kurdish regions on Twitter[7] in the aftermath of the peace process in 2015 reveals important insights about the contested nature of social media. Over 19,000 tweets (excluding retweets) were shared between July 1, 2015, and December 31, 2015, with selected hashtags including—but not limited to—"Dersim is burning" [Dersim yanıyor], "there's fire in Lice" [Licede Yangın Var], and "Cudi Mountain is burning" [Cudi Dağı Yanıyor].[8] Figure 18.4 shows a word cloud with the most frequently used words in these tweets, after removing the hashtags, stop words, and phrases linked to other topics (that is mainly regarding increasing follower numbers on Twitter). The most frequently used words are burning [yanıyor], fire [yangın], there is [var], absent [yok], the PKK, the state [devlet], Lice, Kurd [Kürt], and nature [doğa].

The majority of these tweets aimed to make the forest fires in Şırnak, Diyarbakır, and Dersim visible to the public. Similar to the news articles in *Özgür Gündem* and *Evrensel*, these tweets made references to the forest fires initiated by the armed forces in the 1990s and highlighted the intended nature and ideological motivations of the state's actions against the Kurds. There were also critiques toward the apathy about the forest fires taking place in Kurdistan, underlining that the forest fires or environmental destruction in Turkey's Kurdish regions were not deemed important by the majority of the Turkish public and the mainstream media. The state, as well as the public, was being called to notice these forest fires and to take action to extinguish them and stop the destruction of the Kurdish environment.

The Internet, however, does not only function as a sphere for freedom of speech but also allows its users "to engage in othering and hate" (Ozduzen, Korkut, and Ozduzen 2020, 4). Supporting this argument, there were tweets that blamed the PKK for starting forest fires as an act of terror and argued that the PKK was now using a so-called environmentalist discourse to cover up its crimes. There were also tweets suggesting that even if forest fires were resulting from military operations, they were still legitimate in Turkey's fight against terror.

The War Press during Times of Environmental Destruction

The War Press [Savaş Press] title comes from *Özgür Gündem* in August 2015, months before it was shut down in 2016 by an executive decree. The article described the mainstream media, including *Star, Sabah, Vatan, Akşam, Yeni Şafak, Milliyet,* and *Hürriyet,* as fueling the fight against the Kurds led by the AKP government. The article stated that the media has been acting like a warmonger by turning a blind eye to the government's "arrests, imprisonments, executions, civilian massacres, burning of corpses, burning of forests, declarations of states of emergency, and all kinds of torture, and returning to the familiar narratives of 'dishonorable, villain, disloyal, traitor, ignoble' of the 1990s" (Özgür Gündem 2015).

The argument about the close ties between the government and the media in Turkey is well-grounded. Sözeri and Kurban (2014, 208) highlight that the "vast majority of news outlets [in Turkey] belong to large conglomerates with political and financial ties to the government" which becomes a factor that fosters censorship. This, as a result, creates a type of censorship that includes not only top to bottom but also a form of self-censorship to avoid lawsuits, dismissals, fines, and media outlet closures. The suppression of Turkey's media system, which was already clientelist and politicized, has "deteriorated further during the AKP regime" (Yesil 2016, 138). Media freedoms, as well as human rights, have been in sharp decay in Turkey, which was labeled a "hybrid regime" in the Economist Intelligence Unit Democracy Index in 2019. Yesil (2016, 139) draws attention to the similarities between the AKP's tactics and other authoritarian regimes in the world and argues that the AKP uses "information management" as a strategy.[9]

Kampf and Liebes (2013, 3) define the coverage of war and terror in news as "more complex and varied than in the past," and individuals today "confront an abundance of competing images, frame and narratives, from above and below, from far away and close by, and from involved and uninvolved sources." Despite numerous local accounts on the ground (see Akkuş 2015a; 2015b; 2015c), reports (Halkların Demokratik Partisi 2015; Demokratik Toplum Kongresi 2015), and the positive correlation between conflict and fire evident through data analysis, the state, as well as the mainstream media, failed to cover the escalated conflict from different perspectives. The AKP, which once seemed devoted to overthrowing Kemalism and its military tutelary, replaced it with Erdoganism, and the Kurds have continued to be labeled as domestic threats to the well-being and indivisible unity of Turkey under both ideologies dominating the republic (Dinc 2021, forthcoming). The indifference regarding the environmental destruction in Turkey's Kurdish regions can also be understood through the deeply embedded Turkish nationalism. Although the Turkish nationalist perspective views Turkey as a united whole, the indifference to the environmental destruction in Turkey's Kurdish regions shows that some parts of Turkey can be subjected to destruction and devastation, so long as it supports the security of the nation and the well-being of the state.

Consequently, we see that the Kurdish regions in Turkey have been systematically subjected to ecological destruction through dam projects (Bilgen 2018; Bilgen 2020; Ağar and Böhm 2021), mining activities, and deforestation. Even during times of reduced conflict, environmental destruction continues to take place under banners of developmentalism or securitization. However, the scope and intensity of destruction significantly increases during times of escalated conflict. Particularly with the imposition of states of emergency and military security zones, the destruction of the environment becomes unrestrained, while organizational or civilian action to prevent such destruction is stonewalled. This has been the case in Turkey's Kurdish regions once again, since the end of the peace process in the summer of 2015.

In this chapter, I have focused on the case of forest fires in Dersim and Şırnak, in the immediate aftermath of the peace process between Turkey and the PKK. Using different sets of empirical data, I have shown the positive correlation between fires and conflict in Dersim and Şırnak, and portrayed the coverage—or lack thereof—in both print and social media through examples from selected newspaper coverage and Twitter posts. In doing so, I have argued that the systematic ecological destruction cannot be prevented with an apolitical approach or by viewing ecological destruction in parts of Turkey where Sunni-Turkishness is not the hegemonic power (that is mainly Kurdish, Alevi areas) as a natural or legitimate outcome of "protecting the homeland from division and terrorism." Likewise, the approach of blaming the Kurds or the PKK for the forest fires happening in the non-Kurdish-dominated parts of Turkey has to be changed. Instead, we need a political ecology perspective in Turkey that breaks what Ünlü (2016) defined as the "Turkishness contract" and understands that the environment is equally important in all parts of Turkey (Dinç 2020b). Likewise, it is important to emphasize that such responsibility should be expected not only from the ruling party but also from the opposition parties, as well as local, national, and international environmentalist groups that often fail to draw attention to the destruction of the environment in Turkey's predominately Kurdish regions (Dinç 2020c).

NOTES

1. The data was gathered through remote sensing techniques based on fire data (downloaded from NASA's Fire Information for Resource Management System, FIRMS), satellite images (detected by MODIS satellite sensors and their corresponding algorithms), and geo-referenced global conflict data (downloaded from the Conflict Data Program, UCDP). I would like to thank Lina Eklund and Aiman Shahpurwala for their help in retrieving and analyzing this data.

2. For a study on the effects of conflict on land systems in Iraq, which specifically focuses on the impact of the Turkey-PKK conflict on forested areas in the conflict and fire in the Kurdistan Region of Iraq, see Eklund et. al. 2021.

3. The data provides information on active fires or "hotspots" detected by MODIS satellite sensors and their corresponding algorithms. The algorithm determines if there has been a fire in a 500 x 500 meter area, and then centers the point within that area. The data is provided in point shapefile format with WGS 1984 native projection. Only data classified as Type "0," indicating vegetative fires, were found.

4. Each conflict event in the UCDP dataset is defined as an instance of organized violence with at least one fatality.

5. The expression 'çocuk işi' [it's a snap] was also removed from the word cloud to only keep the words relevant to the forest fires in the regions.

6. Aydın Engin no longer writes in *Cumhuriyet.*

7. Over 19,000 tweets were extracted from Twitter's Advanced Search using Python between July 1, 2015 and December 31, 2015 with the GetOldTweets3 library (https://pypi.org/project/GetOldTweets3/) using specific hashtags about the forest fires in Dersim, Şırnak, and Diyarbakır. Since many of these hashtags are often used together, the data was pre-processed by deleting duplicate tweets. These tweets were also pre-processed using the Natural Language Toolkit (NLTK) library on Python (https://pypi.org/project/nltk/). This process removed any unnecessary punctuation symbols, and Turkish stopwords (that is "bu," "ve," "ya," and "filan"). The author would like to thank Cansu Ozduzen for her help in retrieving and analyzing Twitter data.

8. Full list of hashtags included in the search are: #DersimYaniyor [Dersim is Burning], #DersimYakiliyor [Dersim is being burnt], #DersimeSesVer [Give voice to Dersim], #DersimeSesOl [Be the voice of Dersim], #DersimeSuOl [Be water for Dersim], #DersimeBirDamlaSuOl [Be a drop of water for Dersim], #DersimKöyleriBosaltiliyor [Villages are evacuated in Dersim], #TunceliYaniyor [Tunceli is burning], #LiceYaniyor [Lice is burning], #LiceyeSesVer [Give voice to Lice], #LicedeYanginVar [There is fire in Lice], #LicedeDogaKatliamiVar [The nature is destroyed in Lice], #CudiYaniyor [Cudi is burning], #CudideDogaKatliamiVar [The nature is destroyed in Cudi], #CudiDagiYaniyor [Cudi Mountain is burning], #CudiDagindaKatliamVar [There is a massacre in Cudi Mountain].

9. In the years following the end of the peace process and re-securitization/re-militarization, forest fires continued to take place in Turkey's Kurdish regions. These fires were not widely reported in the mainstream media, and when they were shared on social media, the response from the government and state cadres often highlighted that these were exaggerated or fake news.

REFERENCES

Ağar, Celal Cahit, and Steffen Böhm. 2018. "Towards a Pluralist Labor Geography: Constrained Grassroots Agency and the Socio-Spatial Fix in Dêrsim, Turkey." *Environment and Planning A: Economy and Space* 50, no. 6: 1228–49.

Akkuş, Alper Tolga. 2015a. "Yeşil Gazete #ormanlarımızıyaktırmayacağız Demek Için Amed'de." *Yeşil Gazete* (August). https://yesilgazete.org/yesil-gazete-ormanlarimiziyaktirmayacagiz-demek-icin-amedde/.

Akkuş, Alper Tolga. 2015b. "Yanan Ormanlar ve Yanan Hayvanların Etlerinin Kokusu." *Yeşil Gazete* (August). https://yesilgazete.org/yanan-ormanlar-ve-yanan-hayvanlarin-etlerinin-kokusu/.

Akkuş, Alper Tolga. 2015c. "#ormanlarımızıyaktırmayacağız 3 : Şırnak, Silopi, Görümlü ve Eruh." *Yeşil Gazete* (August). https://yesilgazete.org/ormanlarimiziyaktirmayacagiz-3-sirnak-silopi-gorumlu-ve-eruh/.

ANF News. 2020. "HPG Denies Reports Blaming the PKK for Forest Fires in Turkey." *ANF News*, October 12, 2020. https://anfenglish.com/news/hpg-denies-reports-blaming-the-pkk-for-forest-fires-in-turkey-47197.

Antranik (Yeritsyan). 2012. *Dersim Seyahatname*. Translated by Payline Tomasyan. Istanbul: Aras.

Arslan, Serhat. 2014. "Kürdistan'da Doğa/Ekoloji Katliamı." *Toplum ve Kuram* no. 9: 62–74.

Austin, Jay & Carl Bruch. 2000. *The Environmental Consequences of War: Legal, Economic, and Scientific Perspectives*. Cambridge: Cambridge University Press.

Bilgen, Arda. 2018. "A project of destruction, peace, or techno-science? Untangling the relationship between the Southeastern Anatolia Project (GAP) and the Kurdish question in Turkey." *Middle Eastern Studies*, 54, no. 1: 94–113.

Bilgen, Arda. 2018. "The Southeastern Anatolia Project (GAP) Revisited: The Evolution of GAP over Forty Years." *New Perspectives on Turkey* 58: 125–54.

Christensen, Christian. 2011. "Twitter Revolutions? Addressing Social Media and Dissent." *The Communication Review* (Yverdon, Switzerland) 14, no. 3: 155–57.

Crisis Group Europe. 2016. "The Human Cost of the PKK Conflict in Turkey: The Case of Sur." Briefing 80. Diyarbakır, İstanbul, Brussels: International Crisis Group. https://www.files.ethz.ch/isn/196474/b080-the-human-cost-of-the-pkk-conflict-in-turkey-the-case-of-sur.pdf.

Cûdi, Renas. 2020. "Cûdi'de yeni-kolonyal müdahale ve eko-kırım." Yeni Özgür Politika, April 4, 2020. https://justpaste.it/50tdw.

Daily Sabah. 2020. "Forest Fire Hits Turkey's Hatay a Day after PKK Terror Attack." *Daily Sabah*, October 27, 2020.

Demokratik Toplum Kongresi. 2015. "Orman Yangınları Araştırma, İnceleme ve Gözlem Raporu." https://docplayer.biz.tr/52778763-Orman-yanginlari-arastirma-inceleme-ve-gozlem-raporu.html.

Dicle, Amed. 2017. *2005–2015 Türkiye-PKK GÖRÜŞMELERİ: Kürt Sorununun Çözümüne 'Çözüm Süreci' Operasyonu*. Neuss: Mezopotamya Yayınları.

Dinc, Pinar, Lina Eklund, Aiman Shahpurwala, Ali Mansourian, Augustus Aturinde, and Petter Pilesjö. 2021. "Fighting Insurgency, Ruining the Environment: The Case of Forest Fires in the Dersim Province of Turkey." *Human Ecology*. https://doi.org/10.1007/s10745-021-00243-y.

Dinç, Pınar. 2020a. "Çözüm Süreci Sonrası Dersim'de Orman Yangınları ve Çatışma İlişkisi." *Birikim* 378 (September): 109–120.

Dinç, Pınar. 2020b. "Doğanın imhası, Türklük Sözleşmesi'nin gereği." Interview by Barış Balseçer. *Yeni Özgür Politika*, July 20, 2020. https://www.ozgurpolitika.com/haberi-doganin-imhasi-turkluk-sozlesmesi-nin-geregi-2583.

Dinç, Pınar. 2020c. "Dr. Pınar Dinç: Çatışmalar artınca yangınlar da çoğalıyor." Interview by Şerif Karataş. *Evrensel*, September 20, 2020. https://www.evrensel.net/haber/414492/dr-pinar-dinc-catismalar-artinca-yanginlar-da-cogaliyor.

Dinc, Pinar. [forthcoming] 2021. "Dersim 1937–38: Shifts and Continuities in the State Discourse and Reasoning under Kemalism and Erdoğanism." In *The State of the Kurds in Erdogan's 'New' Turkey*, edited by Nikos Christofis. London: Routledge.

Eklund, Lima, Abdulhakim Abdi, Aiman Shahpurwala, and Pinar Dinc. 2021. "On the Geopolitics of Fire, Conflict and Land in the Kurdistan Region of Iraq." *Remote Sensing* 13, no. 1: 1575. https://www.mdpi.com/2072-4292/13/8/1575.

Etten, Jacob van, Joost Jongerden, Hugo J. de Vos, Annemarie Klaasse, and Esther C. E. van Hoeve. 2008. "Environmental Destruction as a Counterinsurgency Strategy in the Kurdistan Region of Turkey." *Geoforum* 39, no. 5: 1786–97.

Gurses, Mehmet. 2012. "Environmental Consequences of Civil War: Evidence from the Kurdish Conflict in Turkey." *Civil Wars*, 14, no. 2: 254–71.

Halkların Demokratik Partisi. 2015. "Temmuz-Ağustos 2015'te Kürdistan'daki Orman Yangınlarına İlişkin Gözlemler ve Teknik İnceleme Raporu."

Halkların Demokratik Partisi. 2020. "Heyetimiz Hatay'da incelemelerde bulunuyor: Bu yangınları çıkaranların peşini bırakmayacağız." *Halkların Demokratik Partisi*, October 14, 2020. https://www.hdp.org.tr/tr/heyetimiz-hatay-da-incelemelerde-bulunuyor-bu-yanginlari-cikaranlarin-pesini-birakmayacagiz/14728.

Hunt, Stephen E. 2019. "Prospects for Kurdish Ecology Initiatives in Syria and Turkey: Democratic Confederalism and Social Ecology." *Capitalism Nature Socialism* 30, no. 3: 7–26.

Jongerden, Joost, Hugo de Vos, and Jacob van Etten. 2007. "Forest Burning as Counterinsurgency in Turkish-Kurdistan: An Analysis from Space." *The International Journal of Kurdish Studies* 21, nos. 1–2: 1–16.

Kampf, Zohar, and Tamar Liebes. 2013. *Transforming Media Coverage of Violent Conflicts: The New Face of War*. Basingstoke, UK: Palgrave Macmillan.

Khaldarova, Irina, and Mervi Pantti. 2016. "Fake News." *Journalism Practice* 10, no. 7: 891–901.

Kurdish Human Rights Project. 2002. "Internally Displaced Persons: The Kurds in Turkey." London: Human Rights Project.

Nûçe Ciwan. 2020. "Ateşin Çocukları İnsiyatifi: Hatay'daki kutsal ateşi selamlıyoruz." *Nûçe Ciwan*, October 9, 2020. https://www.nuceciwan55.com/2020/10/09/atesin-cocuklari-insiyatifi-hataydaki-kutsal-atesi-selamliyoruz/.

Orhan, Gözde. 2013. "Ekoloji ve Siyaset: Munzur Baraj Projelerine Karşı Toplumsal Direniş Örneği." In *Dersim'i Parantezden Çıkartmak: Dersim Sempozyumu'nun Ardından*, edited by Zeliha Hepkon, Songül Aydın, and Şükrü Aslan. İstanbul: İletişim Yayınları.

Ozduzen, Ozge, and Aidan McGarry. 2020. "Digital Traces of 'Twitter Revolutions': Resistance, Polarization, and Surveillance via Contested Images and Texts of Occupy Gezi." *International Journal of Communication* 14: 2543–63. https://ijoc.org/index.php/ijoc/article/view/12400.

Ozduzen, Ozge, Umut Korkut, and Cansu Ozduzen. 2020. "'Refugees Are Not Welcome': Digital Racism, Online Place-Making and the Evolving Categorization of Syrians in Turkey." *New Media & Society*. https://doi.org/10.1177/1461444820956341.

Özgür Gündem. 2015. "Savaş Press." Özgür Gündem, August 4, 2015.

Sözeri, Ceren, and Dilek Kurban. 2014. "The State of the Journalistic Profession in Turkey." In *Media Policies Revisited*, edited by Evangelia Psychogiopoulou. Basingstoke, UK: Palgrave Macmillan.

TATORT Kurdistan. 2013. *Democratic Autonomy in North Kurdistan*. Translated by Janet Biehl. Porsgrunn: New Compass Press.

Ünlü, Barış. 2016. "The Kurdish Struggle and the Crisis of the Turkishness Contract." *Philosophy & Social Criticism* 42, nos. 4–5: 397–405.

van Bruinessen, Martin. 1994. "Forced Evacuations and Destruction of Villages in Dersim (Tunceli) and Western Bingol, Turkish Kurdistan." Netherlands Kurdistan Society.

Yesil, Bilge. 2016. *Media in New Turkey: The Origins of an Authoritarian Neoliberal State*. Urbana: University of Illinois Press.

Chapter 19

Breaking the Kill Chain

Exposing to Challenge British State and International Corporate Complicity in Turkey's Killer Drone Industry

Ceri Gibbons

The Turkish state's escalating militarism and violations of human rights, especially in the political aftermath of the 2016 attempted coup, are well documented (Çınar and Şirin 2017; Aydin and Avincan 2020; Yilmaz, Caman, and Bashirov 2020; Gurses 2020). For a number of geopolitical reasons, Turkey's military developments continue with minimal criticism. For example, it was only when Turkey ordered a rival Russian missile system, the S-400, that the United States took offence at Turkish actions. The United States punished Turkey by ejecting it from the American international F-35 (Joint-Strike Fighter) program (Targeted News Service 2020), and then imposed sanctions on selected Turkish defense officials (Jakes 2020). The European Union and some of its member-states have also criticized Turkey for its military intervention in Libya, which according to a UN Panel of Experts was among three countries that "routinely" and "sometimes blatantly supplied weapons" in violation of the arms embargo (UN Security Council 2019).

Meanwhile, support and even praise for Turkey's military adventures remains loud and clear from another Western power, the United Kingdom. The UK defence secretary Ben Wallace recently hailed Turkey's deployment of Bayraktar TB2 armed drones in Libya and Nagorno-Karabakh as "a game changer." While Wallace's excited speeches brought great pleasure to Turkish nationalists (*Daily Sabah* 2021), he omitted to mention that Turkey's arms transfers to these war zones violated UN and Organization for Security and Co-operation in Europe arms embargoes that the United Kingdom is legally obligated to uphold. The UK government has also responded to Turkey's violations of international law in the illegal invasions of northeastern Syria

(Talmon 2018; Peters 2018; Marinelli 2018) and cross-border assassinations of PKK fighters in Iraq with either silence or only passing criticism.

The United Kingdom, alongside other Western powers, considered the invasion of Afrîn, and ongoing military incursions into northeastern Syria in 2019, sufficient reason to announce the suspension of new arms sales to Turkey (Warrell and Pitel 2019; Hall and Trew 2019), and revoked some extant licenses related to land-based military equipment, due to Turkish military operations in Syria (Department of Trade 2020). This, however, proved to be merely a cosmetic move. In practice, the gesture amounted to a temporary review, a pause before arms export applications and approvals resumed a few months later in preparation for the next Turkish act of aggression in Nagorno-Karabakh/Artsakh.

The "game-changing" development that seems to have most impressed the British Defence Secretary was not so much a technological advance, as a cost-effectiveness miracle. At an estimated cost of $1–2 million dollars each (Sabbagh 2020), it is claimed by Azeri sources that the Bayraktar TB2 was used to damage or destroy a large proportion of the estimated £1.4 billion in Armenian military hardware lost in the conflict (Mehdiyev 2020). The Bayraktar TB2 is approximately ten times cheaper than the Protector UAV (Chuter 2020), a UAV (Unmanned Aerial Vehicle, popularly known as a "drone") supplied by US and Israeli companies for the over-budget UK Watchkeeper program (Burt 2020). But the UK defence minister's praise for Turkey's killer drones can also be read as deeply self-interested. It was the United Kingdom that provided the key hardware that armed the Bayraktar TB2 in 2014 and it appears that the UK-Turkey killer drone supply chain continues to expand to this day.

Wikileaks documents show Turkey had attempted to purchase armed UAVs from the United States over many years in order to attack the PKK, attempts that were blocked because of American congressional concerns (American Embassy, Ankara. 2010). The suspension of diplomatic relations with Israel, especially in the period after the deadly Israeli attack on Gaza aid ship Mavi Marmara in 2010, that killed nine Turkish civilians, appears to have blocked Turkey's other source of supply of UAVs, Israel (Stratfor 2010).

On July 19, 2012, Turkey finally announced it would develop its own completely indigenous armed drone program (*Hürriyet Daily News* 2012). At this time, however, Turkey did not have the weapons release systems needed to integrate its newly developed laser guided micro-munitions onto its new lightweight UAVs, and secretly remained dependent on the UK arms industry for this critical missing link.

It was the United Kingdom's less scrupulous arms export control regime that finally solved Turkey's armed UAV supply problem. On April 29, 2014, after a two-month license application process, involving cross-departmental scrutiny by the Department for Business, Ministry of Defence, and the Foreign Office, the UK government granted an export license to the Brighton arms company EDO MBM Technology Ltd (EDO) to export a small number of "Hornet bomb racks" to Roketsan in Turkey. These miniature bomb ejector release units had been specifically designed for lightweight munitions for UAVs. When the UK decision was made, Turkey was supposedly engaged in a "democratic opening" intended to reach a political settlement on the Kurdish question. Just how the United Kingdom's supply of key UAV weapon systems to Turkey, long sought for the specific purpose of attacking PKK insurgents, would assist a peace process is difficult to understand.

Dan Sabbagh and Bethan McKernan, writing in *The Guardian,* published evidence of the United Kingdom's secret supplies of Hornet bomb racks to Turkish UAVs on November 27, 2019. Their report (Sabbagh and McKernan 2019), which was based on documents provided by the present author, answered the question of how Turkey, in defiance of an American ban on the export of critical launcher technology had test-fired its first missile from an "indigenously" made drone within three years of announcing the plan to do so. The *Guardian* story was immediately denied on Twitter by Selcuk Bayraktar, co-founder of the privately owned Turkish company Baykar Makina that produces the Bayraktar TB2. He tweeted:

> We never got it from you. It is very expensive (and) it does not work in all cases. We have designed and produced a much more advanced, cost-effective version of our own. You cut your own belly. (@Selcuk, translation from Turkish. Posted on Twitter, November 27, 2019[1]; see also *Daily Sabah* 2019)

Sixteen months after publication, *The Guardian* has yet to receive a legal threat from Selcuk Bayraktar or any of the named companies involved, and the story has been widely quoted by international news media outlets in several languages. It is now clear that *The Guardian*'s finding that EDO's involvement had ended in around 2015 at the development stage was based on the limits of the evidence available in 2019. The bigger story of ongoing UK government-licensed supplies to Turkey with Hornet bomb racks, used in violation of arms embargoes, and attacks that have cost the lives of thousands across the Middle East and North Africa, remains untold. New evidence uncovered since 2019 that is outlined below unveils this bigger story and submits a case that UK-Turkey killer drone supply chain is now open to legal challenge.

THE HORNET BOMB RACK

In 2012, John Eaton, director of design of EDO, presented his company's latest research and development work to the US National Defense Industry

Association (NDIA) Joint Armament Conference. Slide 10 of Eaton's presentation shows a CAD image of a small micro-munition release unit the size of a dollar bill and weighing less than a kilogram.[2] This "WASP station interface unit (SIU)" is the prototype for what would later become the Hornet bomb rack. John Eaton's Powerpoint sets out the research and development challenge the company faced to meet a demand for the next generation of laser-guided micro-weapons weighing less than twenty-five kilograms for the emerging military UAV market.

Shortly afterward in 2012, it was announced that the Defense Industry Executive Committee (SSİK) in Turkey had assigned responsibility for research into the production of the country's first armed unmanned aerial vehicles (UAVs) to Turkish Aerospace Industries (TAI) using the ANKA system as the platform (Warnes 2014). The ANKA had been selected to become Turkey's first armed drone system (Bekdil 2017a).

On February 19, 2014, EDO applied to the Department for Business, Innovation, and Skills (BIS) for its first standard individual export licence (SIEL) to send Hornet bomb racks to Roketsan in Turkey. Rokestan is the largest Turkish-state controlled bomb and missile maker and was tasked with arming the ANKA drone.[3] BIS granted EDO the SIEL on April 29, 2014, for the export of "Hornet bomb Racks," described as "equipment for launching/handling/control equipment for munitions" to be shipped to Roketsan "for testing and evaluation purposes." By its description as "equipment," under the UK Military List rating "ML4.1.b," it is clear that this is the actual hardware, not simply components or technology related to it. According to Department for International Trade (DIT) disclosures under FOIA in 2020, none of these license application documents mention the names of any Turkish UAV even though EDO must have known the purpose for which this equipment was intended when it sent its license application.[4]

By this point in 2014, Roketsan had been working publicly with TAI for over a year on the ANKA weaponization project. The DIT licence database values the first Hornet equipment export on April 29, 2014, at £32,650.00, likely indicating a small number of items being dispatched.[5]

The description of "testing and evaluation purposes" fits the remit of the SSİK-sponsored research into UAV weaponization jointly undertaken by TAI and Roketsan. As Turkey's largest state-controlled bomb and missile prime contractor, Roketsan is capable of producing tube-based launchers for missiles, but evidently did not have the specific expertise required to make miniature bomb release units for guided bombs and munitions. Roketsan's 50lb UAV micro-munition was originally called the SMM—small micro-munition—a name later changed to the Turkish (MAM, or Mini Akıllı Mühimmat). The first MAM developed was called the MAM-L.[6]

With the UK government-approved supply of the Hornet bomb rack, the weaponization of ANKA was all set to proceed at an accelerated pace, but Turkey's armed UAV program then dramatically changed course, and Roketsan did not integrate the MAM-L onto the ANKA in 2014, but only four years later. As a result, the ANKA would not be the first, or even the second, Turkish armed UAV. Why was this? Most likely because on August 28, 2014, Recep Tayyip Erdoğan was sworn into office as the president of the Republic of Turkey.

While the full reasons are still unclear, from August 2014 onward, the state-funded ANKA project was delayed, and Rokestan worked instead with the rival privately funded Bayraktar TB2 project. The UK supplies of Hornet bomb racks to Roketsan were prioritized for use to integrate the MAM-L onto the TB2, instead of the ANKA. By December 2014, EDO had filed patents on its design, perhaps concerned by the unexpected diversion of its technology into the Turkish private sector.[7]

On May 5, 2015, L3Harris announced it would sign a Memorandum of Understanding (MoU) with TÜBİTAK SAGE, the Defense Industries Research and Development Institute of Turkey, for the purpose of "developing and integrating advanced smart carriage and release technology for international military aircraft." As the L3Harris press release put it, "the endeavor will harness each party's complementary skills and experience to produce and market a smart rack with enhanced capabilities for NATO aircraft, including international F-16 fighters and other strategic and unmanned platforms."[8]

A TÜBİTAK SAGE website announcement confirmed that the MoU signing took place on May 7, 2015, at the 12th International Defense Industry Fair (IDEF) Istanbul, but neglected to mention the "unmanned platforms" side of the deal. The article was later hidden from the website (TÜBİTAK SAGE 2015).

A few weeks after the signing on June 29, 2015, EDO applied for a new UK license to export Hornet bomb rack equipment to Rokestan. This time FOIA disclosure shows the license application was "stopped" after six weeks of processing, on August 10, 2015, due to lack of correct documentation.[9]

Around this time, an L3Harris representative answered e-mail questions from the author about the Hornet bomb rack that EDO was promoting in international arms fairs in Paris and New Orleans. Asked for more details about the Hornet, he replied:

> The Hornet is a lightweight, micro pylon RU that is capable of carrying any sub 50 lbs. smart weapon. It was designed by EDO MBM, a subsidiary of Harris Corporation in the UK and is currently in production. . . . For smaller UAVs, the weapon primes are developing smaller stores (e.g., Roketsan's micro weapon at

about 50 lbs . . .). (L3Harris representative based at the Brighton factory, e-mail interview with author 2015)

On September 21, 2015, DIT finally granted EDO two new SIEL licenses for Hornet bomb rack equipment to Roketsan with a value of £84,832—twice the total value of the first 2014 export. Business was growing.[10] In October 2015, EDO filed its accounts for the 2014 financial year. In a statement to investors on the company's "Future Developments," the then managing director, Paul Hills reported the business had "identified a number of potential prospects for financial year 2015" including, "technology led opportunities on the emerging unarmed air platform market. . . . The company is well placed and fully prepared to invest and exploit the targeted growth areas, domestically and internationally."[11]

On November 26, 2015, EDO filed its first European patent on the Hornet bomb rack design. This would have been urgently needed, because one week later, on December 3, 2015, Baykar Makina carried out its first test flight of the Bayraktar TB2 with a Roketsan weapon. A video of the test shows a Roketsan munition carried under the wing of a Bayraktar TB2 in flight above a Turkish landscape (Yusuf Akbaba @ssysfakb. Posted on Twitter, July 10, 2018).[12] For a moment, the light catches a black-and-white mechanism carrying the Roketsan bomb under the wing of the UAV as the aircraft banks to one side. The distinctive exterior design of the Hornet bomb rack is momentarily illuminated, before returning to the shadows.

Two weeks later, on December 17, 2015, Baykar Makina made Turkish national headlines. The company announced the first successful test-firing of a Roketsan weapon from the Bayraktar TB2. At the same time, Selcuk Bayraktar was launched as a national hero, becoming the "godfather" of Turkey's new global killer drone superpower. Five days later, on December 22, 2015, EDO applied for a new export license to export more Hornet bomb rack equipment to Roketsan, worth £186,080, a value greater than all its 2014 export licenses combined. This time it took three months for the UK government to approve the export, but eventually the shipment was approved. By April 19, 2016, EDO had expanded its Hornet bomb rack equipment exports to £468.306 and 2016 saw the UK government granting no less than ten further Hornet bomb rack equipment licenses from EDO to Rokestan, alongside licenses permitting exports of "technology for UAVs" and all for Turkish end use.[13]

In a spectacular wedding in May 2016 attended by most of the Turkish government, with presidents and prime ministers from several neighboring countries plus thousands of guests, in what was clearly promoted as a kind of royal wedding, the national hero Selçuk Bayraktar married Sumeyye Erdoğan and became the President Erdoğan's son-in-law (*Daily Sabah* 2016a). That same month, perhaps less happily for Seleçuk Bayraktar, saw a public disclosure made by Nick Guard, the L3Harris International Business Development Manager. At an arms fair in New Orleans called AUVSI, Guard was promoting EDO's Hornet bomb racks to potential international customers. Guard, perhaps unwisely, divulged to an *IHS Jane's Defence Weekly* journalist that L3Harris had "supplied its Hornet weapons carriage system to Turkey's Baykar Makina for integration on the company's Bayraktar TB2," further confirming that "Harris is looking to set the standard and become the carriage solution of choice" for other Turkish UAV manufacturers, such as TAI and Vestal (Williams 2016).

Following the failed coup attempt in Turkey, in July 2016, Erdoğan imposed emergency powers and purged the military and other institutions of his suspected opponents. This event also gave him the pretext to personally seize the direction of the Turkish defense industry, especially after his referendum victory the following year (Bekdil 2017b). On September 4, 2016, a report in the pro-government Turkish daily newspaper *Daily Sabah* announced that the Bayraktar TB2 had begun patrolling the skies (*Daily Sabah* 2016b). The article reported a "national version" of its rocket launcher that was "previously supplied from abroad." This may have been intended to counter Nick Guard's indiscretion in New Orleans four months earlier. The official narrative of the "national version" was important not only to the Turkish domestic audience who were being presented with the Bayraktar TB2 as an icon of nationalist and indigenous technological achievement, but also to prospective international customers, such as Qatar, Tunisia, Ukraine, and Azerbaijan, who needed to be able to trust that all critical systems were under Turkish control when they signed a purchase agreement.

If the *Daily Sabah* report had been accurate, it would be reasonable to expect to find fewer EDO exports of Hornet bomb racks to Rokestan as the "national version" was developed and produced to replace it. But just weeks later, EDO's license applications were actually increasing in value, alongside UAV-related technology, to over £1.3 million by the end of 2016. As Erdoğan's authoritarian grip tightened, EDO's sales of Hornet bomb racks also increased. By the end of 2017, EDO had made a further £1.4 million in Hornet exports to Roketsan, consisting of components and spare parts. In November 2017, *Defense News* reported that the Karayel produced by Vestel had made its own debut at the Dubai Air Show, as Turkey's second domestically produced armed UAV. Hudai Ozdamar, Roketsan's director for market development, was quoted in the report declaring that Roketsan had now become the domestic supplier to the three Turkish UAV manufacturers, TAI, Baykar, and Vestel. He also revealed the fatal flaw of Turkey's armed drones:

"We try to design all our systems to be self-sufficient. We don't want to be limited by export licensing restrictions (Opall-Rome 2017)." Ozdamer knew that regardless of how small the proportion of a weapons system is dependent on international suppliers, this reliance can still impact the whole system if a critical part, such as a bomb rack, is blocked by a foreign government's export licensing controls.

The United Kingdom is bound under its own laws and international obligations not to export military goods to destinations where they are likely to be diverted to recipients in breach of international law. But neither EDO nor Roketsan acknowledge the use of UK designed and manufactured Hornet bomb racks on Turkish UAVs, and since April 2014, the UK government has deliberately ignored this use. So long as this end use is unacknowledged, UK support for Turkish military expansion and aggression, using armed drones, will no doubt continue.

Apart from the Nick Guard disclosure to *IHS Janes Defence Weekly* in May 2016, L3Harris has made no reference to the use of its Hornet bomb rack on the Bayraktar TB2 in any of its promotional materials, even though this fact could potentially be a major selling point. L3Harris's only mention of the operational use of the Hornet is a vague reference in a 2020 sell sheet: "The Hornet receptacle is now incorporated into two in-service munitions and is being adopted for other users."[14] There can be no doubt that this is a reference to Roketsan's MAM-L and MAM-C micro-munitions used on the TB2 and to be used on the Vestal Karayel and TAI ANKA.

As if to confirm this, the sell sheet even contains a photograph of a Hornet bomb rack installed on an ANKA drone, but does not mention it in the text.

EDO's Hornet bomb rack exports to Turkey reached a new peak in 2018, with a value of nearly £2.5 million by the end of the year. This rise in exports came in the aftermath of the Turkish invasion of Afrîn, which the Turkish state pushed to global media outlets as a shameless sales pitch of the Bayraktar TB2 to potential international customers. President Erdoğan had commissioned a military attack on Afrîn resulting in deaths, injuries, and thousands of displaced people. Using the international media coverage of the invasion, which was supported by drone attack videos, Erdoğan promoted his son-in-law's armed drones to the international market (*Daily Sabah* 2018). By March 2018 Turkey had confirmed sales to its close ally Qatar (Ergöçün 2018). By April 2018, the Trump administration had taken notice of the challenge to US industry's place in the international UAV marketplace and announced a change in its strict controls on American exports. The United States declared its intention to enable American companies to "increase their direct sales to authorized allies and partners" in order to create jobs and "substantial export revenues" (US Department of State 2018). In this way, the US arms giant L3Harris had successfully used its UK subsidiary EDO MBM to circumvent the US congressional ban to arm Turkey's UAVs, and

EDO MBM Hornet Bomb Rack Export licenses 2014-2019 (£s)

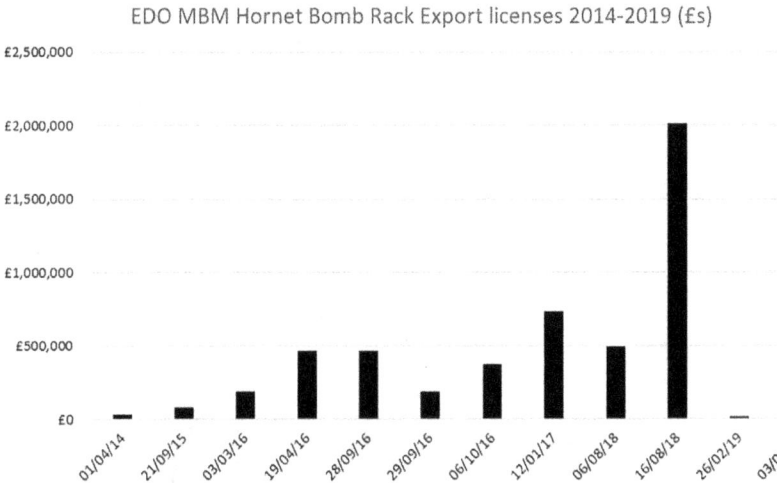

Figure 19.1 EDO MBM Hornet Bomb Rack Export Licenses (value £) 2014–2020. *Source:* © bar chart created by Ceri Gibbons.

thereby opened its US business to the international markets created by this new Turkish threat to US market dominance. Soon, Greece would be ordering MQ-9 Reaper UAVs from the United States, fitted with L3Harris WESCAM sensors to protect its borders from Turkish Bayraktar TB2 UAVs also fitted with L3Harris WESCAM sensors.

By late 2019, Turkey was armed with nearly a hundred Bayraktar TB2s, which it then deployed with devastating effect in its invasion of northeastern Syria in yet more promotional slaughter.

In this context, it is not surprising that EDO's activities are unpopular with many people in the local community in Brighton. This is reflected in a statement in a recent EDO report filed with Companies House: "As a defence company we are unfortunately subject to local protest and as such purposely seek to maintain a low profile within the wider community" (George Gardner, managing director of EDO MBM Technology Ltd, Brighton, UK: a subsidiary of L3Harris Inc, "EDO MBM Technology Limited Report and Financial Statements for the 6 month period ended 3 January 2020").

THE ANCA REPORT

In October 2020 two Bayraktar TB2 drones were shot down over Nagorno-Karabakh/Artsakh by Armenian forces during the escalation of the conflict

Table 19.1   EDO MBM Hornet Bomb Rack Export Licenses (value £) 2014–2020

Date	Export licenses (£s)
01/04/14	£32,650.00
21/09/15	£84,832.00
03/03/16	£186,080.00
19/04/16	£468,306.00
28/09/16	£468,306.00
29/09/16	£190,872.00
06/10/16	£370,893.00
12/01/17	£731,495.00
06/08/18	£496,664.00
16/08/18	£2,012,906.00
26/02/19	£18,101.00
03/04/19	£50,000.00
2020	withheld data

NB These amounts refer to Standard Individual Export Licenses. The table does not include other kinds of licence that may exist but, to date, have not been disclosed. (Table compiled by Ceri Gibbons).

with Azerbaijan. The Armenian National Committee of America (ANCA) published a report online with detailed photographic evidence of the wreckage. This confirmed that Western companies had made some of the drone's components, many of them civilian aviation suppliers reportedly unaware that they had been serving the Turkish drone program. Turkey's claims that it had produced a 100 percent indigenous original design were now demonstrated as plainly false (Sabbagh 2020).

What followed was a highly effective international campaign led by Armenian diaspora groups in Canada, the United States, Germany, and the United Kingdom to challenge the supply of UAV components to Turkey. After legal pressure from these campaigners, the Canadian government, at least temporarily, banned WESCAM's continued supply of wide-spectrum imaging sensors and cameras to Turkey (Defense Brief 2020a; Gallagher 2020). Elsewhere, companies manufacturing products for civilian use ended the supplies of their technologies for Turkish UAVs, including Rotax in Austria (Sevunts 2020), and Garmin in the United States (Defense Brief. 2020b). In the United Kingdom, Armenian diaspora activists produced a petition that attracted thousands of signatures and the Armenian Embassy wrote a letter to persuade Andair Ltd to boycott Turkey's UAV industry. This was a significant victory for campaigners as until that time the Turkish media had presented the United Kingdom as a solid Western backer in contrast to international criticism from the United States and the EU (New Arab Staff 2021).

Figure 19.2 A Damaged UAV Bomb Rack Sits Next to Remnants of Shot Down Bayraktar TB2 Bearing an Azeri flag. The photograph was taken at an Armenian Ministry of Defence facility on October 29, 2020, and published by the Armenian National Committee of America. *Source:* © ANCA.

Perhaps most significantly, the ANCA report provided the first clear evidence of the ongoing operational use of the Hornet Bomb rack on Bayraktar TB2s. Hi-res photographs of the internal mechanism of the UAV's bomb rack, advertised by Baykar as their own original national design, clearly show an identical design to EDO's 2014 online patent drawings.

Figure 19.3   A Closer Side View of a Damaged UAV Bomb Rack from a Shot Down Bayraktar TB2 UAV. The photograph was taken at an Armenian Ministry of Defence facility on October 29, 2020, and published by the Armenian National Committee of America (ANCA). The internal mechanism of the bomb rack is identical to that of a design patented by EDO MBM Technology Ltd in 2014. Patent available online. *Source:* © ANCA.

CONCLUSION

Over the period 2014–2020, EDO exported Hornet bomb racks valued in total at over £5 million to Roketsan. These were then supplied to Baykar Makina, and installed on the Bayraktar TB2s used to violate arms embargoes in Syria, Libya, and Nagorno-Karabakh and to commit possible war crimes.

This new evidence builds the case for a potential legal challenge to UK government licensing decisions concerning EDO's Hornet bomb rack exports to Turkish UAVs. Through such a challenge, we aim to expose more details of covert state-corporate complicity in the supply of critical weapon systems for Turkey's murderous killer drone industry and seek to break the kill chain. Solidarity activists from Armenian and Kurdish solidarity groups are now working together with arms trade activists, human rights litigators, and other researchers to achieve this aim.

NOTES

1. See https://twitter.com/Selcuk/status/1199734594851016704.

2. Powerpoint presentation, "Enabling effective introduction of next generation micro payload capabilities," May 15, 2012. https://ndiastorage.blob.core.usgovcloud api.net/ndia/2012/armaments/Wednesday13668eaton.pdf.

3. Information Rights Unit, Department of International Trade supplied under Freedom of Information request DIT reference: FOI2021/00116 "Revised request EDO MBM Technology Export Licences (Hornet related)," responses to Ceri Gibbons, January 11, 2021, to February 11, 2021. https://www.whatdotheyknow.com /request/revised_request_edo_mbm_technolo#incoming-1722561.

4. Department of International Trade, supplied under Freedom of Information request DIT reference: FOI2020/03094 "EDO MBM Technology Ltd Export licence end users," responses to Ceri Gibbons, June 29, 2020 to October 8, 2020: https:// www.whatdotheyknow.com/request/edo_mbm_technology_ltd_export_li#incoming -1653613.

5. Campaign Against Arms Trade database, "UK export licences applied for military goods to Turkey," https://caat.org.uk/data/exports-uk/licence-list?region=Turkey &date_from=2014-04-29&date_to=2014-04-29.

6. See https://www.roketsan.com.tr/en/product/mam-l-smart-micro-munition/.

7. See patent images and details: https://patents.google.com/patent/US1020 8771B2/en?inventor=Lewendon+James; https://patentimages.storage.googleapis. com/26/58/45/a8342d19fe3592/GB2532969A.pdf; https://patents.google.com/patent /EP3287368A1/en?assignee=edo+mbm&oq=edo+mbm.

8. L3Harris Press release, "Exelis, TUBITAK SAGE to collaborate on development of smart release technology for global customers." L3Harris website, May 5, 2015; https://www.l3harris.com/newsroom/press-release/2015/05/exelis-tubitak-sage -collaborate-development-smart-release-technology.

9. Department of International Trade, supplied under Freedom of Information request DIT reference: FOI2020/03094 "EDO MBM Technology Ltd Export licence end users," responses to Ceri Gibbons, June 29, 2020, to October 8, 2020. https://

www.whatdotheyknow.com/request/edo_mbm_technology_ltd_export_li#incoming -1653613.

10. Department of International Trade, Freedom of Information request reference: Annex A – FOI 2021/00116, "Revised request EDO MBM Technology Export Licences (Hornet related)," response to Ceri Gibbons, February 5, 2021. https://www. whatdotheyknow.com/request/717399/response/1718597/attach/html/4/00116%20 Annex%20A.pdf.html, cross-referenced to Campaign Against Arms Trade, "UK export licences applied for Grenades, bombs, missiles, countermeasures including launching/handling/control equipment for munitions to Turkey," https://caat.org.uk/data/ exports-uk/licence-list?region=Turkey&rating=ML4&item=launching%2Fhandling %2Fcontrol+equipment+for+munitions&date_from=2015-09-21&date_to=2015.

11. "EDO MBM Technology Limited: Report and Financial Statements for the year ended 31 December 2014," accessible at: https://beta.companieshouse.gov.uk/ company/00402684/filing-history/MzEzMjc1NDc0M2FkaXF6a2N4/document.

12. See: https://twitter.com/ssysfakb/status/1016587040585117696.

13. Information Rights Unit, Department of International Trade supplied under Freedom of Information request DIT reference: FOI2020/02389 "UAV related export licenses to Turkey - EDO MBM," responses to Abdullah Hall, May 27, 2020 to June 24, 2020, https://www.whatdotheyknow.com/request/uav_related_export_licenses _to_t_2#incoming-1590971.

14. L3Harris, Wasp and Hornet Carriage and Release Units sell sheet: https:// www.l3harris.com/sites/default/files/2020-09/l3harris-wasp-and-hornet-carriage-and -release-units-sell-sheet-sas.pdf.

REFERENCES

American Embassy, Ankara. 2010. "Secretary Gates' Turkey Bilateral Visit: Scenesetter." January 26, 2010. Public Library of US Diplomacy on Wikileaks. https://wikileaks.org/plusd/cables/10ANKARA126_a.html.

Aydin, Hasan, and Koksal Avincan. 2020. "Intellectual Crimes and Serious Violation of Human Rights in Turkey: A Narrative Inquiry." *The International Journal of Human Rights* 24, no. 8: 1127–1155. https://doi.org/10.1080/13642987.2020.1713108.

Bekdil, Burak Ege. 2017a. "Turkey's First Armed Anka Drone to Be Delivered in 2017." *Defense News*, January 17, 2017. https://www.defensenews.com/air/2017 /01/17/turkeys-first-armed-anka-drone-to-be-delivered-in-2017/.

Bekdil, Burak Ege. 2017b. "Turkey's Erdogan Decrees Sweeping Defense Procurement Takeover." *Defense News*, December 27, 2017. https://www. defensenews.com/industry/2017/12/27/turkeys-erdogan-decrees-sweeping-defense -procurement-takeover/.

Burt, Peter. 2020. "Watchdog Reports Continuing Problems with Protector and Watchkeeper Drone Programmes. August 18, 2020. https://dronewars.net/2020/08/ 18/watchdog-reports-continuing-problems-with-protector-and-watchkeeper-drone -programmes/#more-13346.

Center for the National Interest. 2020. "Britain's Tempest Sixth-Generation Fighter Could Blow Away the F-35." *States News Service*. March 20, 2020, Nexis.

Chuter, Andrew. 2020. "UK Orders First Three Protector Drones from General Atomics." *Defense News*, July 15, 2020. https://www.defensenews.com/global/europe/2020/07/15/uk-orders-first-three-protector-drones-from-general-atomics/#:~:text=An%20MoD%20spokesperson%20denied%20that,180%20million%2C%20or%20%24230%20million.

Çınar, Özgür H., and Tolga Şirin. 2017. "Turkey's Human Rights Agenda." *Research and Policy on Turkey* 2, no. 2: 133–143. https://doi.org/10.1080/23760818.2017.1350354.

Daily Sabah. 2016a. "Erdoğan's Daughter Sümeyye Ties the Knot with Selçuk Bayraktar in Istanbul." *Daily Sabah*, May 14, 2016. https://www.dailysabah.com/politics/2016/05/14/erdogans-daughter-sumeyye-ties-the-knot-with-selcuk-bayraktar-in-istanbul.

Daily Sabah. 2016b. "Turkey's Domestically-made Armed Drone Starts Patrolling the Skies." *Daily Sabah*, September 4, 2016. https://www.dailysabah.com/turkey/2016/09/04/turkeys-domestically-made-armed-drone-starts-patrolling-the-skies.

Daily Sabah. 2018. "Turkey's Bayraktar TB2 Drones Enable Swift, Precise Victory Against YPG/PKK in Syria's Afrin." *Daily Sabah*, April 19, 2018. https://www.dailysabah.com/war-on-terror/2018/04/19/turkeys-bayraktar-tb2-drones-enable-swift-precise-victory-against-ypgpkk-in-syrias-afrin/.

Daily Sabah. 2019. "Turkish Drone Maker Baykar Dismisses Report it Owes Success to UK Technology." *Daily Sabah*, last modified December 20, 2019. https://www.dailysabah.com/defense/2019/11/27/turkish-drone-maker-baykar-dismisses-report-it-owes-success-to-uk-technology.

Daily Sabah. 2021. "Turkey's Defense Industry not far from Becoming Trendsetter." *Daily Sabah*, January 21, 2021. https://www.dailysabah.com/business/defense/turkeys-defense-industry-not-far-from-becoming-trendsetter.

Defense Brief. 2020a. "Canada Halts Defense Exports to Turkey in Wake of Nagorno-Karabakh Clash." *Defense Brief* website. October 6, 2020. https://defbrief.com/2020/10/06/canada-halts-defense-exports-to-turkey-in-wake-of-nagorno-karabakh-clash/.

Defense Brief. 2020b. "Garmin Condemns Use of its Products in Turkish Combat UAS." *Defense Brief* website. November 6, 2020. https://defbrief.com/2020/11/06/garmin-condemns-use-of-its-products-in-turkish-combat-uas/.

Department of Trade. [UK]. 2020. Strategic Export Controls: Licensing Statistics, October to December 2019: Statistical Commentary. *GOV.UK* website. April 15, 2020. https://assets.publishing.service.gov.uk/government/uploads/system/uploads/attachment_data/file/959058/strategic-export-controls-commentary-1-October-31-December-2019.pdf.

Gallagher, Kelsey. 2020. "Killer Optics: Exports of WESCAM Sensors to Turkey—A Litmus Test of Canada's Compliance with the Arms Trade Treaty." Ploughshares Special Report (September 2020). Waterloo, ON: Project Ploughshares. https://ploughshares.ca/wp-content/uploads/2020/09/TurkeyWESCAMReportSept.2020.pdf.

Ergöçün, Gökhan. 2018. "Turkey's Baykar to Export Armed UAVs to Qatar." *Anadolu Agency* website. March 14, 2018. https://www.aa.com.tr/en/middle-east/turkeys-baykar-to-export-armed-uavs-to-qatar/1088587.

Gurses, Mehmet. 2020. "The Evolving Kurdish Question in Turkey." *Middle East Critique* 29, no. 3: 307–318. https://doi.org/10.1080/19436149.2020.1770448.

Hall, Richard, and Bel Trew. 2019. "Syria War: UK Suspends New Arms Sales to Turkey Amid International Outcry Over Offensive Against Kurds; Suspension Follows Other EU Countries Halting Sales. *The Independent.* October 15, 2019, Nexis.

Hürriyet Daily News. 2012. "Turkey Set to Produce its Own Armed UAVs." Hürriyet Daily News, July 19, 2012. https://www.hurriyetdailynews.com/turkey-set-to-produce-its-own-armed-uavs-25830.

Marinelli, Stefano. 2018. "The Use of Force of Turkey in Rojava after the Capture of Afrin. Consequences for International Law and for the Syrian Conflict." *International Law Blog,* March 26, 2018. https://internationallaw.blog/2018/03/26/the-use-of-force-of-turkey-in-rojava-after-the-capture-of-afrin-consequences-for-international-law-and-for-the-syrian-conflict/.

Jakes, Lara. 2020. "U.S. Imposes Sanctions on Turkey Over 2017 Purchase of Russian Missile Defenses." *New York Times,* December 14, 2020. https://www.nytimes.com/2020/12/14/us/politics/trump-turkey-missile-defense-sanctions.html.

Mehdiyev, Mushvig. 2020. "Armenia's Military Equipment Loss in Recent Karabakh War Stands at $4 Billion, Exceeding State Budget." *Caspian News*, December 6, 2020. https://caspiannews.com/news-detail/armenias-military-equipment-loss-in-recent-karabakh-war-stands-at-4-billion-exceeding-state-budget-2020-12-6-0/.

New Arab Staff. 2021. "UK Company Nixes Exports to Turkey over Nagorno-Karabakh Drone Use." *The New Arab* website. January 14, 2021. https://english.alaraby.co.uk/english/news/2021/1/14/uk-company-cancels-turkey-exports-over-karabakh-drone-use.

Opall-Rome, Barbara. 2017. "Turkey's Newest Armed Drone Makes Debut at Dubai Airshow." *Defense News*, November 15, 2017. https://www.defensenews.com/digital-show-dailies/dubai-air-show/2017/11/15/turkeys-newest-armed-drone-makes-debut-at-dubai-airshow/.

Peters, Anne. 2018. "The Turkish Operation in Afrin (Syria) and the Silence of the Lambs." *Blog of the European Journal of International Law,* January 13, 2018. https://www.ejiltalk.org/the-turkish-operation-in-afrin-syria-and-the-silence-of-the-lambs/.

Targeted News Service. 2020. *"Congressional Research Service Report: 'Turkey: Background & U.S. Relations.'"* November 10, 2020, Nexis.

Reuters. 2019. "Erdogan Says It's Unacceptable that Turkey Can't Have Nuclear Weapons." *Reuters*, September 4, 2019. https://www.reuters.com/article/us-turkey-nuclear-erdogan-idUSKCN1VP2QN.

Sabbagh, Dan. 2020. "UK Wants New Drones in Wake of Azerbaijan Military Success." *The Guardian*, December 29, 2020. https://www.theguardian.com/world/2020/dec/29/uk-defence-secretary-hails-azerbaijans-use-of-drones-in-conflict.

Sabbagh, Dan, and Bethan McKernan. 2019. "Revealed: How UK technology Fuelled Turkey's Rise to Global Drone Power." *The Guardian*, November 27, 2019. https ://www.theguardian.com/news/2019/nov/27/revealed-uk-technology-turkey-rise-global-drone-power.

Sevunts, Levon. 2020. "Bombardier Recreational Products Suspends Delivery of Aircraft Engines Used on Military Drones." *CBC* website. October 25, 2020. https://www.cbc.ca/news/politics/turkey-armenia-azerbaijan-drones-bombardier-1 .5775350.

Stratfor. 2010. "The Troubled Acquisition of Israeli UAVs." E-mail June 22, 2010. The Global Intelligence Files on Wikileaks. https://wikileaks.org/gifiles/docs/13 /1331116_turkey-the-troubled-acquisition-of-israeli-uavs-.html.

Talmon, Stefan. 2018. "Difficulties in Assessing the Illegality of the Turkish Intervention in Syria." *German Practice in International Law,* January 26, 2018. https:// gpil.jura.uni-bonn.de/2018/01/difficulties-assessing-illegality-turkish-intervention -syria/.

TÜBİTAK SAGE. 2015. "TÜBİTAK SAGE ve EXELIS Karşılıklı Anlayış Belgesi İmzaladı." [TÜBİTAK SAGE and EXELIS Signed a Certificate of Mutual Understanding]. http://www.sage.tubitak.gov.tr/tr/haber/tubitak-sage-ve-exelis-karsilikli-anlayis-belgesi-imzaladi (site discontinued). Accessible via Wayback Machine, March 8, 2021. https://web.archive.org/web/20150919032936/; http:// www.sage.tubitak.gov.tr/tr/haber/tubitak-sage-ve-exelis-karsilikli-anlayis-belgesi -imzaladi.

UN Security Council. 2019. Final Report of the Panel of Experts on Libya Established Pursuant to Security Council Resolution 1973 (2011). S/2019/91419-188162/376. (December 9, 2019). https://undocs.org/S/2019/914.

U.S. Department of State. 2018. "Briefing on Updated Conventional Arms Transfer Policy and Unmanned Aerial Systems UAS Export Policy." https://2017-2021.state .gov/briefing-on-updated-conventional-arms-transfer-policy-and-unmanned-aerial -systems-uas-export-policy/index.html.

Warnes, Alan. 2014. "TAI Exhibits Attack Helicopter and Anka UAV." *Ain Online*, July 13, 2014. https://www.ainonline.com/aviation-news/aerospace/2014-07-13/ tai-exhibits-attack-helicopter-and-anka-uav.

Warrell, Helen, and Laura Pitel. 2019. "UK Suspends all New Arms sales to Turkey." *FT.com.* October 15, 2019, Nexis.

Williams, Huw. 2016. "Xponential 2016: Hornet carriage system equips Bayraktar UAV." *IHS Jane's 360* website, May 9, 2016. https://www.janes.com/article/ 60084/xponential-2016-hornet-carriage-system-equips-bayraktar-uav. Removed from website, but accessed via Wayback Machine, March 6, 2021. https://web. archive.org/web/20160510134607/https://www.janes.com/article/60084/xponential-2016-hornet-carriage-system-equips-bayraktar-uav.

Yilmaz, Ihsan, Mehmet Efe Caman, and Galib Bashirov. 2020. "How an Islamist Party Managed to Legitimate its Authoritarianization in the Eyes of the Secularist Opposition: The Case of Turkey." *Democratization* 27, no. 2: 265–82. https://doi .org/10.1080/13510347.2019.1679772.

Part VI

CONCLUSIONS

Chapter 20

"To Plant the Tree of Tomorrow"

Seeding and Spiraling Ecologically Aware Democratic Autonomy beyond the Kurdish Freedom Movement

Stephen E. Hunt

In recent decades, several social movements have adopted similar aims to the Kurdish ecology initiatives, seeking coherent and viable alternatives to both unsustainable neoliberalism and state capitalism. This chapter considers instances where democratic autonomy, together with social ecology, is being scaled across from, and to, the Kurdish freedom movement (KFM), to blossom in other regions. It will consider instances of solidarity between networks sympathetic to democratic autonomy and ecological justice internationally and the potential for these to connect and converge in more expansive planetary transformation. Support for radical democracy has appeared in such diverse situations as the new municipalism that emerged in Barcelona following the Indignados movement (Russell 2019), as an alternative paradigm to address the Israeli-Palestinian conflict, the South African Shack Dwellers' Movement (Anderson 2018), in the Kabyle Commune of Barbacha (Tarlacrea 2018) in North Africa, and in elements of Fiji's dual-power system (Hall 2019). The following briefly identifies several scenarios where experiments in democratic autonomy are unfolding. I shall foreground a different theme in each: the Zapatistas (challenging extractivism through democratic autonomy); Cherán (security); Brazil's Landless Workers' Movement or MST (pedagogy); La Via Campesina (gender parity in an internationalist network); Mapuche (indigeneity); possibilities in the Middle East (conflict resolution); Cooperation Jackson (ecologically sustainable solidarity economy). Also noteworthy are coordinating bodies, such as Symbiosis, which are emerging to promote principles of democratic autonomy and social ecology. I will conclude with a consideration of some of the challenges and opportunities available to the development of such models.

To observe that the struggles of the Zapatistas, KFM, and those of other present-day sites of contestation, such as West Papua, Brazil, and Mapuche lands in Chile and Argentina, share similar desires for regional autonomy and common ecological aspirations is to make no bold claim.[1] They are re-energized campaigns to defend the commons from enclosure and to maintain autonomy and cultural integrity against demands for land, natural resources, and exploitable labor in the face of colonizing forces that have long undertaken practices leading to the annihilation and assimilation of indigenous peoples. There are correspondences both in the recent history of resistance in Kurdistan and Chiapas (Cooperativa Integral Catalana 2015) and in the flourishing of revolutionary imaginaries. Sometimes the lines of continuity between these separate struggles remain parallel, in other cases, there are points of convergence through mutual recognition and solidarity. To identify these strong correlations is not to downplay the culturally distinctive and historically specific experiences of each (Gambetti 2009, 75 fn 94), yet comparison helps us to understand developing affinities and conceptual links. In this context, there have been mutually reinforcing gestures of recognition and solidarity, powerful signals of hope, and waymarks toward building international capacity to confront the regimes of securitization and unaccountable power that imperil humanity. Acknowledgment of the potential for a confederation of authentic democratic initiatives, moreover, is even more essential since Western mainstream media, and even supposedly "progressive" discourses, have largely marginalized such resistance.

ZAPATISTAS

All watersheds are arbitrary, yet January 1, 1994, is a strong contender to be the symbolic date on which, territorially, the present experiments in democratic autonomy began. On this day, the Zapatista Army of National Liberation (EZLN) launched their insurgency in several towns across the Chiapas region of southeastern Mexico, declaring what would be known as "autonomous communities" (Marcos 2001, 227). While the Zedillo administration's ratification of the North American Free Trade Agreement (NAFTA) was the immediate provocation for the Zapatistas' uprising, the underlying issues were grounded in historic racism and indigenous exclusion, settler colonialism, dispossession, impoverishment, and land rights conflict. Abolishing the land tenure protections afforded by the Revolutionary Agrarian Law of 1915, in the interests of free trade, was effectively to revert to the precarity that prevailed in the early twentieth century. The old cause of land and freedom was dramatically reignited.

The Zapatista rebel territories have participatory decision-making structures configured through community assemblies, autonomous rebel municipalities, and five regional bodies, concerned with coordination and equitable distribution, known as *caracoles* (Starr, Martínez-Torres, and Rosset 2010). The suggestively poetic natural history metaphor of the caracol is variously translated from the Spanish as "conch" or "snail." Just as various are interpretations of the meaning such spiraling shells hold for the Zapatistas. Gaston Bachelard, the French phenomenologist, devoted much consideration to the expansive meaning of shells' "transcendental geometry," one of humanity's oldest and most fertile symbols, in his celebrated *Poetics of Space* (1994, chap. 5). In Subcomandante Marcos's account, the notion of the caracol was based on a tale where it is claimed that early men and women had caracol-shaped hearts, prompting himself and other Zapatistas to designate the five carcoles with fantastical names during a wild swimming trip (Marcos 2003). He was sensitive to the richly allusive political connotations of the non-gladiatorial shell-dwelling creatures as absurdist banner carriers for the revolution. For Pablo González Casanova, the symbolism of the conch resonates since it is about amplifying communities previously unheard, projecting their voices by producing a sound that travels far away (Casanova 2005, 79 fn 1). For John Ross, the conch or snail evokes a similar calling together of society, with the spiraling structure of the carapace of either also representing a connection between the Zapatistas' internationalist commitment to both the inner domain of local democracy and the outside world (2005). The microcosm and macrocosm are a great weight to carry, and so for Rebecca Solnit, the Zapatistas' revolutionary snails, recognized as "one of its principal symbols," also brings to mind the humbler necessity for progress to be patient and "grounded" (2008). The deliberative nature of consensus decision-making is a slow and sometimes arduous process, at its best embodying the virtues of patience, in marked contrast to the hypermobility of neoliberalism. Certainly, land snails are more prevalent in Zapatista artwork, than marine volutes, with masked mollusks beautifully crisscrossing the heartlands of Chiapas and the re-envisioned landscapes of a reconstructed world. And what better way to call to mind the idea of homeland when the snail and its home are synonymous? This image of self-containment, therefore, recalls the notion of autonomy, a concept beloved of both the Zapatista and Kurdish struggles.

Both the EZLN and PKK undertook far-reaching processes of critical re-evaluation, fundamentally re-examining state socialist approaches to political power. The failure of vanguardism and top-down regimes to achieve emancipation, the necessity for women's empowerment and liberation to achieve full social revolution, and the need to protect the natural environment against monoculture and extractivism have led the Zapatistas and KFM to pursue similar pathways. Subcomandante Marcos advocates a transformation in power relationships, arguing that "the problem of power is not a question of who rules, but of who exercises power" (2001, 46). Such words, formally embedded in the project to institute dispersed power in community, municipal, and regional assemblies in Chiapas, sound close to social ecologist Murray Bookchin's communalist principles. Nevertheless, there is little evidence that the Zapatistas have been influenced by Bookchin's ideas. It is also believed that fellow social ecologists failed to engage Bookchin's interest in the Zapatista movement, even though it emerged more than a decade before the end of his life. This contrasts with the profound influence of Bookchin's ideas upon Abdullah Öcalan and other Kurdish thinkers as they reappraised their political philosophy at the outset of the new millennium. In recent years, however, members of the Institute of Social Ecology (ISE), such as Brian Tokar and Daniel Chodorkoff, have referenced and acclaimed the Zapatista movement's progress. The ISE's Eleanor Finley has explicitly noted the importance of Zapatista decision-making through democratic assemblies, in which the movement has revived indigenous processes whereby communities and regional populations are represented through "hundreds of autonomous *municipios* and five regional capitals called *caracoles* (snails) whose spirals symbolize the joining of villages" (Finlay 2017).

There are further affinities between the EZLN and the KFM on the family tree of radical social movements. PKK members were aware of developments in Mexico and the Zapatistas' new political direction. They are believed to have intensively analyzed and discussed the implications of the experiments in direct democracy during a two-day conference in 1998 (McGuire 2019, unpaginated).[2] Öcalan took into account the Zapatista model when rethinking Kurdish praxis during his early years in prison (Gambetti 2009, 75 fn 94). Furthermore, Marlies Casier highlights Kurdish activists' role in the anti-globalization movement, which emerged in the late 1990s after the Zapatista uprising and proliferated during the following decade (2015, 139). Known in activist circles as a "movement of movements," this developed internationally around several high-profile social forums, perhaps modeled on the earlier Zapatista *Encuentro* for Humanity and against Neoliberalism, notably the 2001 World Social Forum first held at Porto Alegre, Brazil, and the European Social Forums from 2002 onward. A key development was the formation of a Mesopotamia Social Forum (MSF) which hosted international social forums in Diyarbakır in 2009 and 2011. This represented a significant coalescence between pro-Kurdish activism and international social movements since, according to Casier, it attracted participants from across the Middle East and Europe, while the Zapatistas "featured as a reference point throughout" (2015, 145). Such reciprocity broadened the base of discussions about democratic autonomy. Issues of environmental degradation also featured prominently in these events, again facilitating exchanges of ideas between ecological activists of various persuasions, with debates about social ecology already an integral part of the discourse (Romão 2009; Biehl 2011). The 2011 MSF featured an ecology tent (ANF News 2011), while the Mesopotamia Ecology Movement (MEM) emerged from discussions at the Ecology Forum that was part of this event (Ayboğa 2015). The prominence of ecological concerns at these gatherings reinforced social ecology's premise that ecology is integral to political struggle, not a single-issue add-on. More recently, this has been born out by the convergence of social justice and ecological strands of the anti-globalization movement in the climate justice movement.

Underlying conflicts in Chiapas and Kurdistan is the imperative to exploit the natural environment through extractivism. The Zapatista stronghold, the abundant and biodiverse Lacandon jungle, is the largest expanse of rainforest surviving in Mexico and home to several endangered species (Howard 1998, 359). The Chiapas region is also a treasury of abundant natural resources that attracts corporate bounty hunters eager for oil and gas reserves, cattle lots and cash crops, timber, and uranium. The exploitation of such resources and commodities for export, together with substantial logging operations, requires extensive road-building and other infrastructure. Furthermore, hydroelectric schemes on the Grijalva River have displaced thousands; as in Turkey's Kurdish regions, hydro-hegemonic projects continue to be proposed and constructed (and resisted) (Howard 1998, 368).[3] Opportunities for huge profits for foreign investors abound. Subcomandante Marcos sardonically notes that the exchange value of such commodities includes such riches as "ecological destruction, agricultural plunder, hyperinflation, alcoholism, prostitution, and poverty" (2001, 23). The Kurdish region is similarly blessed with oil and gas reserves, forests, and plentiful mineral resources, for which local communities enjoy "benefits" from neoliberal investment. In Chiapas and Kurdistan alike, raw materials are for the most part extracted for value-added processing elsewhere. Such ecological injustice, therefore, is at the heart of conflicts in both regions.

Ecological justice is integral to the political resistance in Chiapas and other parts of the Americas where democratic autonomy is developing. A core demand of the First Declaration of the Lacandon Jungle (January 2, 1994), therefore, was to "suspend the robbery of our natural resources in the areas controlled by the EZLN" (Marcos 2001, 15). When the Zapatistas controlled mountain areas in Chiapas, they took measures to protect the forests, banning the cutting down of trees and hunting of wild animals in 1994–1995 (Marcos 2001, 243). Conservation of natural resources was a key part of the discussion in the 1996 San Andres Accords (Marcos 2001, 143 fn). This was reflected in the Second Declaration of La Realidad for Humanity and against

Neoliberalism, which included assertions "against the destruction of the environment" and aspirations "for the defense and protection of the environment" (Marcos 2001, 124), proclaimed the same year. Philip Howard argues that the Zapatistas' "most enduring claim" is "freedom and relief from the escalating environmental scarcities that have been impoverishing their communities" (Howard 1998, 357). Any indigenous people's assertions of local autonomy, however, are considered dangerous. Marcos was aware that the oppressive circumstances that communities in Chiapas were facing, demonstrated most extremely by the notorious Acteal Massacre of 1997, were of a kind with situations whereby paramilitaries were used to terrorize dissident populations elsewhere, such as "the war of extermination in Guatemala. Or Vietnam? Or Kurdistan?" (Marcos 2001, 143). Shockingly, 148 land and environmental protectors were killed in recorded incidences in Latin America in 2019. This is an increasing number, with sixty-four killings in Columbia (twenty-four in 2018), twenty-four in Brazil (twenty in 2018), eighteen in Mexico (fourteen in 2018), fourteen in Honduras (four in 2018), and twelve in Guatemala (sixteen in 2018), and many others across the continent and worldwide including forty-three murders in the Philippines (Statista.com 2019 and Statista.com 2020). This underscores the perceived threat that environmental protection poses to powerful adversaries and the high cost of activism to defend the natural world.

Given this emphasis on the protection of the natural world, it is unsurprising that Marcos uses a further ecological metaphor to evoke the EZLN's political project:

> To plant the tree of tomorrow, that is what we want. [. . .] The tree of tomorrow is a space where everyone is, where the other knows and respects the other others . . . it is a place with democracy, liberty, and justice: that is the tree of tomorrow. (Marcos 2001, 282)

DEMOCRATIC AUTONOMY IN CENTRAL AND SOUTH AMERICA

In Cherán, Michoacán state, in central-western Mexico, a community uprising and declaration of self-government has been directly and explicitly linked to forest preservation. Here, environmental protection and democratic autonomy have joined in a practical and inspiring way. As in Chiapas, policies of economic liberalization intensified the drive to extract profits from the exploitation of natural resources in indigenous commons, resulting, for example, in accelerated loss of tree cover. The 2011 uprising occurred when around a hundred of the town's citizens, with broad community support, confronted loggers illegally taking timber from depleted communal forests. This action was a direct challenge to a triad of powerful interested parties in alleged collusion: the criminal cartels running logging operations; local politicians; state law enforcement and security forces. All were directly or indirectly profiting from the exploitation of the forests. Given the failure of the mayor and police to protect the community from the cartels, and indeed their alleged intervention in favor of the latter (Zizumbo Colunga 2015, 56), all three were expelled from the municipality. The democratic governance that has since administered the town draws upon the indigenous Purépecha people's traditional forms of government, enshrined in a Council of Elders. This has served as a form of organizational resilience, implemented to ensure that decision-makers are directly accountable to neighborhood assemblies. A member of the autonomous movement in Cherán explained that political parties were abolished to prevent infiltration and dominance by criminal drug trafficking syndicates, believed to be working in collaboration with mainstream politicians. He explained, "We knew that if there was one leader, the crime cartel would extort them. They would either buy them or murder them" (TV Cherán 2018). Autonomous structures of accountability have therefore enabled the scrutiny of chosen administrators and better protected them from intimidation.

Consideration of Cherán's case illustrates both the robustness of such experiments in democratic autonomy, but also the challenges in extending them beyond local circumstances. Cherán has endured and thrived since 2011, preventing the return of criminal cartels and political parties. Like the Zapatistas, the people of Cherán networked effectively, appealing to a regional, national, and even international constituency of support to attain political leverage and to prevent a return of direct state control. In contrast to experiments in Turkey's Kurdish municipalities, the Mexican Supreme Court formally recognized Cherán's autonomous status in 2014, reflecting the contrasting constitutional frameworks and principles going back to the foundations of the Mexican (1917) and Turkish (1921) Republics. Community-defense units called "ronda comunitaria" have prevented incursions from cartels and dealt with anti-social behavior, while criminal activity has markedly reduced. The activities of these militias are embedded in the community by being integrated into "fogatas" or "Campfire barricades," set up as security checkpoints, but also functioning as "core meeting points and the basis for the collective decision making process," since they directly feed into the neighborhood assemblies (El Enemigo Común 2018). Forest patrols ("guardabosques") have forced criminal cartels out of the forests, and an ambitious reforestation project has been underway for several years, alongside other ecological achievements, including pioneering efforts to make Cherán a zero-waste town (Covarrubias 2019; Pressly 2016).

As in Kurdish regions and Chiapas, claims for ecological justice are closely linked to struggles against what Cortés Calderón identifies as the "extractivism process," according to David Harvey's notion of "accumulation by dispossession," undertaken by the state and private corporations, often in complicity with criminal enterprises (Cortés Calderón 2018, 63). While, to date, Cherán has enjoyed fragile yet positive outcomes, successfully maintaining security and enhanced environmental sustainability, prevailing levels of community solidarity are not readily replicated elsewhere. Indeed, attempts to follow the model in the neighboring town of Nathuatzen, demographically similar and facing comparable threats to well-being and livelihoods from cartel-based illegal logging operations, have been more precarious. In primary research, Daniel Zizumbo Colunga demonstrates significantly diminished community orientation in Nathuatzen, when compared to Cherán, resulting in less capacity to mobilize effectively and in divisive local politics (Zizumbo Colunga 2015, 72–94; Lopez 2020). Furthermore, the direction of the wider autonomous movement, the Auto Defensa Movement (founded in Ayutla, 2013), is as yet unclear. There are fears that cartels have infiltrated, corrupted, and appropriated these armed groups, initially created for the purpose of community defense, with the likelihood that combatants have become embroiled in gang wars and trafficking rather than effective forces of liberation.

Nonetheless, awareness of mutual struggles for authentic democratic autonomy, gender equality, and ecological sustainability at the local and regional level are being voiced in intercontinental expressions of international solidarity. In late 2019, for example, members of Kongreya Star, representing activist women in northeastern Syria, recorded a video message to a gathering of over 3,500 women hosted by Zapatistas in Chiapas (Women Defend Rojava 2020). Zapatista women and other delegates also declared solidarity with women in northeastern Syria while displaying a "Defend Rojava" banner, due to Turkish state forces' and proxy militias' ongoing attacks (@ RISEUP4ROJAVA on Twitter, January 3, 2020). Such connections reflect mutual recognition of the challenges and aspirations shared across time and space, despite different contexts. The Lobo de Mar Collective created a striking mural—"Las Compas"—at a Zapatista Art Festival, showing a Kurdish heval from the YPJ sitting side by side with a compañera from the EZLN, in front of a militia woman from the CNT/FAI (Spanish anarcho-syndicalist union), a juxtaposition saluting a longer tradition of resistance dating back to the Spanish Revolution (Mallett-Outtrim 2016). Several commentators (e.g., Stanchev 2015) have noted the affinities and the centrality of women in such revolutions. While the municipalities of Chiapas and Cherán have Kurdish counterparts, several large grassroots networks and organizations are also experimenting with, structures of democratic autonomy, incorporating ecological ideas, across the Earth.

One of the most successful social movements in South America, Brazil's formidable Movimento dos Trabalhadores Sem Terra (MST) or Landless Workers' Movement, founded in 1985, also cherishes principles of grass-roots democracy, internationalism, feminism (Martínez-Torres and Rosset 2010, 159), and ecological sustainability comparable to those of the KFM and the EZLN (Magdoff and Williams 2017, 299–300). Like Cherán, the MST is organized without a single identifiable leader, both in keeping with its egalitarian political ideology and to build accountability and resilience into its structures, making it difficult to destroy the movement by either the bribery of high-powered individuals effectively acting as clients of big business, or their imprisonment or assassination. The MST embeds attention to relationships with the living world in its commitments to equitable access to land, sustainable agroecology, food sovereignty, and cooperative ownership. Radical pedagogy is at the heart of the MST's aspirations, with well-developed educational programs based on instilling competencies in critical challenge that could be the basis for useful exchanges of ideas with the academies in Rojava. Rural settlement schools have elements of national educational curricula, while "content connected to the student's own reality is also valued, such as agroecology, healthy and poison-free food and values such as cooperation, solidarity and cultural appreciation."[4] David Meek's study of emancipationist educational programs, derived from the ideas of Paulo Freire and others, finds that "both agroecology and transformative education are central to the MST's ideology and struggle" (2015, 2). Using a political ecology of education framework, Meek examines the challenges that the MST faces in its efforts to facilitate deep understanding of structures of oppression and creative means for transformation. The curriculum of his particular case study takes a problem-based approach to the conflicts and contradictions within contested claims for land use in a critical reflective way, conducting primary field studies intended to dislocate previous assumptions, integrated with intensive field research to, in Meek's words, "advance counter-hegemonic forms of production" (5). Teachers and students are encouraged to act in the process of co-creating knowledge and seek pathways for radical change in keeping with Freire's recognition that "the reality of oppression not as a closed world from which there is no exit, but as a limiting situation which they can transform" (Freire 1996, 31).

There have been some exchanges between the MST and Kurdish and Turkish delegates, with international delegates at the MST's 2015 *III International Congress on Dilemmas of Humanity* signing a solidarity call condemning Turkish hostilities against Kurdish populations (Yeniay, Harvey, and Aysu 2015). Jose, a Colombian delegate at the 2019 2nd Youth Conference of the Middle East, insightfully commented upon the common struggles and need for reciprocity between internationalists with aspirations for radical democracy, concluding that "we need to apply the methods used by the Kurdish struggle and struggles in South America to actually start a revolutionary movement that is there for the people and not only for some abstract ideas" ("Jose" 2019).

The MST was also a founder member of La Via Campesina (LVC/ International Peasants' Movement), an extensive network founded in 1993 that is intent upon combining radical democracy with agroecology and food sovereignty. Originally emerging from a congress in Nicaragua, LVC is now a truly international alliance. LVC is an impressive attempt to organize a democratic confederation of millions of land workers, with rotating representatives and the equal representation of women, on a model comparable to the Kurdish movement's structures of co-chairs (Menser 2008, 30–31; Martínez-Torres and Rosset 2010, 157, 159, 162; Imperial 2019). The existence of such well-developed structures and an ambition to create and maintain a constructive alternative model of participatory democracy, alongside gender equality and ecological sustainability, make LVC an obvious network for affinity and solidarity with the KFM. Some links have been forged in recent years.[5] Ercan Ayboğa of the MEM confirmed positive links between the movements were underway in 2017, when

> Via Campesina held a big conference in the Basque Country. A friend from our ecology movement (MEM) was there as an observer. There are many cooperatives in Rojava, so the agricultural cooperatives of Rojava have applied to become to become members of Campesina. Campesina wants to send a delegation to Rojava. I also met the MST in Amed last year, when they stayed for two or three days. So, they are very aware of what's going on in Kurdistan. (Skype interview with Stephen Hunt, August 18, 2017)

Given LVC's origins and continuing stronghold in Latin and Central America, many other long-running mobilizations there have also aligned with the movement. The indigenous Mapuche peoples in Chile and Argentina have worked reciprocally within LVC to maintain access to, and protect the integrity of, their seed heritage and to support agroecology. Transnational corporations and large landowners have increasingly appropriated land for intensive monocultural agriculture and infrastructure projects, such as hydroelectric schemes. Mapuche groups affiliated to LVC, such as the UST-MNCI movement, have organized to resist the dispossession of common ancestral lands (Imperial 2019, 245). Deforestation and the cultivation of monocultures of pine or eucalyptus have accompanied such land grabbing. This is a recent episode of the so-called Columbian Exchange, in which indigenous flora and horticulture have been replaced, raising concerns about the loss of local ecosystems, that have evolved to be resilient within local geographical and meteorological conditions. Mapuche groups and their allies are endeavoring to observe appropriately situated land cultivation practices, using resilient crop varieties that respect the terroir, developed to thrive in local geographical conditions to maintain traditional livelihoods and protect cultural diversity. Mapuche campaigns for human rights and ecological justice have led to some convergence and solidarity with the environmental movement and powerful Chilean student movement (Torres-Salinas et al. 2016; Leone and Ponce 2019).

Scientific researcher Jeremy Anbleyth-Evans, who has been working with the Mapuche people, reports that

> struggles in Wallmapu, include the Coordinadora Arauco Malleco (CAM), who have been carrying out non-violent "monkey wrenching" of deforestation equipment. Their efforts are based on the logic that the state has permitted the dispossession of medicinal and food rich forests, and the destruction of sacred groves, where the guardian spirits dwell. These sacred forests have been cut down and given to corporations such as Forestal Arauco, owned by the Angelini family during the Pinochet dictatorship. CAM has put forward its political project based upon the rejection of Western categorisation, such as the concept of the nation-state. Its autonomist project is based on the complete territorial recuperation of land Wallmapu lost during the Pacification of Araucania in 1883. Additionally, in the marine environment, the Mapuche Huilliches in Los Lagos Coordinadora Willy Werken Wichan have been working to realise marine conservation plans to limit the impacts of industrial salmon aquaculture and overfishing. Their approaches to conservation, are based on participatory democracy, or Trawün, and the celebration of Itrofil Mogen, biodiversity as sacred. (Jeremy Anbleyth-Evans, e-mail to Stephen Hunt, July 6, 2020)

State military incursions and the criminalization of Mapuche attempts to defend ancestral lands (known to them as "Wallmapu"), from what they regard as neo-colonialism and extractivism, has led them, and other Chileans and Argentinians, to recognize a common cause with the KFM (Villanueva 2018). The Committee of Solidarity with Kurdistan, Chilean, and Wallmapu Region has been set up to study and promote ideas of democratic autonomy and jineolojî within Chile. The Internationalist Commune of Rojava has also sent a statement of solidarity to Mapuche campaigners (Internationalist Commune of Rojava 2017). Such solidarity underlines the common recognition that for indigenous and precarious communities across the planet, resistance against neoliberal expansion is a struggle for the survival of their unique cultures and for the protection of the lands upon which they depend. A decline in the irreplaceable knowledge of local ecosystems and linguistic diversity is a loss to all humanity.

DEMOCRATIC AUTONOMY AND
CONFLICT IN THE MIDDLE EAST

While the relevance of the model of democratic confederalism nearer to the Kurdish homelands in the fractured Middle East is clear, its value and promise are not, at least yet, well-established outside of northeastern Syria. Abdullah Öcalan presents his vision of a paradigm shift to multi-ethnic democratic confederalism as a means to resolve nation-state conflicts across the Middle East. Intractable conflicts, sectarianism, and nationalist rivalries keep the peoples of the region divided—to the (questionable) benefit of external global actors—and are an indictment of systems of injustice that have been endemic during the era of capitalist modernity. Öcalan's inclusive program of direct democracy within federal structures, therefore, seems an attractive and compelling alternative. It is a proposition, to reframe ongoing hostilities and thereby advance conflict resolution, to further processes of decolonization, and to unlock the potential for renewal in the region, thus enabling formerly embattled communities to better achieve a positive cultural realization. The application of such an approach to the Israel-Palestine conflict, for example, is explicit in Öcalan's writings (2013, 40–44). To date, democratic confederalism remains a marginal and minority interest in the conflict, but grassroots autonomous structures have precedents in the Palestinian struggle (Haifawi [2015]), and forms of confederalism have received some notice and sympathy from Israeli commentators (Scheindlin and Waxman 2017) as a means to reorient regional dynamics. A small number of internationalists of Palestinian heritage have been participating in the Revolution in Rojava which has helped to build relationships with Arabic affiliates (Soz 2017). Rojava has also received support from within Israel, where there were pro-Kurdish demonstrations during the 2019 Turkish attacks on Rojava and, more ambivalently from the Kurdish perspective, from the Israeli state (4kurdistan. 2019; Vardi 2019). The emancipatory approach of democratic confederalism, embracing the pillars of gender liberation and ecological sustainability, has also attracted some attention in Lebanon, at least within communities of Kurdish heritage which have recognized the model's relevance as an alternative to prevailing political instability and corruption (Maxwell 2017).

In this multi-ethnic northeastern Syria, the autonomous political structures have extended beyond majority Kurdish communities to include Arabs, Syriac Christians, Turkmens, Yazidis, Chechen, and Armenians. Ecological aspects of the Revolution too have extended to cities in AANES with majority Arabic populations. Notably, Raqqa has started to come to terms with the trauma of Islamic State occupation and violence by including mass tree-planting in its reconstruction plans, clearing landmines, and reclaiming parks and green spaces that had reportedly functioned as mass graves during the era of terror (ANF News 2020). At present, there are multi-ethnic civil councils with co-chairs in principal cities without Kurdish majorities under the Syrian Democratic Council of AANES, including Raqqa, Al-Tabqa, Manbij, and Deir ez-Zor.

Powerful evidence that progressive experiments can arise in the most deprived and precarious circumstances is the fine example of the communalist Makhmur Refugee Camp, near Erbil in the KRI. The refugee camp, established in around 1998 for people displaced during the conflict in Turkey, has operated within autonomous democratic structures that predate Rojava. Makhmur been considered as a "laboratory" of ideas in practice. It is run autonomously following the principles of democratic confederalism with direct democracy, gender parity, and cooperatives. It also maintains a people's defense unit, necessary for security since ISIS attacked when they held territory in the area and there is an ongoing Turkish military threat, with planes targeting Makhmur with airstrikes in 2020 (Çiçek 2020) and drone attacks in 2021. The settlement also endures in the context of a KRG-imposed economic embargo and limited support from United Nations agencies. Despite such external hostility and the inhospitable climate with its searing heat and arid terrain, residents have grown food and cultivated vegetation for shelter and to improve air quality over many years, to the extent that it has been described as an orchard, or even as an oasis (Democracy Communalist 2020; ANF News 2018).

DEMOCRATIC AUTONOMY IN THE WEST

Viyan Qereçox, an internationalist volunteer, reflecting on the perspectives heard during recent experiences in Rojava, was impressed by ambitious expressions that are rarer in social movements in the West:

> "So, global democratic confederalism, how are we going to get there?" They are really thinking in these huge terms, while still coping with the day-to-day stuff, but having the courage, ambition, and political clarity to know what is necessary, to really take it seriously and to take themselves seriously, and to say, "We are revolutionaries and we are connected to social movements all around the world. [. . .] How do we as people in northeast Syria contribute towards the project of global democratic confederalism, because it is not enough to have it here, we need to have it everywhere for it to be sustainable everywhere." So that was incredibly inspiring and strengthening, and I think we need to do with a little bit more big picture thinking in the UK. ([Qereçox] 2020)

So, what are the opportunities for scaling across such projects internationally? In the West, activists from the Kurdish diaspora and their allies are generating networks, projects, and initiatives that seek to support the Kurdish movement's adaptation of social ecology and other compatible movements in the Global South. These take three complementary approaches and forms. First, and ultimately perhaps the most efficacious, is to create communal structures rooted in principles of democratic autonomy and confederalism, gender equality and anti-hierarchical practice, and ecological sustainability. Second, direct solidarity within the Kurdish region, from joining the struggle through the international brigades fighting with the Kurdish militias, in which cause several lives have been given, to practical work in the cooperatives, health and welfare work, education, or involvement in ecological projects, such as Make Rojava Green Again. Beyond Kurdistan, practical solidarity initiatives are proving effective and valuable, such as raising funds and skills sharing to support specific projects, for example, Co-operation in Mesopotamia's "Water for Rojava" project (part of the UK-based Solidarity Economy Association and Kiptik's mutual aid projects to enable the design and provision of ecological stoves and water systems in Chiapas.[6] Third, there are efforts to raise awareness of progressive Kurdish ideas through campaigns, demonstrations, and direct action, through activist discussion groups and webinars set up to learn and share ideas and to consider what can appropriately be adapted to other contexts, and through academic publications and conferences.

Cooperation Jackson in Mississippi is an advanced example of an ambitious eco-socialist project in the United States. Launched on May Day 2014, Cooperation Jackson draws upon rich traditions of Black radicalism. The cooperative network's Jackson-Kush Plan aspires to meld a solidarity economy, underpinned by principles of ecological sustainability, with policies of democratic governance. Co-founder Kali Akuno stated:

> We aim to help create ecologically regenerative methods of production, that is, engage in techniques of production, distribution, consumption, and recycling that replenish our ecosystems rather than just extract from them. ("Stir to Action." Powerpoint presentation on Cooperation Jackson, May 21–23, 2019)

Unsurprisingly, Cooperative Jackson members have taken an interest in developments in northeastern Syria and regularly demonstrate solidarity with the Kurdish struggle. While the initiative in Mississippi emerged autonomously, during a discussion with Ercan Ayboğa of the MEM, Akuno observed that they had independently reached a similar political position and practice in a way that was not coincidental but rather much to do with the conditions and impact of global capitalism (Jackson Rising 2018). Cooperation Jackson takes a strategic approach with a requirement for new members to contract in with a philosophical and practical commitment to the tenets of solidarity economy. These principles are integrated with an identification of what are currently considered to be specific local threats to community livelihood and the environment in a historical and international context.

There have been many attempts to coordinate anti-capitalism and global justice internationally; some of them achieving large and powerful mobilizations. Unfortunately, among the cyberspace junk floating in the internet's pluriverse is the detritus of many discontinued eco-socialist initiatives, including People's Global Action, the social forums, Reclaim the Streets, Occupy, and Indymedia to name but a few of the more prominent movements that did not outlast historical moments. Fortunately, eco-socialist thinking is also continually revitalized by ongoing initiatives. Kurdish groups have made efforts to work with larger mobilizations. London Kurdish Solidarity, for example, ran workshops at the national Extinction Rebellion action in London 2019 (Anderson 2019), and activists in Rojava have made links with Fridays for a Future (Make Rojava Green Again 2020). The ISE, started by Murray Bookchin in Vermont in 1974, represents enduring commitment to synthesize theory and practice, the India-based Radical Ecology Democracy forum offers a refreshingly internationalist perspective that includes sustained interest in the Kurdish initiatives, while the recently launched Symbiosis,[7] citing the "Kurdish freedom struggle" among its key inspirations, has explicitly set itself the ambition of creating a federation of communities and municipalities to build "democracy and ecological society from the ground up" across North America. The Transnational Institute of Social Ecology (TRISE) is a Pan-European network that works with social movements to promote the ideas of social ecology in the academic sphere. The Network for an Alternative Quest has organized several conferences in Hamburg to explore capitalist modernity and alternatives with attention to Kurdish achievements. While undertaking invaluable work, it must be conceded that these networks are specialist and marginal, with low visibility even in left, libertarian, and environmentalist networks, and certainly within wider public discourse including the established media, social media, and education.

This is not to diminish their consequence or effectiveness, yet raises the question of how such networks can help build sufficient capacity and post-capitalist infrastructure necessary to scale across as an effective counter-power. In addition to their intrinsic importance as forums for discussion and the development of ideas, however, they have two vital roles. First, projects such as Cooperation Jackson demonstrate a capacity to intervene impactfully in the cultural and economic life of a specific locality. Second, networks for democratic autonomy have a significant role in seeding critical awareness and ideas for viable alternatives, thereby contributing toward emerging counter-hegemonic social movements' development. These must refine their praxis if they are to transcend the limitations of previous mobilizations. Current mass movements, such as Extinction Rebellion, Black Lives Matter, and youth campaigns against climate change, for example, operate according to fluid (sometimes contested) framings, interests, and priorities. To successfully enact the profound social transformation necessary to realize their objectives, they should positively affirm ecological justice and develop their critique of hierarchy and class and their understanding of the systemic roots of oppression, given the intimate connections between extractivism, colonialism, patriarchy, and racism. They must recognize and resist the inevitable media demonization and state criminalization that confront any challenge to the entrenched interests of capital and the state. They should be alert to any dilution of their demands or appropriation of their cause. They should seek to avoid factionalism and to attain a powerful convergence of interests beyond parallel single-issue campaigns. Finally, moreover, social movements should aim to transcend the confines of defensive battles, to project a constructive vision of the future. Citing the work of David Harvey and Ernst Bloch, Ruth Levitas insightfully argues that it is necessary to envision a future that is viable and desirable, sufficiently open so as not to be determined by a rigid template, but also sufficiently closed that it is not empty of content beyond an unelaborated hope for better times to come (Levitas 2013, 123–26). Activists must at once be aware of the weight of the historic and structural forces that constrain contemporary realities, yet in a volatile present, be resistant to foreclosure and open to possibilities for liberation. The KFM's model of democratic confederalism goes some way to advancing such a paradigm where adaptable practices complement coherent structures.

It is not sufficient to survey the existence of isolated instances of ecologically aware experiments in democratic autonomy as a glimmer of hope in a deteriorating global situation if they are not sustained and more widely actualized. Yet this phenomenon represents a fascinating dialectic in which liberatory ideas, from such insurgent thinkers in the West as Murray Bookchin, Maria Mies, and Immanuel Wallenstein, have been taken up and put into practice by the Kurdish movement, where they have been radically recontextualized in a concrete form that has in turn been a touch-paper to ignite experiments such as the municipalist Barcelona En Comú. Barcelona councilor, Laura Pérez, for example, explicitly states that they are "trying to follow this path" (Barcelona En Comú 2019, 22), as she pays tribute to the Kurdish communalist movement. Such developments signpost potential to achieve the "coherent strategic convergence" (Pritchard and Worth 2016, 11) that is necessary to unite left-wing emancipatory traditions around shared politics of non-domination. At a moment of ecological crisis, this conception of non-domination should embrace social justice and also be extended to the human relationship with the rest of the biosphere. Broadly this would entail a synthesis of what we already know, informed by historic insights and coherent thinking, and contingent upon fast-moving political and technological developments. Effective means to build capacity and resilience: land, infrastructure, harmony, and knowledge (embedded in cultural heritage, also open source) based on a flexible platform of achievable demands, using a diversity of strategies appropriate to prevailing social and political circumstances. These social and ecological imperatives must converge and spiral if we are to plant the tree of tomorrow.

NOTES

1. Gullistan Yarkin (2015, 38) notes that the EZLN, MST, and Via Campesina, like the PKK, all abandoned Marxist national liberation strategies and state socialist objectives.

2. Ercan Ayboğa also cites the Zapatistas as a direct and significant influence on the Kurdish concept of democratic confederalism (Ayboğa and Biehl 2011).

3. Communal landholders ("ejidatarios") from the indigenous Zoque community have continued to resist the construction of the Chicoasén II Hydro Project, with campaigns including hunger strikes, leading the Mexican state to detain opponents of the scheme (SIPAZ 2015; Hobson 2016, 300–02).

4. There are 2,000 rural schools set up on occupied settlements, at the instigation of the MST, quilombolas (descendants of enslaved peoples) and other rural movements, although administrated by local municipalities or the state. Concerningly, Jair Bolsonaro's far-right, pro-agribusiness government is threatening the schools' future, and indeed to criminalize the MST (Friends of the MST 2019).

5. In Qamishlo, Rojava in 2014, for example, members of the Union of Kurdish Students in Syria marked LVC's international day of peasant's struggle with a campaign to promote organic agriculture and oppose the spread of genetically modified organisms (La Via Campesina 2014).

6. See the following websites: https://mesopotamia.coop/; https://www.solidarity economy.coop/; and https://www.kiptik.org/projects.html.

7. See https://www.symbiosis-revolution.org/.

REFERENCES

4kurdistan. 2019. "The People of Israel Stand in Solidarity with Kurdistan Syria." October 13, 2019. Video. https://youtu.be/wSFR50HOQuk.

Akuno, Kali, and Ajamu Nangwaya. 2017. *Jackson Rising: The Struggle for Economic Democracy and Black Self-Determination in Jackson,* Mississippi. [Cantley, QC]: Daraja Press.

Anderson, Tom. 2019. "Last Day of Marble Arch Occupation Spent Discussing a Revolution with Ecology at its Centre." *The Canary,* April 26, 2019. https://www.the canary.co/global/world-analysis/2019/04/26/last-day-of-marble-arch-occupation -spent-discussing-a-revolution-with-ecology-at-its-centre/.

Anderson, Tyler. "Why Libertarian Municipalism is more Needed Today than ever Before." *Uneven Earth* website. November 25, 2018. http://unevenearth.org/2018/11/why-libertarian-municipalism-is-more-needed-today-than-ever-before/.

ANF News. 2011. "All set for Mesopotamia Social Forum in Diyarbakýr." *ANF News*, September 18, 2011. https://anfenglishmobile.com/features/all-set-for-mesopotamia-social-forum-in-diyarbakyr-4071.

ANF News. 2018. "Makhmur Refugee Camp: From a Desert to an Orchard." *ANF News*, July 23, 2018). https://anfenglish.com/news/makhmur-refugee-camp-from-a-desert-to-an-orchard-28442.

ANF News. 2020. "100,000 Olive Trees Planted in Raqqa." *ANF News*, February 7, 2020. https://anfenglish.com/news/100-000-olive-trees-planted-in-raqqa-41436.

Ayboğa, Ercan. 2015. "Restructuring Mesopotamia Ecology Movement: Strengthening the Change in Ecological Awareness!" working paper, *Academia.edu* (August 5, 2015): https://www.academia.edu/18468207/Mesopotamian_Ecology_Movement.

Ayboğa, Ercan, and Janet Biehl. 2011. "Kurdish Communalism." (Ercan Ayboğa interviewed by Janet Biehl). *New Compass* blog. April 16 and September 20, 2011. http://new-compass.net/article/kurdish-communalism.

Bachelard, Gaston. 1994. *The Poetics of Space* [1958]. Translated by Maria Jolas in 1964. Boston, MA: Beacon Press.

Barcelona En Comú (ed.). 2019. *Fearless Cities: A Guide to the Global Municipalist Movement*. Oxford: New Internationalist.

Biehl, Janet. 2011. "Report from the Mesopotamian Social Forum." *New Compass* website. May 13, 2011. http://new-compass.net/node/265.

Casanova, Pablo González. 2005. "The Zapatista 'Caracoles': Networks of Resistance and Autonomy." *Socialism and Democracy* 19, no. 3: 79–92. https://doi.org/10.1080/08854300500257963.

Casier, Marlies. 2015. "Beyond Kurdistan?: The Mesopotamia Social Forum and the Appropriation and Re-imagination of Mesopotamia by the Kurdish Movement." In *The Kurdish Issue in Turkey: A Spatial Perspective*, edited by Zeynep Gambetti and Joost Jongerden, 136–54. Abingdon, UK: Routledge.

Çiçek, Meral. 2020. "Meral Çiçek, of the Kurdish Women's Relations Office, on the Bombing of Makhmur," Video interview with Jeff Miley, *Peace in Kurdistan* website, June 18, 2020. https://www.peaceinkurdistancampaign.com/meral-cicek-of-the-kurdish-womens-relations-office-on-the-bombing-of-makhmur/.

Cooperativa Integral Catalana. 2015. "From Chiapas to Rojava—More than Just Coincidences." *Cooperativa Integral Catalana* website. February 9, 2015. https://cooperativa.cat/from-chiapas-to-rojava-more-than-just-coincidences/.

Cortés Calderón, Sofía Valeria. 2018. "Security, Justice, and Forest Protection! An Indigenous Community's Fight Against Illegal Deforestation and Organised Crime in Cherán, Mexico." In *Green Crime in Mexico: A Collection of Case Studies*, edited by Ines Arroyo-Quiroz and Tanya Wyatt, 63–73, Palgrave Studies in Green Criminology. [London]: Palgrave Macmillan.

Covarrubias, Jennifer Gonzalez. 2019. "A Mexican Indigenous Town's Environmental Revolt." *Phys.org* website. December 21, 2019. https://phys.org/news/2019-12-mexican-indigenous-town-environmental-revolt.html.

Democracy Communalist. 2020. "Maxmûr, Blooming in the Desert." January 15, 2020. Video. https://www.youtube.com/watch?v=NG8NzsaEsm0.

El Enemigo Común 2018. "Celebrating Seven Years of Self-Governance in Cherán, Michoacán." *Intercontinental Cry* website. August 15, 2018. https://intercontine ntalcry.org/celebrating-seven-years-of-self-governance-in-cheran-michoacan/.

Finlay, Eleanor. 2017. "Reason, Creativity and Freedom: The Communalist Model." *ROAR*. February 11, 2017: https://roarmag.org/essays/communalism-bookchin-direct-democracy/.

Freire, Paulo. 1996. *The Pedagogy of the Oppressed* [1970], rev. ed. translated by Myra Bergman Ramos. London: Penguin.

Friends of the MST. 2019. "Special Report: Brazil Education in the Countryside" *Friends of the MST* website. https://www.mstbrazil.org/content/special-report-brazil-education-countryside.

Gambetti, Zeynep. 2009. "Politics of Place/Space: The Spatial Dynamics of the Kurdish and Zapatistas Movements." *New Perspectives on Turkey* no. 41 (Fall): 43–87. https://doi.org/10.1017/S0896634600005379.

Hall, Glenn. 2019. "Forms of Freedom: Dual Power in Fiji." *Harbinger: A Journal of Social Ecology* 4, no. 1 (Winter). https://harbinger-journal.com/issue-1/forms-of-freedom/.

Haifawi, Yoav. [2015]. "Democratic Confederalism and the Palestinian Experience." *The Region*, December 13, 2017. https://theregion.org/article/12189-democratic-confederalism-and-the-palestinian-experience.

Hobson, Bernadette. 2016. "The Clean Energy Poverty Market: The Case of Chiapas, Mexico." In *Water Planet: The Culture, Politics, Economics, and Sustainability of Water*, edited by Camille Gaskin-Reyes, 300–302. Santa Barbara, CA: ABC-CLIO.

Howard, Philip. 1998. "The History of Ecological Marginalization in Chiapas." *Environmental History* 3, no. 3 (July): 357–377. https://doi.org/10.2307/3985184.

Imperial, Miranda. 2019. "New Materialist Feminist Ecological Practices: La Via Campesina and Activist Environmental Work." *Social Sciences* 8, no. 8 (August): 235–250. https://doi.org/10.3390/socsci8080235.

Internationalist Commune of Rojava. 2017. "From Kurdistan to Wallmapu—Solidarity with the Mapuche!" *Internationalist Commune* website. November 12, 2017. https://internationalistcommune.com/kurdistan-wallmapu-solidarity-mapuche/.

Jackson Rising. 2018. "The Rojava Revolution and Democratic Confederalism: A Cooperation Jackson dialogue with Ercan Ayboga." March 24, 2018. Video. https ://www.youtube.com/watch?v=vsSni5urAMM.

"Jose." 2019. "Interview: Rojava, Venezuela and South America—Between Imperialism and Revolutions," *Internationalist Commune* website. May 6, 2019. https://internationalistcommune.com/rojava_venezuela/.

Klein, Hilary. 2015. *Compañeras: Zapatista Women's Stories*. Oakland: Seven Stories Press.

La Via Campesina. 2014. "Syria (Qamishlo, Al-Hasaka) Distributing Flyers on Organic Farming and Natural Seeds." *La Via Campesina* website. April 17, 2014.

https://viacampesina.org/en/event/april-17-syria-qamishlo-al-hasaka-distributing
-flyers-on-organic-farming-and-natural-seeds/.

Levitas. Ruth. 2013 *Utopia as Method: The Imaginary Reconstitution of Society.*
Basingstoke, UK: Palgrave Macmillan.

Leone, Miguel and Camila Ponce. 2019. "New Solidarity for the Mapuche Indigenous
Movement in Today's Chile." *OpenDemocracy* website. January 16, 2019. https
://www.opendemocracy.net/en/democraciaabierta/new-solidarity-for-mapuche-
indigenous-movement-in-today-s-chile/.

Lopez, Oscar. 2020. "A Town Torn Apart: Mexico's Indigenous Communities Fight
for Autonomy." *Reuters*, January 2, 2020. https://uk.reuters.com/article/us-mexico-
indigenous-cities-feature-trfn/a-town-torn-apart-mexicos-indigenous-communities
-fight-for-autonomy-idUSKBN1Z10M9.

Magdoff, Fred, and Chris Williams. 2017. *Creating an Ecological Society: Toward a
Revolutionary Transformation.* New York: Monthly Review Press.

Make Rojava Green Again. 2020. "25th September Global Climate Strike: Greetings
from Rojava to Fridays For Future Youth of the World!" *Make Rojava Green
Again* website. September 18, 2020. https://makerojavagreenagain.org/2020/09
/18/25th-september-global-climate-strike-greetings-from-rojava-to-fridays-for-
future-youth-of-the-world/.

Mallett-Outtrim, Ryan. 2016. "The Zapatista's CompArte Art Festival in Images."
New Internationalist website. August 16, 2016. https://newint.org/features/web
-exclusive/2016/08/15/zapatista-art-festival.

Marcos, Subcomandante Insurgente (Rafael Sebastián Guillén Vicente). 2001. *Our
Word is Our Weapon: Selected Writings*, edited by Juana Ponce de León. New
York: Seven Stories Press.

Marcos (Subcomandante Insurgente). 2003. "Each Caracol now had a Name
Assigned." July 2003, *The Struggle* website. http://struggle.ws/mexico/ezln/2003
/marcos/caracolJULY.html.

Martínez-Torres, María Elena, and Peter M. Rosset. 2010. "La Vía Campesina: The
Birth and Evolution of a Transnational Social Movement." *Journal of Peasant
Studies*, 37, no. 1: 149–75. https://doi.org/10.1080/03066150903498804.

Maxwell, Clare, 2017. "The Kurds of Lebanon: Identity, Activism and Ideology."
New Eastern Politics. January 30, 2017. https://www.neweasternpolitics.com/the
-kurds-of-lebanon-identity-activism-and-ideology-by-clare-maxwell/.

McGuire, Emily. 2019. "Timeline: Kurdish Resistance in Turkey and Syria
1847-2019." *Women Defend Rojava* website. https://womendefendrojava.net/wp-
content/uploads/2019/10/Timeline-Kurdish-Resistance-in-Turkey-and-Syria-
Emily-McGuire-w-Addendum-1_compressed.pdf.

Meek, David. 2015. "Learning as Territoriality: The Political Ecology of Education
in the Brazilian Landless Workers' Movement." *Journal of Peasant Studies* 2
(January 20, 2015): 1179–1200. https://doi.org/10.1080/03066150.2014.978299.

Menser, Michael. 2008. "Transnational Participatory Democracy in Action: The Case
of La Via Campesina." *Journal of Social Philosophy* 39, no. 1 (Spring): 20–41.
https://doi.org/10.1111/j.1467-9833.2007.00409.x.

Öcalan, Abdullah. 2017. *The Political Thought of Abdullah Öcalan: Kurdistan, Women's Revolution and Democratic Confederalism.* Translated by Havin Guneser. London: Pluto and International Initiative.

Prichard, Alex, and Owen Worth. 2016. "Left-wing Convergence: An Introduction." *Capital and Class* 40, no. 1: 3–17, https://doi.org/10.1177/0309816815624370.

Pressly. Linda. 2016. "Mexico—The Town That Said, 'No'!" *Crossing Continents*, broadcast by BBC Radio 4. December 26, 2016. https://www.bbc.co.uk/sounds/play/b085xpzk.

[Qereçox], Viyan. 2020. "Democratic Confederalism: Learning from Grassroots Democracy," Kurdistan Solidarity Network webinar. May 19, 2020. Video. https://www.facebook.com/watch/?v=868009160376437.

Romão, João. 2009. "Mesopotamia Social Forum (MSF) in Diyarbakir." *Transform Europe!* Website. October 12, 2009. https://www.transform-network.net/cs/blog/article/mesopotamia-social-forum-msf-in-diyarbakir/.

Ross, John. 2005. "Celebrating the Caracoles: Step by Step, The Zapatistas Advance on the Horizon." *Humboldt Journal of Social Relations* 29, no .1: 39–46. https://www.jstor.org/stable/23263124.

Russell, Bertie. 2019. "Beyond the Local Trap: The New Municipalism and the Rise of the Fearless Cities." *Antipode* 51, no. 3 (June): 989–1010. https://doi.org/10.1111/anti.12520.

Scheindlin, Dahlia, and Dov Waxman. 2016. "Confederalism: A Third Way for Israel-Palestine." *The Washington Quarterly* 39, no. 1 (Spring 2016): 83–94. http://dx.doi.org/10.1080/0163660X.2016.1170482.

SIPAZ. 2015. "Chiapas: Ejidatarios from Chicoasén Initiate Hunger-Strike Against Chicoasén II Dam-Project." *SIPAZ Blog.* November 19, 2015. https://sipazen.wordpress.com/2015/11/19/chiapas-ejidatarios-from-chicoasen-initiate-hunger-strike-against-chicoasen-ii-dam-project/.

Solnit, Rebecca. 2008. "Revolution of the Snails: Encounters with the Zapatistas." *Common Dreams* website. January 16, 2008. https://www.commondreams.org/views/2008/01/16/revolution-snails-encounters-zapatistas.

Soz, Baz. [2017]. "Interview with a Palestinian Internationalist [Baz Soz] in Rojava." *Internationalist Commune* website. (July 4, 2018. [first pub. in German in *Lower Class Magazine*, June 14, 2017]. https://internationalistcommune.com/better-than-a-one-or-two-state-solution-would-be-a-no-state-solution-a-palestinian-internationalist-in-rojava/.

Stanchev, Peter. 2015. "From Chiapas to Rojava: Seas Divide us, Autonomy Binds us," *ROAR* (February 17, 2015), https://roarmag.org/essays/chiapas-rojava-zapatista-kurds/.

Starr, Amory, María Elena Martínez-Torres, and Peter Rosset. 2010. "Participatory Democracy in Action: Practices of the Zapatistas and the Movimento Sem Terra." *Latin American Perspectives* 38, no. 1: 102–19. https://doi.org/10.1177/0094582X10384214.

Statista.com. 2019. "Number of Land Activists and Environmental Defenders Murdered in Selected Countries in Latin America in 2018." July 2019. https://www.statista.com/statistics/884020/number-activists-murdered-latin-america-country/.

Statista.com. 2020. "Number of Land and Environmental Protectors Killed Worldwide in 2019, by Country. July 2020. https://www.statista.com/statistics/563444/worldwide-killings-of-land-and-environmental-defenders-by-country/.

Tarlacrea, Matouf. 2018. "Other Rojavas: Echoes from the Free Commune of Barbacha: Chronicling an Autonomous Uprising in Northern Africa." Translated. from 2014 French edition by Michael Disnivic and Habiba Dhirem-Kasper. [s.l.]: Crimethinc.

Torres-Salinas, Robinson et al. 2016. "Forestry Development, Water Scarcity, and the Mapuche Protest for Environmental Justice in Chile." *Ambiente e Sociedade* 19, no. 1 (January 1, 2016): 121–44. https://doi.org/10.1590/1809-4422asoc150134r1v1912016.

TV Cherán. 2018. "Celebrating Seven Years of Self Governance in Cherán, Michoacán." August 2, 2018. Video. https://www.youtube.com/watch?v=oDspeScGZrQ.

Vardi, Sahar. 2019. "The Hypocrisy of Israel's Alliance with the Kurds." *+972 Magazine*. October 18, 2019. https://www.972mag.com/israel-kurds-turkey-arms-exports/.

Villanueva, Pilar. 2018. "From Rojava to the Mapuche Struggle: The Kurdish Revolutionary Seed Spreads in Latin America." *Toward Freedom* website. September 12, 2018. https://towardfreedom.org/story/archives/americas/from-rojava-to-the-mapuche-struggle-the-kurdish-revolutionary-seed-spreads-in-latin-america/.

Women Defend Rojava. "Rojava Women's Movement - Message to Zapatista Women." December 27, 2020. Video. https://www.youtube.com/watch?v=NmdRbfHkP60.

Yarkin, Gullistan. 2015. "The Ideological Transformation of the PKK Regarding the Political Economy of the Kurdish Region in Turkey." *Kurdish Studies* 3, no. 1(May): 26–46. https://kurdishstudies.net/journal/ks/article/download/390/383/393.

Yeniay, Özlem, David Harvey, and Abdullah Aysu, "Solidarity Call from Brazil's Landless Workers' Movement for Turkey." *Bianet* website. August 5, 2015. http://bianet.org/english/print/166580-solidarity-call-from-brazil-s-landless-workers-movement-for-turkey.

Zizumbo Colunga, Daniel. 2015. "Taking the Law into Our Hands: Trust, Social Capital and Vigilante Justice." PhD diss., Vanderbilt University. https://etd.library.vanderbilt.edu/etd-11202015-202702.

Chapter 21

Concluding Reflections on the Kurdish Ecology Initiatives

Stephen E. Hunt

This collection acknowledges the immense challenges that face the Kurdish freedom movement and its ecological aspirations. These concerns cannot be disentangled from the present conflict and fluid political situation. At the time of writing, Turkish state forces and their proxies continue to attack northeastern Syria. Relations between the AANES and Nechirvan Barzani's Kurdistan Regional Government in Iraq, which has economic and political ties with the Turkish state, have also become increasingly and dangerously fractious, particularly since the territorial defeat of ISIS as a common enemy. Western nations are reluctant to check the military adventurism of NATO ally Turkey, while Western corporations continue their long-standing arms sales to the Middle East (Feinstein 2012), an issue Ceri Gibbons demonstrates in the current collection. Within Turkey, repression in internal matters is no more likely to be reversed in the foreseeable future than aggression in external affairs. The Biden Presidency in the United States may take a less sympathetic stance toward Recep Tayyip Erdoğan's ambitions, following its predecessor's betrayal of the Syrian Democratic Forces after their defeat of ISIS. Perceived strategic priorities and the interests of global capitalism, however, will unquestionably prevail. While Russian and Iranian forces continue to bolster the Assad regime in Syria, the AANES will continue to share a hostile boundary with areas governed by its old Ba'athist adversary to the South. In Iran too, there is no sign that the oppression of Kurdish populations, including those engaged in the ecological activism that Allan Hassaniyan's chapter analyzes, will abate. In such contexts, the pursuit of an ecological agenda, and the development of popular assemblies and a solidarity economy face formidable obstacles within the AANES, and remain for the most part aspirational beyond. Added to this deteriorating security situation, disruptions such as the coronavirus pandemic have slowed progress for "Make Rojava Green Again" and other practical ecological projects.

Particular planetary and regional threats to ecological well-being are outlined in the introduction. While compiling this book, research detailing yet another symbolic watershed in the intensification of anthropogenic impact was announced: the startling claim that human-made artifacts are overtaking natural materials in terms of mass on the Earth's surface (Elhacham et al. 2020). This follows concerns about the loss of living biomass, estimated to have reduced by 11 percent since 1900 (Schramskia et al. 2015). The principal objective of the present collection has not been, however, to add to the ever-expanding literature about the Earth's plight. It has aimed rather to understand the project for ecological solidarity within the principles and structures of democratic confederalism and to consider what can be learned from this approach. The Kurdish movement is idealistic and ambitious in its aspirations, but at the same time aware of the challenges it faces and limitations to its progress. The AANES's assessment, for example, is candid:

> In the field of ecology, the political system has set itself admirable goals which it has only gone a small way towards fulfilling. Many projects have been set up, particularly in the field of agricultural cooperatives, but basic structural developments in waste and water treatment, fuel and energy and building standards have yet to be made. The Autonomous Administration has made crucial steps in identifying areas for development and international partners, but often lacks the necessary institutional capacity, funding and expertise to take action. (Rojava Information Center 2019, 55)

The first point to note is that the foregoing chapters have demonstrated that Kurdish ecology is distinct from mainstream Western liberal environmental campaigns. It is a form of political ecology, specifically social ecology, and as such integrated within the wider framework of democratic confederalism. The Kurdish paradigm uses the theory of social ecology to understand ecological destruction through deep historical roots and the systemic problems currently arising from the dominance of capitalism and statism. Öcalan applied Fernand Braudel's concept of the *longue durée* to consider the impact of seminal civilization in Mesopotamian city states[1] upon subsequent human development and domination of the natural world. Öcalan has been particularly concerned with the position of women. As Hammy observes in the present collection, Öcalan prioritized critique of patriarchy as a foundational form of hierarchical domination with damaging social consequences. It follows that women's liberation is regarded as central to social revolution: jineolojî is at the heart of the KFM's projects. In keeping with this emphasis, women play a central role, for example through the democratic assemblies, cooperatives, and famously in the self-defense units. In addition to acknowledging the reproductive roles and unpaid domestic work that have been harnessed and exploited so effectively through the capitalist division of labor, the Kurdish perspective also recognizes the role of women in producing the world's wealth. Women account for a major proportion of labor to meet core social needs, outnumbering men as healthcare workers and teachers, and making a significant, although unquantified, contribution to food production (OECD 2019; Katsarova 2020; Doss et al. 2018). The capitalist economy fails to reflect such facts in the equitable distribution of profits, while, globally, women are politically underrepresented. If, as social ecologists believe, the ecological crisis has its origins in structural domination and inequality, the pillars of ecological sustainability and women's liberation are complementary.

Central to the holistic paradigm promoted by the Kurdish movement, the Zapatistas, and compatible initiatives, such as Cooperation Jackson, is the principle that social justice and ecology are mutually reinforcing liberation struggles. The movements for democratic confederalism are based on the notion that extractivism is an ideological expression of colonial thinking as well as an ecologically destructive and unsustainable practice. Critics with a social ecological outlook reject conventional justifications that make the case that the destruction of the living world is imperative for economic growth, and therefore deemed a social good. They counter that relentless damage to the living world is destructive of real social and economic value, with sell-offs of natural resources typically enriching and empowering privileged beneficiaries while negatively impacting the quality of life and displacing other communities and nonhuman species. Unsustainable ecological destruction is at the expense of the resilience and livelihoods of others in the present, and in generations to come. What David Harvey (2004) terms "accumulation by dispossession" is accomplished through land grabs with attendant displacements of human communities, forest clearances, and despoliation of local ecologies. Relatedly, it is instructive that Silvia Federici's influential work *Caliban and the Witch* (2014, 14) links enclosure with witch hunts (and other violent acts in the war against women) and slavery, as inseparable forms of primitive accumulation, and that equally call for an inseparable response in the present day, as historically. Today, such social traumas as environmental degradation, the oppression of women, and exploitative working conditions are negative counterparts to the three pillars of democratic confederalism. Democratic confederal principles directly contest these injustices with political claims for ecological well-being, women's liberation, inclusive multi-ethnic participatory democracy with a solidarity economy. It is no coincidence that some of the most committed opponents of ongoing processes of primitive accumulation, within the anti-globalization movement, have emerged in regions such as the Chiapas. Here the Zapatistas are aware of the state violence and economic chicanery that exploits, dispossesses, and displaces communities that suffer poverty and increasing debts to those that they supply with expanding quantities of timber, fossil fuels, and foodstuffs. Equally, populations in central Africa states are rewarded with conflict and habitat destruction for their bounty of natural resources of critical demand, such as rare earth minerals, due to patterns of uneven development endemic to advanced capitalism. It is understood therefore that isolated attempts to create build alternative political programs would be unsuccessful (Akuno and Nangwaya 2017, 33). Single-issue approaches can easily lead to reactive and isolated mobilizations. The West's social movements have much to learn from the perceptions of indigenous communities on the frontlines of ecological destruction through colonialism and extractivism, who have achieved enduring relationships with the natural environment (Imperial 2019, 250; Torrez 2011, 50). Connecting struggles against racism and colonialism with direct ecological solidarity, therefore, is a means to integrate aspirations to build a movement of movements better able to resist both social injustice and unsustainable capitalist modes of production.

So ecological solidarity calls for a particular kind of international solidarity, one that is direct, builds connected networks and structures, and is not solely symbolic. This is the journey that the Kurdish movement has been navigating as it attempts to resolve and move beyond tensions between ethno-nationalist bonding over a patriotic love of the sacred land, evoked in national liberation struggles for independence, and socialism's internationalist aspirations. Furthermore, democratic confederalism is an expansive project to create a "commune of communes." This is critically distinct from globalization or varieties of internationalism that entail the sacrifice of the local for the good of the whole. Social ecology is after all a vision of interdependence. Timothy Morton's concept of *subscendence*, a theoretical underpinning in *Humankind: Solidarity with Nonhuman People* (2017, chapter 3), may be helpful here. Morton introduces a radical reversal in his invitation to reconceive the conventional idea that the "whole is greater than the sum of its parts," asking us instead to entertain the notion that the parts can be greater than the whole. He provocatively and counterintuitively reframes the relationship between the whole and the particular. Morton suggests that the restricted ontology of the abstract whole can be less than the proliferating, multidimensional, and infinitely complex aspects of the particular. There is an ecological danger in thinking of organisms as replaceable, disposable, and expendable so long as the whole endures. This awareness is in harmony with the political understanding that the imposition of the unitary state has violent logic and consequences for the civitas and cultural diversity. The call for solidarity across and beyond humanity, therefore, chimes well with the insights of democratic confederalism, a form of interdependence in the realm of human politics and social ecology. Social ecologist John P. Clark similarly invites us to adopt an internationalist planetary consciousness that does not obliterate local ecosystems or cultural specificity in the cause of an abstract totality. He enjoins us to embrace a "universal particularity" that rejects exclusive parochialism and xenophobia. Clark proposes instead that we follow the example of French regionalist thinkers in "authentic reaching out to others beyond one's one own region" (2017, 224). Fellow social ecologist Brian Tokar, therefore, calls for the development of profoundly inclusive "Confederations of democratic communities and regions" to "develop new continental and global institutions" (2019). In short, the aspirations of the Kurdish political and ecological initiatives have implications far wider than the Kurdish region. Kurdish activist and revolutionary Besime Conca appeals, "Come join us, and let's protect this revolution together, because this is not just a Kurdish revolution, this is an internationalist revolution for all" (Conca 2020).

Which leads us to consider questions of counterpower: how to scale across such experiments and build capacity? "Resist and build" are the watchwords of Cooperation Jackson. Yet, despite many impressive mobilizations, statements, and initiatives, there have been limits to the practical international solidarity that has been achieved since the emergence of the global justice movement during the 1990s. Since, as has been suggested, ecology is the science of interdependence, it follows that ecological resilience—or the lack of—matters for the prospects of all life on Earth, including, of course, within that biosphere, all human lives. Expanding ecological solidarity, therefore, is not just desirable but essential. In addition to building political networks and structures, what follows identifies capacity building in four related areas: political and economic infrastructure, media, education, and the value of the Kurdish paradigm in the development of a social imaginary.

The chapter "Plant the Tree of Tomorrow," surveys movements with affinities to democratic confederalism and social ecology which could potentially develop alongside, beyond, and in solidarity with, the Kurdish experiments. Despite formidable obstacles, one of the most hopeful and enlightened aspects of the administration in Rojava is its inclusive approach to capacity building. The example of democratic confederalism has won over sizable Arab populations, especially since Kurdish, Arab, and Syriac militias fought together as the Syrian Democratic Forces to defeat ISIS with US support. In the AANES, districts with majorities of Arab and other ethnic populations are now practicing democratic confederalism, notably including liberated Raqqa, the one-time ISIS capital. Perhaps the most extraordinary development has been recent efforts to rehabilitate mostly young, low-ranking ISIS affiliates in the notorious Al-Hawl camp. Here there have been initiatives to educate jihadist youth with ideas about direct democracy and gender equality, to counter indoctrination undertaken during the so-called "Caliphate" (Rojava Information Center, 2020). Furthermore, as Ercan Aboğa's chapter in the present book shows, the Kurdish movement has been making serious efforts to embrace ecology within its democratic structures and solidarity economy, even though, to date, this aspect has been less prominent than other parts of the political program.

For social ecological principles to be successfully adopted for humans and other species alike, therefore, they need to be both embedded in the practice of the developing political and economic infrastructure, and fully integrated into the alternative post-capitalist paradigm. As this happens, ideas, such as food sovereignty, restoration ecology, and liberatory technology, become part of the collective mentality. The ubiquitous and intensifying systemic crises of capitalism create a vacuum for envisioning alternatives. Viyan Qereçox, an internationalist volunteer with experience of the ecological initiatives, reflects upon the utopian thinking grounded in radical pragmatism found in Rojava.

> Having some kind of ideology that ties us together allows us to hold the contradictions within our strategy and actions, which is absolutely crucial in terms of fighting climate change. We work in a reality in which it's impossible to fully embody our ecological values in the way that we live, and getting overly fixated on this more lifestylist approach to sustainability cuts off a lot of possibilities to organize on a more collective and fundamental level. In Rojava, the ecological aspect of the Revolution has faced countless challenges and is riddled with contradictions. Even though the movement is committed to sustainability, much of it runs off the profits of fossil fuel extraction, the lack of infrastructure means that people burn trash and dump waste, and the embargo means that more sustainable technology is incredibly hard to access. Sometimes decisions need to be made in which a more ecological approach comes into contradiction with a more practical-term approach. ("Democratic Confederalism: Learning from Grassroots Democracy," Scottish Solidarity with Kurdistan webinar, May 19, 2020)

Nevertheless, Qereçox reports that progress is being made since practitioners are building capacity in a spirit of experimentation and scaling across "pockets of deep practice" and adaptability.

The foregoing chapters have given the ecological aspects of the Kurdish movement, the attention and the respect afforded by critical evaluation that they deserve. To expand such conversations, communication is a priority. It is necessary to seek inventive ways to share knowledge and exchange ideas through multiple forms of media. Mainstream media marginalization has long been a major obstacle to public discussion about the kind of alternatives that the Kurdish and Zapatista movements propose and the oppression they face. The big news stories relating to the Kurdish movement, such as the Kurdish and allied militias' major role in the defeat of ISIS, are rarely aired or printed in the West. While the KFM is right to highlight the undoubted similarities between the South African and Kurdish liberation struggles, and parallels between the long-term imprisonment of Nelson Mandela on Robben Island and that of Abdullah Öcalan on İmralı Island, there remains a wide gap in the media attention devoted to the two issues. From 2020, the coronavirus pandemic severely disrupted face-to-face networking to advance international solidarity. Yet within activist circles, the proliferation of Kurdish webinars facilitated deep discussion and exchanges about opportunities and threats. Digital forums, for example, the excellent Co-operation in Mesopotamia project, have also helped fund-raising for practical international solidarity to aid social and environmental infrastructure. Digital technologies are affording opportunities to learn and participate that are inclusive and enable participation without the time and expense incurred in attending physical seminars and conferences. Social media helps to address these deficiencies in information communication, although pro-KFM Twitter and Facebook accounts are regularly suspended. Face-to-face communication and alternative print and broadcast media remain vital to energize public debate. Above all, challenging marginalization and the uncritical repetition of the Turkish state's misrepresentation of the Kurdish movement as separatist and terroristic are urgent priorities for raising awareness of Kurdish aspirations, including the ecological efforts.

Education is key to ensuring that social ecology's place within the paradigm of democratic confederalism is understood, and its implications realized in practice. The Kurdish educational academies' radical pedagogy encourages both collective learning and an autodidactic spirit among movement followers. As we have seen, ideas from social ecology furnish the Kurdish movement with a philosophy of intellectual and practical transformation, being a kind of permaculture for diverse but sustainable ideas. The present collection shows, for example, the relevance of the ideas of Bookchin and Öcalan for rethinking the human relationship to ecosystems. Paulo Freire's ideas are influential too, for instance, in shaping the curricula of the new Rojava University (Fox 2018). The Brazilian educator's proposals to decolonize power relations are linked to ecological well-being. Freire looks to reconfigure education so that it rejects inculcating habits of objectification and control leading to "necrophily" (the impulse to convert organic life into non-living matter), and instead nurtures "biophily," oriented toward creative life force (1996, 58). Above all, the KFM calls for a revolution of ideas and mentality. Rigorous processes of self-reflection and critical evaluation evident in practices such as *Tekmîl* (defined as undertaking critical self-assessment with love and respect) and study groups have multiple benefits: nurturing the culture of collaborative learning through political participation that underpins self-education and outreach; confronting anti-social traits rather than excluding perpetrators; helping to create a dynamic and highly motivating social imaginary. Social ecology tells us that *the* ecological crisis is a crisis of ideas, in which for the natural world to flourish, it is necessary to decolonize the mentality of domination and exploitation as a terrain, expressed, for example, in the contradictory capitalist idea that increased consumerism is beneficial for the economy.

Social commentator Aaron Bastani cites the "absence of collective imagination" as a "most pressing crisis" (2019, 21), alongside the ecological crisis as the leading threat to the well-being of humanity and the rest of the biosphere. Clark, further emphasizes the urgent need for post-capitalist dissidents to undertake the work of "critical and dialectical analysis of the social imaginary" to understand the deterministic functioning of the current state and corporate attainment of "imaginary supremacy" by defining reality in the minds of its subjects. He makes the case for creating a "liberatory social imaginary" (2019, chapter 16). The post-capitalist imaginary being developed within the Kurdish freedom movement calls for, and is enacting, a profound shift in social priorities, a model of eco-socialism where the well-being of society and the biosphere are privileged over the needs of capital. The movement operates on the premise that, ultimately, such a transformation cannot be purchased, or attained by technological invention alone, but must be created through critical consciousness and collective endeavor.

Democratic confederalism represents an advanced attempt to work together to put such principles into practice with outcomes that may be partial and imperfect, yet are nonetheless hopeful. To date, the KFM has successfully endured in the face of state hostility, and communities in northeastern Syria have built alliances and combined courage, self-discipline, and vision to defeat ISIS, one of the most aggressive terror organizations of recent years. Alongside successful social defense, the Kurdish movement has continued to develop the infrastructure of the solidarity economy, participatory assemblies, free academies, and ecological initiatives. In this collection, we have traveled far and glimpsed many possibilities. The contributors have taken us on a journey of discovery to visit the women's eco-village at Jinwar, the bostans of Amed, the assemblies of the Mesopotamia Ecology Movement, the autonomous gardens of the Makhmur Refugee Camp, and to the women's cooperatives and communal markets of northeastern Syria. We have considered the inestimable challenges and messy complexities that have confronted social movements and ecological activists in situations as varied as the campaigns against dam-building at Hasankeyf and Dersim, for urban green space at Gezi Park, for the protection of the natural world in Rojhelat, and on the frontlines of the Hevsel Gardens resistance. In each case, it has been necessary to navigate with nuance tensions between utopian impulses and pragmaticism with efforts to defend social achievements, experiment, and identify achievable transitional measures. There are practical obstacles such as diversifying food production where there has been a legacy of monoculture and providing sustainable livelihoods and decent standards of living for all in the context of economic embargo, oil dependency, and increasing climate change. There are political challenges in attempting to implement participatory democracy in the context of repressive state apparatus and to find legal expression beyond the constraining framework of statist discourses. Moreover, the complicity of Western imperialism is visible in the historic legacy of imposed borders in the Middle East, the dominance of fossil-fuel interests, and the destabilizing influence of the arms industry.

Ecological Solidarity and the Kurdish Freedom Movement has aimed to understand what is distinct about the movement for a Western, English-speaking audience. It has rejected Eurocentric assumptions that "green" issues are a cultural import from the environmental organizations in the West, or even straightforwardly an outlook taken from the political ecology of thinkers, such as Murray Bookchin. This work seeks to inform conversations about the potentially transformational aspects of Kurdish ecology, considering what can be learned and what forms of practical mutual aid are appropriate, within mutually enriching exchanges. Inevitably, there are further aspects of Kurdish ecology and environmentalism to explore such as the activities of Iraqi civil society ecological initiatives, for example the Save the Tigris and Iraqi Marshes Campaign and Humat Dijlah, and the environmental record of the KRG in Iraq; deeper analysis of the way that ecological concerns feature in the education program; the impacts of conflict and development upon the Kurdish region's natural history; the extent and way that predominantly Kurdish communities at home and in the diaspora interpret and represent ecological ideas within their media, art, film, and other cultural manifestations; further analysis of notice given to ecology by international Kurdish solidarity groups and of Kurdish campaigns within Western environmental groups.

The progress that the Kurdish movement has achieved in realizing democratic confederalism to date has been made coherent through a collective commitment to an alternative framework or paradigm. This aims to unite and bond the different components of participatory democracy, gender liberation, and ecology. As we have seen, the presence of what John P. Clark has termed a "liberatory social imaginary" enhances and deepens the sense that it is within humanity's collective grasp to re-envision and positively transform this world. While another world is impossible, since this is the only planet we have, we have seen through the Kurdish example that another ecology movement is possible, one for inspiration and with which to make connections, not for simple imitation or appropriation. With similar principles of ecological solidarity, perhaps, other more just, sustainable, and harmonious outcomes may be possible for this world.

NOTE

1. The first rudimentary state forms emerged around 5,000 years ago in areas that today have substantial Kurdish populations, the Tigris and Euphrates Valley (Scott 2017, 7).

REFERENCES

Akuno, Kali and Ajamu Nangwaya. 2017. *Jackson Rising: The Struggle for Economic Democracy and Black Self-Determination* in *Jackson, Mississippi.* [Montréal]: Daraja Press.

Bastani, Aaron. 2019. *Fully Automated Luxury Communism.* London: Verso.

Clark, John P. 2019. *Between Earth and Empire: From the Necrocene to the Beloved Community.* Oakland: PM Press.

Conca, Besime. 2020. "Defending the Rojava Revolution." Besime Conca interviewed by Ashish Kothari. *Radical Ecology Democracy* website. June 24, 2020. https://www.radicalecologicaldemocracy.org/redweb-conversations-series-defending-the-rojava-model/.

Doss, Cheryl, Ruth Meinzen-Dick, Agnes Quisumbin, and Sophie Theis. 2018. "Women in Agriculture: Four Myths." *Global Food Security* 16 (March 2018): 69–74. https://doi.org/10.1016/j.gfs.2017.10.001.

Elhacham, Emily, Liad Ben-Uri, Jonathan Grozovski, Yinon M. Bar-On and Ron Milo. 2020. "Global Human-made Mass Exceeds all Living Biomass." *Nature* (December 9, 2020). https://doi.org/10.1038/s41586-020-3010-5.

Federici, Silvia. 2014. *Caliban and the Witch: Women, the Body and Primitive Accumulation.* 2nd rev. ed. Brooklyn, NY: Autonomedia.

Feinstein, Andrew. 2012. *The Shadow World: Inside the Global Arms Trade.* London: Penguin.

Fox, Edward. 2018. "A New University, Born in the Chaos of War." *Al-Fanar Media* website. August 22, 2018. https://www.al-fanarmedia.org/2018/08/a-new-university-born-in-the-fog-of-war/.

Freire, Paulo. 1996. *The Pedagogy of the Oppressed* [1970]. New revised edition translated by Myra Bergman Ramos. London: Penguin.

Harvey, David. 2004. "The 'New' Imperialism: Dispossession by Accumulation." *Socialist Register* 40: 63–87. https://socialistregister.com/index.php/srv/article/view/5811.

Imperial, Miranda. 2019. "New Materialist Feminist Ecological Practices: La Via Campesina and Activist Environmental Work." *Social Sciences* 8, no. 8 (August): 235–50. https://doi.org/10.3390/socsci8080235.

Katsarova, Ivana. 2020. "Teaching: A Woman's World." *European Parliament Research Service* website, March 2020. https://www.europarl.europa.eu/RegData/etudes/ATAG/2020/646191/EPRS_ATA(2020)646191_EN.pdf.

Morton, Timothy. 2017. *Humankind: Solidarity with Non-Human People*. London: Verso.

OECD. 2019. "Women Are Well-Represented in Health and Long-Term Care Professions, but Often in Jobs with Poor Working Conditions." *OECD* website, March 2019. https://www.oecd.org/gender/data/women-are-well-represented-in-health-and-long-term-care-professions-but-often-in-jobs-with-poor-working-conditions.htm.

Qereçox, Viyan. 2019. "Lessons for the UK from Rojava: Political Culture, Ideology, Democracy." *Komun Academy* website (November 17, 2019): https://komun-academy.com/2019/11/17/lessons-for-the-uk-from-rojava-political-culture-ideology-democracy/.

Rojava Information Center. 2019. *Beyond the Frontlines: The Building of the Democratic System in North and East Syria*, version 4 (December 19, 2020): https://rojavainformationcenter.com/storage/2019/12/Beyond-the-frontlines-The-building-of-the-democratic-system-in-North-and-East-Syria-Report-Rojava-Information-Center-December-2019-V4.pdf.

Rojava Information Center. 2020. *Hidden Battlefields: Rehabilitating ISIS Affiliates and Building a Democratic Culture in Their Former Territories* (December): https://rojavainformationcenter.com/storage/2020/12/RIC_HiddenBattlefields_-DEC2020.pdf.

Schramskia, John R., David K. Gattiea, and James H. Brow. 2015. "Human Domination of the Biosphere: Rapid Discharge of the Earth-Space Battery Foretells the Future of Humankind." *Proceedings of the National Academy of Sciences* 112, no. 31 (August 4, 2015): 9511–17. https://doi.org/10.1073/pnas.1508353112.

Scott, James C. 2017. *Against the Grain: A Deep History of the Earliest States*. New Haven: Yale University Press.

Tokar, Brian. 2019. "The Promise and Pitfalls of Localism," Opener for GTI Forum Think Globally, Act Locally? *Great Transition Initiative* website (August): https://greattransition.org/gti-forum/global-local.

Torrez, Faustino. 2011. "La Via Campesina: Peasant-led agrarian reform and food sovereignty." *Development* 54, no. 1: 49–54. https://doi.org/10.1057/dev.2010.96.

Glossary

The naming of territory within Kurdish areas is complicated and contested so contributors have used variant terms. This text has mostly regularized nomenclature on the following principles. Within current national borders, parts of Kurdistan have been referred to by identifying the administrating state which is more familiar to an international readership. For example, "Turkey's Kurdish areas," or "Iran's Kurdish region," but "AANES," or "Rojava," to reflect the prevailing administration within that area. The Kurdish names for selected majority Kurdish areas are supplied below. For specific centers of population, within Kurdish majority areas, the preference has been to use Kurdish names, for example, the older Kurdish "Dêrsim" rather than the Turkish appellation of "Tunceli."

AANES: Autonomous Administration of North and East Syria. So named since 2018 (previously Democratic Federation of Northern Syria). Originally, and still popularly, known as Rojava, incorporating the cantons of Kobanî, Afrîn, and Cizîrê, the current territory is a more extensive area, retitled to reflect the region's multi-ethnic character. However, Afrîn canton has been under occupation since the Turkish invasion of 2018, and other areas liberated from Islamic State by the Syrian Democratic Forces (militias allied to AANES) also subsequently became part of the Turkish occupation of northern Syria following offensives in 2019.

AKP (Adalet ve Kalkınma Partisi): The Turkish Justice and Development Party, a right-wing political party, currently the party of government led by its founder, President Recep Tayyip Erdoğan.

Amed: Turkish name, Diyarbakir.

Bakûr: Turkey's Kurdish region.

Başûr: South Kurdistan/Kurdistan Region of Iraq.

Daesh: See ISIS.

Democratic Confederalism: Political system based on Abdullah Öcalan's Declaration of Democratic Confederalism in Kurdistan (2005). Democratic confederalism rests on the three pillars of "grass-roots based democracy," an "ecological model of society," and "the liberation struggle of women."

DTK/KCD (Turkish: Demokratik Toplum Kongresi; Kurdish: Kongreya Civaka Demokratik): Democratic Society Congress, an organization and coordinating body set up to promote and implement the structures of democratic confederalism in Turkey's Kurdish areas.

Ecological solidarity: A synthesis at once informed by insights from the diversity and evolutionary force of the science of ecology, and fired by traditions of solidarity, justice, and mutual aid from emancipatory social movements. The dynamics of interdependence characterize both. In keeping with the philosophy of social ecology, ecological solidarity calls for profound social transformation to respect such interdependence and to address the destructive human relationship with the living world.

HADEP (Halkın Demokrasi Partisi): People's Democracy Party, Pro-Kurdish party banned by the Turkish state in 2003.

HDK (Halkların Demokratik Kongresi): People's Democratic Congress (Turkey).

HDP (Halkların Demokratik Partisi): People's Democratic Party (Turkey).

ISIS: Islamic State of Iraq and Syria, often called "Daesh."

KCK (Koma Civakên Kurdistan): Kurdish Communities Union, umbrella organization representing over-arching structure of the Kurdish Freedom Movement, tasked with coordinating ideas of democratic confederalism among political parties, including PKK and PYD, across borders.

KRI: Kurdistan Region of Iraq. Nechirvan Barzani's Kurdistan Democratic Party currently runs the Kurdistan Regional Government (KRG).

Kurdish freedom movement (KFM): Broad movement based on the principles of democratic confederalism, direct democracy, gender equality, ecological sustainability, and solidarity economy. This body of ideas has its most expansive formulation in the philosophy of Abdullah Öcalan, who has developed them while reading a broad range of political thought during his imprisonment on İmralı Island. The KFM emerged in its present form during the transformation of the Kurdish struggle following the end of the Cold War and a gradual rejection of Marxist-Leninist notions of state socialism. The KFM's principles are being implemented in practice in AANES and have particular support within Kurdish majority areas of southeastern Turkey, as well as across Kurdish areas in the Middle East, and also within the Kurdish diaspora and the international solidarity movement among left-wing and anarchist groups, feminists, and especially social ecologists and eco-socialists. Similar principles have been adopted by the Zapatistas in Mexico and by international organizations such as La Via Campesina.

MEM: Mesopotamia Ecology Movement.

MRGA: Make Rojava Green Again.

PKK (Partiya Karkerên Kurdistan): Kurdistan Workers' Party.

PYD (Partiya Yekîtiya Demokrat): Democratic Union Party (Syria).

Raa Haqi (Kırmancki) / **Rêya Heqî** (Kurmanci): Kurdish Alevi religion and wisdom tradition, literally meaning the "path of the eternal truth." There are variant spellings in both Kırmancki and Kurmanci.

Rojava: Most common term for predominantly Kurdish area of northeastern Syria, now officially AANES (see above).

Rojhelat: Predominantly Kurdish area within the state and jurisdiction of Iran, known as eastern Kurdistan.

TEV-DEM (Tevgêra Cîvaka Demokratîk): The Movement for a Democratic Society. Umbrella organization for civil structures and groups in ANNES.

YPG/YPJ: People's Protection Units—predominantly Kurdish militia in northeastern Syria, now integrated into the Syrian Democratic Forces. The YPJ refers to the women's brigades.

Index

Note: Page locators in italics refer to figures.

affinity groups, x, 4, 17
Afrîn, xxiv, xxv, 90, 117, 134, 280, 286
agriculture, xxiii, xxvii–xxviii, xxix, 48,
50, 52–54, 77, 78, 85, 87–90, 101–
12, 115–30, 139, 143–44, 153–55,
219, 308; agricultural cooperatives,
121, 123, 143, 144n5, 154–55, 219,
308, 322; Agricultural Revolution,
xxxiii, 153; agricultural workers,
106–7, 112n6, 118, 130n3, 130n4;
agroecology, xi, xix, xxviii, *92*,
115, 120–22, 124, 126, 129, 307–8;
Ba'athist policy on, *92*, 116–17, 155;
conflict affecting, xxiii, xxiv, xxvi,
xxvii, 230, 265; dams affecting,
xxvii; diversification of, 119, 126,
219; industrial agriculture, xxvi,
xxix, 90, 116–17, 155, 308; in NE
Syria/Rojava, 115–17, 119, 121, 123,
130n2; urban agriculture, 101–12;
women in, 144–45, 154–55, 158,
170, 185. *See also* food sovereignty;
monoculture; organic growing
agroecology. *See* agriculture
air pollution, xxi–xxii, xxiv, xxx, 311
AKP (Adalet ve Kalkınma Partisi/
Justice and Development Party,
Turkey), xxviii, xxxi, 51, 84, 175,
177–80, 182, 186, 187, 193, 196,
198, 199, 230, 238n9, 263, 265,
271–73
Akuno, Kali, 312
Alevism: ancestors, 226–27, 229,
233, 235, 238n8, 245, 247, 249;
animals, respect for, xiii, xx,
228–30, 243, 247, 250, 256, 257;
ikrar (sacred oath, institutions),
228, 237; Kurmanc ethnic identity,
227, 238n3, 243–52, 257, 258n1;
massacre and genocide of Kurdish
Alevis, xiv, 226–27, 230, 233,
237, 238n8; *Ocax/Ocaks* system,
227, 229, 232, 233, 237–39, 244,
247–48; sacred places (*jiares*), xiii,
xx, xxx, 225–37, 243, 245–48, 250;
tradition, reimagination of, 227,
232–37; tribes (*aşirets*), 227–29,
233, 236, 237n1, 238n3, 244, 245,
247–50, 254–55, 257–58; water,
reverence for, xiii, 228, 245, 247–
55, 257. *See also Rêya Heqî/Raa
Haqi* faith
alienation, x, xxxiii, 28, 34, 50, 79, 80,
88, 136, 153

About the Editor and Contributors

ABOUT THE EDITOR

Stephen E. Hunt is a writer, university academic support coordinator, ecological activist, and radical historian. His PhD in English literature was awarded for a research thesis on Romantic ecology at the University of the West of England. He continues to research within the environmental humanities. He was a founder member of the Bristol Kurdish Solidarity Network, part of the Kurdish Solidarity Network. He is a member of the Association for the Study of Literature and the Environment (ASLE-UKI) and on the executive committee of the Angela Carter Society. He has written for publications including *Capitalism Nature Socialism*, *Green Letters*, and *Environment and History*. Stephen's books include *The Revolutionary Urbanism of Street Farm: Eco-Anarchism, Architecture and Alternative Technology in the 1970s* (2014); *Strikers, Hobblers, Conchies and Reds: A Radical History of Bristol 1880–1939* (co-authored, 2014); and *Angela Carter's 'Provincial Bohemia': The Counterculture in 1960's–1970's Bristol and Bath* (2020).

ABOUT THE CONTRIBUTORS

Marlene A. Payva Almonte holds a PhD from the School of Law and Social Justice of the University of Liverpool. Her doctoral research explores the relationship between human rights and climate change and identifies potential areas of reform in the field of international human rights law in light of the climate crisis. Her research interests include human rights law, climate change law, environmental law, international relations, and international humanitarian law. She is a lawyer and also holds a master's of laws (LLM)

347

in advanced studies in public international law from Leiden University, a master's in international relations (MSc) from the Barcelona Institute of International Studies, and a bachelor of laws from the University San Martin de Porres.

Azize Aslan has a PhD from the Instituto de Ciencias Sociales y Humanidades Alfonso Vélez Pliego, Benemérita Universidad Autónoma de Puebla, Mexico. She is currently working on the autonomy of Rojava and has written extensively on the organization of the communal economy. She is an activist in the Kurdish women's movement and has actively participated in the Democratic Economy project in Bakur.

Ercan Ayboğa is an environmental engineer and has been politically engaged in ecology struggles related to Kurdistan since his involvement in the Initiative to Keep Hasankeyf Alive against the construction of the Ilisu Dam in North (Turkish) Kurdistan. Ercan is also one of the cofounders of the Mesopotamia Ecology Movement, established in 2012. His activism has been undertaken from Germany and North Kurdistan. He traveled to Rojava on two occasions for research as co-author of the book *Revolution in Rojava* (2015), where he had also a focus on the ecology of that liberated territory. Ercan regularly writes articles for newspapers and magazines.

Fabiana Cioni (1970, Livorno) is an architect and activist and has been traveling in Kurdistan since 2005. Between 2000 and 2007, she worked as planner in Tuscany. She later quitted this job for ethical incompatibility with the public sector. Since 2008, she has been a high-school teacher. She is currently a PhD Student at the IUAV University of Venice.

John P. Clark is an eco-communitarian anarchist writer, activist, and educator in New Orleans. His most recent book is *Between Earth and Empire: From the Necrocene to the Beloved Community* and he is at work on a book on dialectical social ecology. He is coordinator of La Terre Institute for Community and Ecology, professor emeritus of philosophy at Loyola University and a member of the Education and Research Workers' Union of the Industrial Workers of the World (IWW).

Isabel David is a political scientist and assistant professor at the Institute of Social and Political Sciences and research fellow at the Orient Institute, Universidade de Lisboa (University of Lisbon), Portugal. She has published in reputed journals and publishers like Rowman & Littlefield, Palgrave Macmillan, Bloomsbury, Routledge, Lexington, Amsterdam University Press, ABC-CLIO, *British Journal of Middle Eastern Studies, Turkish*

Studies, Journal of Civil Society, Journal of Contemporary European Studies, Mediterranean Quarterly. She is co-editor (with Kumru F. Toktamis) of the recently created "Culture, Society and Political Economy in Turkey" book series with Peter Lang Academic Publishers.

Dilşa Deniz is a socio-cultural Kurdish anthropologist and is presently visiting scholar at Harvard University Divinity School. She was dismissed from her position in Turkey, in February 2016 after signing a peace petition. Her research focuses on gender, and cultural, political and religious practices in relation to Kurdish Alevis, Alevi geography, myths of Alevism in Kurdish communities, particularly in the city of Dersim, an ancient urban center for Kurdish Alevism in Anatolia.

She worked extensively as an activist and organizer in the women's movement in Turkey. She holds a PhD in Social Anthropology and has published articles and book chapters, as well as her monograph.

Pinar Dinc is a researcher at the Department of Physical Geography and Ecosystem Science and an affiliated researcher at the Centre for Advanced Middle Eastern Studies, Lund University. She received her PhD degree in political science at the Department of Government, London School of Economics in January 2017. Her PhD dissertation, *Collective Memory and Competition over Identity in a Conflict Zone: The Case of Dersim*, explores the causes and mechanisms of on-going competition over the nature of national identity through a case study of Dersim in the Turkish Republic. She worked on the Democratic Federation of Northern Syria (Rojava Cantons) as a Swedish Institute fellow at the Center for Middle Eastern Studies (CMES) at Lund University between September 2017 and 2018. In October 2018, she started as a Marie Curie Individual Fellow at CMES with the FIRE (Fighting Insurgency Ruining Environment) project, which focuses on conflict and the environment in the Middle East. She is currently leading the "Turkey Beyond Borders: Critical Voices, New Perspectives" project, which is funded by the Swedish Institute. She also works on a FORMAS project called "Societal impacts of climate stress: An integrated assessment of drought, vulnerability and conflict in Syria" at Lund University.

Laurent Dissard is an anthropologist of Turkey and the Middle East. He is assistant researcher in the interdisciplinary laboratory ITEM of the University of Pau and research affiliate of the EHESS research center CETOBaC in Paris. He received his PhD in 2011 at UC Berkeley, Andrew W. Mellon postdoc in 2013 from the Wolf Humanities Center at the University of Pennsylvania, and a Junior Research Fellowship in 2015 from the Institute of Advanced Studies of University College London (UCL). He has carried out ethnographic and archaeological fieldwork between 2001 and 2015 in Eastern Turkey, in an effort to critically examine the following themes: heritage politics and the rise of the tourism industry; the relationship between infrastructural development and nationalism; social justice and cultural, religious, and linguistic recognition. He is currently finishing a first book titled *Submerged: Archaeological Rescue and Historical Erasure in Eastern Turkey* (Stanford University Press), on the ways the past, rather than being discovered or erased, is instead "submerged," that is in a constant back-and-forth between absence and presence.

Ceri Gibbons is a political activist, investigative researcher, and writer working on issues of human rights, national security, and arms trade complicity in war crimes. He has worked with UK direct action campaign groups, Smash EDO, Campaign against Arms Trade, and the international legal charity, Reprieve.

Ahmet Kerim Gültekin (PhD) is an anthropologist. He was an assistant professor at the Sociology Department of Munzur University (Dersim, Turkey). He was dismissed via a decree-law in January 2017, issued by AKP (Justice and Development Party) without any juridical process due to signing the peace declaration of "Academics for Peace." The petition was protesting AKP's militarist oppressions and civil rights crimes against Kurdish civilians. Gültekin, shortly after, was sentenced to eight years of imprisonment due to his political and academic activities regarding Kurdish Alevi studies. He currently lives in Germany.

He received his doctorate in ethnology from Ankara University (2013), where he also did his MA (2007) and BA (2004) in the Department of Social Anthropology and Ethnology. His areas of interest are mainly Kurdish Alevism Studies (especially Dersim Studies), anthropology of religion, sacred places, and ethnography (methodological approaches on oral history and collective memory). He has published many articles and book chapters. He is the author of the books *Tunceli'de Kutsal Mekân Kültü* (Sacred Place Cults in Tunceli) 2004, *Tunceli'de Sünni Olmak* (Being Sunni at Tunceli) 2010, *Kutsal Mekanın Yeniden Üretimi* (The Reproduction of Sacred Place) 2020 and co-editor of *Kurdish Alevis and the Case of Dersim* 2019.

Cihad Hammy was born in Kobanî in 1991. He studied English literature at Damascus University but left his studies because of the war in Syria. He currently resides in Germany and studies English-American studies and philosophy at Hamburg University.

Allan Hassaniyan is a lecturer in Middle East Studies, at the College of Social Sciences and International Studies, at the University of Exeter.

Nicholas Hildyard works with The Corner House, a UK solidarity and mutual learning group. He has worked on water justice issues in the Kurdish region since the early 2000s, notably in opposition to the Ilisu Dam. He is the author of *Licenced Larceny: Infrastructure, Financial Extraction and the Global South* (2016).

Domenico Patassini (1949, Marostica) is a planner and professor of evaluation cultures at the Iuav University of Venice (Italy). He is a member of the scientific committee of the International PhD Programme at the School of Doctorate Studies and leads scientific research on spatial analysis, assessment, and management. He has been involved in urban projects, plans, and educational programs in many African countries since the beginning of the 1980s. Results of studies and research have been published in national and international journals and books.

Thomas James Phillips is a senior lecturer in law at Liverpool John Moores University. His research interests include Kurdish studies, public international law, and human rights law. In 2019, Thomas completed his PhD thesis on the right of self-determination and the Kurds in Turkey, and he has published a number of journal articles, book chapters, and opinion pieces on that topic. He has also traveled to the Kurdistan regions of Iraq and Turkey on numerous occasions to monitor elections and trials, as well as to speak at academic conferences. Thomas is also an active supporter of the "Peace in Kurdistan Campaign" and the "Kurdistan Solidarity Campaign."

Michel P. Pimbert is professor of agroecology and food politics at Coventry University and the director of the Centre for Agroecology, Water and Resilience in the United Kingdom. Over the last thirty-five years, he has done participatory research with indigenous peoples, peasant farmers, and activist scholars to promote food sovereignty, sustainable agriculture, nature conservation, and human rights. From 2013 to 2017, he was a member of the High Level Panel of Experts on Food Security and Nutrition (HLPE) of the Committee on World Food Security (CFS) at the UN Food and Agriculture Organization. His latest co-authored book is titled *Agroecology Now! Transformations Towards More Just and Sustainable Food Systems*.

Clémence Scalbert-Yücel is senior lecturer in Kurdish Studies at the University of Exeter. She has published on Kurdish literature, Kurdish language policies and heritage, including her monograph *Engagement, Langue. et Littérature. Le champ littéraire kurde en Turquie: 1980–2000* (2014). Her new research deals with rural transformations in Kurdistan. She is a member of the editorial board of the *European Journal of Turkish Studies*.

Engin Sustam lives and works in Paris. He is an academician and independent curator, completed his undergraduate studies in Mimar Sinan Fine Arts University in 2000 and had his master's degree in the same faculty (2002). He then gained another master's degree (DEA, 2005) and his PhD (2012) degree in EHESS, Paris, in sociology. He is currently an associate researcher at IFEA Istanbul, Cetobac of EHESS Paris, InCite of University of Geneva and has lectured at Paris 8 St. Denis University. He has worked as a visiting scholar at Queen Mary University of London, the University of Geneva, EHESS, ENS, Paris 8 St. Denis University, FMSH, among others. Sustam will be also a guest researcher at UDK Berlin. He is the author of the books *Kurdish Art and Subalternity, The Emergence of a Subjective and Creative Production Space Between Violence and Resistance in Turkey* (2016) and *Kırılgan Sapmalar: Sokak Mukavemetleri ve Yeni Başkaldırılar* ("Unexpected Insurgency: New Spaces for Global Uprisings") (2020, and to be translated into French). He is an art critic for contemporary art exhibitions on micro-politics, ecology, subculture, gender, and memory. His research and publications focus on postcolonialism, Kurdish cultural studies, social ecology, social movement, art theory, global uprising, sovereignty, violence and counter-violence, antifascist music, and nationalism and racism. Sustam is also the founder of Kargeh Berlin, Paris, and London (Kargeh: Workshop-Art collective) based on cultural art association.

Kumru Toktamış is an associate professor at the Pratt Institute Brooklyn and the coordinator of the cultural studies minor at the Department of Social Sciences and Cultural Studies. She is the co-editor of *Everywhere Taksim*, with Isabel David, where she contributed an article entitled "Invoking and Evoking Nationhood as Contentious Democratization" (2015). Her recent publications include two articles on the Imrali Talks "A Peace That Wasn't: Friends, Foes and Contentious re-Entrenchment of Kurdish Politics in Turkey," in *Turkish Studies* (2018) and "(Im)possibility of negotiating peace 2005–2015 peace/reconciliation talks between the Turkish government and Kurdish Politicians," in *Journal of Balkan and Near Eastern Studies* (2019), and one about the de-democratization of the AKP regime ("Now There is, Now There is Not: The Disappearing Silent Revolution of AKP as Re-entrenchment" in *British Journal of Middle Eastern Studies*).

Federico Venturini is a research associate at the University of Udine (Italy). His current research focuses on zero waste and sustainable tourism. In 2016, he earned his PhD at the University of Leeds. Focusing on the experiences in Rio de Janeiro between 2013 and 2014, in his research he explored the relations between contemporary cities and urban social movements, utilizing participatory/militant research approaches and through the lens of social ecology.

He has been a member of the Advisory Board of the Transnational Institute of Social Ecology, and also the International İmralı Peace Delegation, organized by the EU Turkey Civic Commission. He co-edited, with Thomas Jeffrey Miley, the book *Your Freedom and Mine: Abdullah Ocalan and the Kurdish Question in Erdogan's Turkey*, and with Emet Degirmenci and Inés Morales, the volume *Social Ecology and the Right to the City: Towards Ecological and Democratic Cities*.

www.ingramcontent.com/pod-product-compliance
Lightning Source LLC
Chambersburg PA
CBHW050624280326
41932CB00015B/2519